KU-247-478

Risk Analysis and Society

An Interdisciplinary Characterization of the Field

Edited by

TIMOTHY McDANIELS
University of British Columbia

MITCHELL J. SMALL
Carnegie Mellon University

Published in association with
the Society for Risk Analysis

CAMBRIDGE
UNIVERSITY PRESS

PUBLISHED BY THE PRESS SYNDICATE OF THE UNIVERSITY OF CAMBRIDGE
The Pitt Building, Trumpington Street, Cambridge, United Kingdom

CAMBRIDGE UNIVERSITY PRESS
The Edinburgh Building, Cambridge CB2 2RU, UK
40 West 20th Street, New York, NY 10011-4211, USA
477 Williamstown Road, Port Melbourne, VIC 3207, Australia
Ruiz de Alarcón 13, 28014 Madrid, Spain
Dock House, The Waterfront, Cape Town 8001, South Africa

http://www.cambridge.org

© Cambridge University Press 2004

This book is in copyright. Subject to statutory exception
and to the provisions of relevant collective licensing agreements,
no reproduction of any part may take place without
the written permission of Cambridge University Press.

First published 2004

Printed in the United States of America

Typeface Palatino 10/12 pt.   *System* LATEX 2_ε [TB]

A catalog record for this book is available from the British Library.

Library of Congress Cataloging in Publication Data

Risk analysis and society : an interdisciplinary characterization of the field / edited by
Timothy McDaniels, Mitchell J. Small.
    p.   cm.
Includes bibliographical references and index.
ISBN 0-521-82556-3 – ISBN 0-521-53263-9 (pb.)
1. Technology – Risk assessment.   2. Risk assessment – Social aspects.
I. McDaniels, Timothy, 1948–   II. Small, Mitchell J.

T174.5.R553   2003
363.1'02–dc21   2003046104

ISBN 0 521 82556 3 hardback
ISBN 0 521 53263 9 paperback

6000612257

Risk Analysis and Society
An Interdisciplinary Characterization of the Field

Risk Analysis and Society characterizes the state of the art and science of risk analysis. It views risk analysis as an important basis for informed debate, policy decisions, and governance regarding risk issues within societies. Its twelve chapters were designed and commissioned to provide interdisciplinary insights about the fundamental issues in risk analysis for the beginning of a new century. The chapter authors are some of the leading researchers in the broad fields that provide the basis for risk analysis, including the social, natural, medical, engineering, and physical sciences. Its chapters address a wide range of issues, including new perspectives on uncertainty and variability analysis, exposure analysis and the role of precaution, environmental risk and justice, risk valuation and citizen involvement, extreme events, the role of efficiency in risk management, and the assessment and governance of transboundary and global risks. The book was used as a starting point for discussions at the 2003 First World Congress on Risk, held in Brussels. Individual chapters were commissioned for a "Risk and Governance" Symposium held by the Society for Risk Analysis in June 2000.

Timothy McDaniels is a professor at the University of British Columbia, where he directs the Eco-Risk Research Unit and teaches in three graduate programs. He is a specialist in decision analysis, value elicitation, policy analysis, citizen involvement, and the social and judgmental aspects of risk management in environmental and human health contexts. He has served on advisory panels for the EPA SAB, for the U.S. National Science Foundation, and for Health Canada.

Mitchell J. Small is the H. John Heinz III Professor of Environmental Engineering at Carnegie Mellon University. Not only is he currently chair of the EPA SAB Environmental Models Subcommitee, a councilor for the Society for Risk Analysis, and a feature columnist for the *Journal of Industrial Ecology*, but he also serves as an associate editor for *Environmental Science and Technology*.

WITHDRAWN

Contents

PART THREE. NEW APPROACHES AND NEEDS FOR RISK
MANAGEMENT

Contributors

Vicki M. Bier
Department of Industrial Engineering
University of Wisconsin – Madison
Madison, WI **Ann Bruce**

ESRC Centre for Social and Economic Research on Innovation
    in Genomics
University of Edinburgh
Edinburgh, Scotland

Robin Cantor
Principal and Managing Director
LECG, LLC
Washington, DC

Alison C. Cullen
The Daniel J. Evans School of Public Affairs
University of Washington
Seattle, WA

Mary R. English
Energy, Environment and Resources Center
University of Tennessee
Knoxville, TN

Scott Ferson
Applied Biomathematics
Seatauket, NY

John D. Graham
Department of Policy and Decision Sciences
Center for Risk Analysis
Harvard School of Public Health
Boston, MA

Robin S. Gregory
Decision Research
University of British Columbia
Vancouver, Canada

Yacov Y. Haimes
Department of Systems and Information Engineering
University of Virginia
Charlottesville, VA

Dale Hattis
Center for Technology, Policy, and Industrial Development
George Perkins Marsh Institute
Clark University
Worcester, MA

Saburo Ikeda
Tsukuba University
Tsukuba, Ibaraki, Japan

Per-Olov Johansson
Department of Economics
Stockholm School of Economics
Stockholm, Sweden

Michinori Kabuto
National Institute for Environmental Studies
Tsukuba, Ibaraki, Japan

James H. Lambert
Department of Systems and Information Engineering
University of Virginia
Charlottesville, VA

Timothy L. McDaniels
School of Community and Regional Planning,
Institute for Resources, Environment, and Sustainability
University of British Columbia
Vancouver, Canada

Junko Nakanishi
Institute of Environmental Science and Technology
Yokohoma University
Yokohoma, Japan

Ortwin Renn
Center of Technology Assessment
Stuttgart, Germany

Lorenz R. Rhomberg
Gradient Corporation
Cambridge, MA

Mitchell J. Small
Departments of Civil Engineering & Environmental Engineering and
    Public Policy
Carnegie Mellon University
Pittsburgh, PA

Joyce Tait
ESRC Centre for Social and Economic Research on Innovation
    in Genomics
University of Edinburgh
Edinburgh, Scotland

Iwao Uchiyama
National Institute for Public Health
Tokyo, Japan

Rae Zimmerman
Robert F. Wagner Graduate School of Public Service
New York University
New York, NY

1

Introduction – Risk Analysis and Society

An Interdisciplinary Characterization of the Field

Timothy L. McDaniels and Mitchell J. Small

1. RISK ANALYSIS AND SOCIETY

Being alive means seeking opportunities and taking risks. For people living in modern society at the beginning of the twenty-first century, being alive means grappling with a complex and growing array of risks to the well-being of humans and the natural environment. It also means increasing concern for the how these risks are understood, characterized, and managed. Hence, we have the human dread of and fascination for risk and the increasingly important role of risk analysis within societies.

Since the beginning of human development, risks to health and well-being have led to adaptive responses that open paths for change. When neolithic family groups shared knowledge and resources for combating hunger, thirst, climate, or outside attack, they were trying to manage risks they faced. Jared Diamond's recent book, *Guns, Germs, and Steel*, presents the complex and fundamental decisions faced by hunter-gatherers when considering whether to adopt food production in place of their traditional foraging way of life (Diamond, 1999). Issues of uncertainty, value trade-offs, community knowledge, outside expertise, ethical dilemmas, and the imposition of risks by others were all part of those choices.

Risk management has been a fundamental motivation for development of social and governance structures over the last 10,000 years. The onset of agricultural production brought increasing population and permanent settlements. Concentrated population in turn led to greater risks of drought, famine, and conquest by others. Settlements thus created the need for infrastructures for managing these risks, such as water supply, food storage, and defenses. Large-scale construction in turn required specialization of labor and governance that could harness the collective resources needed for early societal risk management efforts. Without the risks to life and limb faced by individuals, societies would not have developed as they have.

How concepts of risk analysis originated, evolved, and became formalized provides one of the most compelling stories of the history of human thought. A comprehensive paper by Covello and Mumpower (1985) provides an historical perspective on risk analysis and risk management, starting with a group of decision consultants called the Asipu in the Tigris-Euphrates valley of 3200 B.C. That review considers early developments in probability, epidemiology, insurance, and legislation regarding societal risk analysis. It draws several distinctions about shifts in risk analysis and management from early to modern times. Peter Bernstein's remarkable book, *Against the Gods*, presents a chronology of thinking and understanding about risk, beginning with the development of number systems, games of chance and probability, and then tracing the development of economic thought about risk. In lively and accessible terms, he explores the great works on decision theory of the mid- and latter twentieth century, including von Neumann and Morgenstern (1944) and the writing of Tversky, Kahneman, and their colleagues. Bernstein's overall focus is on our understanding and management of financial risk. Yet his theme of risk analysis and decision theory as a means of reframing our conception of uncertainty (no longer simply fate but rather acting "against the Gods") is important for the history of risk analysis in all domains (Bernstein, 1998).

Analyzing and managing societal risks to health, safety, and environmental quality have become dominant themes in the social and natural sciences. Throughout the last century, economists have characterized the entire rationale for government as based in the support for collective efforts that cannot be accomplished through private markets. Most of those collective endeavors involve responses to risks: national defense, natural hazards, public health and safety, environmental protection, social infrastructure, and so forth. Sociologists such as Ulrich Beck (1992) see risk as an organizing principle for understanding the structure and functional relationships of modern societies. Engineers have for years addressed the risk of failure in their designs using safety factors and standards-based approaches. More recently, they have developed and applied probabilistic tools for explicit consideration of risk-cost and risk-risk trade-offs in areas such as dam, transportation, and product safety. Health scientists have seen an explosion in the number and complexity of health issues that require organized societal responses, and with it a demand for informed analysis to guide and tailor these programs. Toxicologists see their models put to use within a growing number of health, welfare, and ecological contexts, with health and environmental risks now considered fundamental elements of infrastructure, product, and regulatory design. All these disciplines have grappled with the role that uncertainty and precaution should play in managing risks.

2. A BRIEF HISTORICAL PERSPECTIVE ON THE DEVELOPMENT OF RISK ANALYSIS

A number of authors have provided accounts tracing the history and modern development of risk analysis as a field of study (Covello and Mumpower, 1985; Graham, 1995; Paustenbach, 1995; Rechard, 1999; Bedford and Cooke, 2001). The history by Rechard (1999), while focusing upon the implications for nuclear power and radioactive waste management, provides a particularly broad overview of the various disciplines that have contributed to the modern state of the field. Building upon the insights of Cumming (1981) and Ruckelshaus (1983), he notes that "risk assessment is not a distinct branch of science; instead it is a . . . 'hybrid discipline,' in which the current state of scientific and technological knowledge is made accessible to society as input to risk management decisions." Rechard traces early developments in probability theory, medicine, environmental health, chemical toxicology, reliability analysis, health and safety regulation, and risk perception and communication. Among the more recent key milestones (including a few that we have added) are:

- In 1924 Lotka, a U.S. physicist, speculates that, based on 1920 coal use, industrial activity will double atmospheric CO_2 in 500 years (http://www.environmentaldefense.org/pubs/FactSheets/d_GWFact.html, accessed January 3, 2002);
- In 1926 Muller discovers that X-rays induce genetic mutations in fruit flies 1,500 times more quickly than normal (http://www.dnacenter.com/geneticshistory.html, accessed January 2, 2002);
- The International Commission of Radiation Protection (ICRP) is established in 1928 in Sweden;
- The United Kingdom specifies a 99.999% reliability for 1-hour flying time of commercial aircraft in 1939;
- von Neumann and Morgenstern publish the *Theory of Games and Economic Behavior* in 1944;
- Monte Carlo methods are first applied in 1947 for diffusion of neutrons through fissile material (Metropolis and Ulam, 1949);
- In 1949 Callendar, a British scientist, speculatively links the estimated 10% increase of atmospheric CO_2 between 1850 and 1940 with the observed warming of northern Europe and North America that began in the 1880s (http://www.environmentaldefense.org/pubs/FactSheets/d_GWFact.html, accessed January 3, 2002);
- The U.S. Food and Drug Administration (FDA) adopts in 1954 a $100\times$ factor of safety for hazardous chemicals to determine an allowable daily intake (Dourson and Stara, 1983; Goldstein, 1990);
- Fault-tree methods developed in 1961 at Bell Labs for U.S. Air Force to evaluate Minuteman missile launch safety (http://www.safeware-eng.com/pubs/SafAnTooReq.shtml, accessed January 3, 2002);

- Starr presents a risk-cost-benefit analysis for nuclear power plants in 1969 (Starr, 1969);
- In 1972 Berg creates the first recombinant DNA molecule and the first successful DNA cloning experiment is performed in California;
- Ames test developed in 1973 to identify chemicals that damage DNA, in order to identify possibly carcinogenic substances (http://www. dnacenter.com/geneticshistory.html, accessed January 2, 2002);
- Kahneman and Tversky publish "Subjective Probability: A Judgment of Representativeness" in 1972 and "On the Psychology of Prediction" in 1973 (subsequently *Judgment under Uncertainty: Heuristics and Biases* with Slovic in 1982);
- Crutzen (1974) and Molina and Rowland (1974) identify key factors affecting depletion of stratospheric ozone;
- The U.S. Nuclear Regulatory Commission publishes the first probabilistic risk assessment for reactor safety in 1975 (Rasmussen et al., 1975; APS, 1975);
- Turner (1975) summarizes mathematical structure of single- and multi-hit dose-response toxicity models;
- The U.S. Environmental Protection Agency (EPA) issues its first formal guidelines for cancer risk assessments in 1976 (U.S. EPA, 1976; see also, Crump et al. 1976; Albert, Train, and Anderson, 1977; IRLG, 1979);
- Page (1978) identifies the character of risks requiring a precautionary approach;
- Crouch and Wilson (1979) examine interspecies comparisons of carcinogenic potency;
- Mackay (1979) proposes fugacity-based method for multimedia environmental modeling;
- Kaplan and Garrick (1981) characterize risk in terms of outcome scenarios, their consequences, and their probability of occurrence;
- The U.S. National Academy of Sciences issues 1983 study on *Risk Assessment in the Federal Government: Managing the Process* (NRC, 1983);
- Additional U.S. National Research Council studies (NRC, 1989; 1996) and a key Presidential/Congressional Commission Report on Risk Assessment and Risk Management (1997) emphasize critical roles for risk communication and social, deliberative processes in guiding risk assessment and risk management activities;
- Montreal Protocol on Stratospheric Ozone ratified in 1987;
- In 1988 the Intergovernmental Panel on Climate Change (IPCC) is founded and issues its first reports in 1990 (IPCC, 1990a, b, c; see http:// www.ipcc.ch/pub/reports.htm);
- International Life Sciences Institute Working Group presents conceptual framework for pathogenic microbial risk assessment for human disease in 1996 (ILSI, 1996);

- Haimes (1999) outlines an important role for risk analysis to address emerging threats to critical infrastructure, including cyber sabotage and terrorism;
- The completion of the Human Genome Project is announced in 2000 (the complete map of the human genome is published in 2001 in the journals *Science* and *Nature*) (http://www.dnacenter.com/geneticshistory.html, accessed January 2, 2002);
- The European Union Environment Commission publishes in 2000 a Communication[1] on the use of the precautionary principle in analyzing risk for environmental and health issues;
- In May 2001, the Convention on Persistent Organic Pollutants (POPs) is signed in Stockholm, Sweden; and
- In November 2001, Advanced Cell Technology of Massachusetts clones first human cells (http://detnews.com/2001/health/0111/26/a01-352254.htm, accessed January 2, 2002).

Clearly, the continued growth of research and applications addressing issues in risk analysis, and their extension to include a broad spectrum of scientific, social, and political perspectives, have been motivated by scientific and technological advances as well as societal needs. The chapters in this volume attempt to characterize the current landscape of risk analysis and to explore the frontiers of risk research and application. In so doing, we recognize that the science behind risk analysis draws upon a wide range of fundamental disciplines, with contributions from different applied fields and new applications emerging at a rapid pace. Table 1.1 provides one representation of the intellectual foundations and adaptations of scientific knowledge that support risk analysis. Clearly, as one moves from the basic disciplines of physics, chemistry, biology, mathematics, logic, and philosophy into the more applied disciplines of environmental science and engineering, medicine, public health, reliability engineering, and systems analysis and on to the social, behavioral, and policy sciences, widely divergent knowledge and insights must be tapped and integrated to solve real problems. This is not easy work, but it is exciting.

The question that we now face is this: To what extent has risk analysis evolved into a defined discipline? Such an *inter*disciplinary evolution can surely entail the benefits of providing a common set of tools and knowledge for addressing complex, multifaceted risk problems. However, it could also lead to a narrowing of the *multi*disciplinary vision of risk analysis that now provides much of its vigor and excitement. By demanding contributions from different disciplinary perspectives, how can we ensure a rich exchange of ideas from the most sound and advanced sources for each? By bringing together in this volume various perspectives on the current state of the social, engineering, health, and ecological risk sciences, we hope to motivate further thinking on this question, but not to resolve it.

TABLE 1.1. *The Disciplinary Foundations of Risk Analysis*

Risk Analysis

Public & Private Infrastructure Security & Safety	National & International Law and Governance	Public Participation

Pollution Prevention & Remediation	Sustainable Development	Business, Finance & Investment	Risk Perception and Communication

Medicine & Public Health		Reliability Engineering and System Analysis	Behavioral and Decision Sciences	Political Science

| | | | | History | |

Health Physics	Toxicology	Epidemiology			Psychology		Ethics

Environmental Science & Engineering

Pharmacology & Physiology	Ecology	Statistics	Economics		Law

Earth Sciences

PHYSICS	CHEMISTRY	BIOLOGY	MATHEMATICS	LOGIC	PHILOSOPHY

We suspect that pushes to integrate and synthesize will continue to be met by pulls to focus and specialize within both traditional disciplines and new, spin-off fields of study. For example, the areas of exposure assessment, infrastructure security, global change science, information systems, and biotechnology all could be seen as spin-offs of risk analysis. This dual process of integration and spin-off is healthy. These pulls and pushes can lead to new and deeper knowledge, as well as more insightful and pertinent solutions to the many pressing challenges that require effective applications of the risk sciences. We hope that this volume provides fodder and motivation for researchers and practitioners who need both to "delve deeper" and to "reach wider" to solve their problems.

3. RATIONALE AND GENESIS FOR THIS COLLECTION

While much attention gets focused on specific risks at specific times, or an array of risks as seen from one discipline, relatively little attention has been paid to comprehensive interdisciplinary perspectives on the relationship between risk issues and the broader societies in which they exist. Even less attention has been paid to exploring issues of the state of development and practice regarding risk issues from these different perspectives.

This book provides a characterization of the state of knowledge, research, and practice in the key technical and social disciplines that contribute to risk analysis. Our emphasis is largely on risk analysis as applied to health, safety, and environment questions, although the book also has relevance for diverse topics ranging from the protection of critical infrastructures such as computer systems to insurance for natural and man-made hazards. These disciplines address the performance of engineered systems, human health and the environment, probabilistic assessment, risk perception and communication, economic valuation of outcomes, and social and political mechanisms and institutions for risk management.

Over three days in June 2000, the Society for Risk Analysis (SRA) held an international symposium at Airlie House, outside Washington, DC. One purpose of the symposium was to begin the process of assessing the current state of risk analysis from many different disciplinary perspectives. A second purpose was to foster informed discussion that considers the state of risk analysis and its contributions to governance in various parts of the world. A third purpose was to lay the groundwork for a series of world congresses on risk issues.

As part of that effort, the U.S. National Science Foundation provided support for a series of commissioned papers by some of the world's leading experts on risk issues. Topics were nominated and refined by the organizing committee, and selected authors were commissioned to prepare chapters that would be unusual in terms of the breadth of perspectives they addressed. The vision was to sponsor a series of papers that offered historical, social, technical, and policy-oriented insights about key aspects of how risk analysis contributes to governance. Initial drafts served as the starting point for group discussions at the symposium and afterwards. All the chapters were subsequently peer reviewed and underwent a series of revisions.

The ten papers resulting from this process together comprise an informed viewpoint on the recent history, current state, and future outlook for the field of risk analysis, and its contributions to private and public decision making. The chapters place particular emphasis on risk analysis within the context of national and international governance. Here the term "risk analysis" collectively refers to risk assessment, risk management, and risk communication. The authors of these chapters include prominent risk scholars from North America, Europe, and Asia, including several past presidents of the SRA.

4. CONTENTS OF THIS BOOK

The papers are presented in three major groupings. The first set addresses the fundamental character of risk, including its inherent variability in natural and engineered systems (Chapter 2 by Dale Hattis); the relationship among system components and whole systems in biological dose-response (Chapter 3, Lorenz Rhomberg); the character and characterization of rare and extreme events (Chapter 4 by Vicki Bier and coauthors); and the social elements of equity and justice that are critical components of risk issues (Chapter 5 by Mary English).

The second section of the book explores advances in methods for risk assessment and analysis. These include a paper by Alison Cullen and Mitchell Small on qualitative and quantitative methods for uncertainty analysis (Chapter 6); an examination of methods for valuing risk by Robin Gregory (Chapter 7); and an assessment of methods for cost-benefit and cost-effectiveness analysis by John Graham, Per-Olov Johansson, and Junko Nakanishi (Chapter 8).

This chapter provides a transition to the final section of the book addressing approaches and needs for risk management. Here Ortwin Renn explores new methods for promoting public input and participation in risk management decisions (Chapter 9). Joyce Tait and Ann Bruce examine institutions for addressing multinational and global risks (Chapter 10), and Michinori Kabuto, Saburo Ikeda, and Iwao Uchiyama provide insights on the special challenges of managing both traditional and newly emerging risks in the developing nations of Asia (Chapter 11).

The concluding Chapter 12 is written by Rae Zimmerman and Robin Cantor, two of the key organizers of the Airlie House symposium that provided the impetus for this book. In their chapter, the authors pull together integrative themes from the papers and the discussion that occurred at the meetings as these chapters were developed.

This collection does not cover all issues in the diverse and growing field of risk analysis.[2] It does however attempt to highlight the key elements at the forefront of risk theory and application that will most influence directions in the field in coming years. The collection shows how risk analysis has evolved from the largely technical disciplines of systems reliability and health sciences to encompass the full range of political, legal, economic, and social considerations that must be addressed when understanding technical systems and their role in society.

It is hoped that the chapters in this book will form the intellectual basis for future world congresses on risk, and for the emergence of an integrated, multidisciplinary interpretation of risk analysis that could be endorsed by the many professional societies devoted to risk issues. With this book, we seek to move beyond a series of single-discipline perspectives regarding risk issues to an interdisciplinary and multidisciplinary integration of perspectives on risk. Real-world problems do not respect disciplinary boundaries. They require integration and the ability to find understanding through the exploration of linkages, multiple structures, and multiple perspectives.

Notes

1. It indicates that measures based on the precautionary principle should be "[p]roportionate to the chosen level of protection; non-discriminatory in their application; consistent with similar measures already taken; based on an examination of the potential benefits and costs of action (or lack of it); subject to review in light of new scientific data; and capable of assigning the burden of proof for producing a more comprehensive risk assessment."

2. The events of September 11, 2001, have made it clear that if risk analysis is to be useful for society, it must be able to address new, unexpected, and even unimagined threats. While the chapters in this book were commissioned and completed before the terrorist attacks of September 11, the concepts and applied perspectives presented here are applicable to understanding and eventually managing risks of terrorism. As a starting point for understanding the role of risk analysis in addressing terrorism, we can turn to the writing of several of the past presidents of the Society for Risk Analysis. The society's journal *Risk Analysis* asked several of the former presidents of the society to prepare short perspective pieces on the potential application of risk analysis to managing terrorist threats. These short papers were published in the June 2002 edition of the journal, in a special issue titled: *Assessing the Risks of Terrorism: A Special Collection of Perspectives Articles by Former Presidents of the Society for Risk Analysis.* The themes and key points raised in those papers provide some perspectives on what risk analysis has to offer, and what key questions must be addressed, in providing analytical insight into decisions about managing terrorist threats.

References

Albert, R. E., Train R., and Anderson, E., 1977. Rationale developed by EPA for the assessment of carcinogenic risks. *Journal of the National Cancer Institute*, 58(5): 1537–41.

American Physical Society (APS). 1975. Report to the American Physical Society by the study group on light-water reactor safety. *Reviews Modern Physics*, 47, Supplement 1: S1–S124.

Beck, U. 1992. *Risk Society: Towards a New Modernity*. Sage, London.

Bedford, T., and Cooke, R., 2001. *Probabilistic Risk Analysis: Foundations and Methods*. Cambridge University Press, Cambridge, UK.

Bernstein, P. J. 1998. *Against the Gods: The Remarkable Story of Risk*. Wiley, New York.

Covello, V., and Mumpower, J. 1985. Risk analysis and risk management: An historical perspective. *Risk Analysis*, 5(2): 103–20.

Crouch, E. A. C., and Wilson, R. 1979. Interspecies comparison of carcinogenic potency. *Journal of Toxicology and Environmental Health*, 5: 1095–1118.

Crump, K. S., Hoel, D. G., Langley, C. H., and Peto, R. 1976. Fundamental carcinogenic processes and their implications for low dose risk assessment. *Cancer Research*, 36(9 Pt. 1): 2973–9.

Crutzen, P. J. 1974. Estimates of possible variations in total ozone due to natural causes and human activities. *Ambio*, 3(6): 201–10.

Cumming, R. B. 1981. Editorial: Is risk assessment a science? *Risk Analysis*, 1(1): 1–3.

Diamond, J. 1999. *Guns, Germs and Steel: The Fates of Human Societies*. Norton, New York.

Dourson, M. L., and Stara, J. F. 1983. Regulatory history and experimental support of uncertainty (safety) factors. *Regulatory Toxicology & Pharmacology*, 3: 224–38.

Goldstein, B. D. 1990. The problem with the margin of safety: Toward the concept of protection. *Risk Analysis*, 10(1): 7–10.

Graham, J. D. 1995. Historical perspective on risk assessment in the federal government. *Toxicology*, 102(1/2): 29–52.

Haimes, Y. Y. 1999. Editorial: The role of the Society for Risk Analysis in the emerging threats to critical infrastructures. *Risk Analysis*, 19(2): 153–7.

Interagency Regulatory Liaison Group (IRLG) Work Group on Risk Assessment. 1979. Scientific bases for identification of potential carcinogens and estimation of risks. *Journal of the National Cancer Institute*, 63(1, July): 241–68.

Intergovernmental Panel on Climate Change (IPCC). 1990a. *Scientific Assessment of Climate Change – Report of Working Group I*, J. T. Houghton, G. J. Jenkins, and J. J. Ephraums, Eds. Cambridge University Press, Cambridge, UK.

Intergovernmental Panel on Climate Change (IPCC). 1990b. *Impacts Assessment of Climate Change – Report of Working Group II*, W. J. McG. Tegart, G. W. Sheldon, and D. C. Griffiths, Eds. Australian Government Publishing, Canberra, Australia.

Intergovernmental Panel on Climate Change (IPCC). 1990c. *The IPCC Response Strategies – Report of Working Group III*. Island Press, Covelo, CA.

International Life Sciences Institute (ILSI). 1996. A conceptual framework to assess the risks of human disease following exposure to pathogens. *Risk Analysis*, 16(6): 841–8.

Kahneman, D., and Tversky, A. 1972. Subjective probability: A judgment of representativeness. *Cognitive Psychology*, 3: 430–54.

Kahneman, D., and Tversky, A. 1973. On the psychology of prediction. *Psychological Review*, 80(4): 237–51.

Kahneman, D., Solvic, P., and Tversky, A., Eds. 1982. *Judgment under Uncertainty: Heuristics and Biases*. Cambridge University Press, Cambridge, UK.

Kaplan, S., and Garrick, B. J. 1981. On the quantitative definition of risk. *Risk Analysis*, 1(1): 11–27.

Mackay, D. 1979. Finding fugacity feasible. *Environmental Science & Technology*, 13: 1218–23.

Molina, M. J., and Rowland, F. S. 1974. Stratospheric sink for chlorofluoromethanes: Chlorine atom-catalyzed destruction of ozone. *Nature*, 249: 810–12.

Page, T. 1978. A generic view of toxic chemicals and similar risks, *Ecology Law Quarterly*, 7: 207–44.

Paustenbach, D. J. 1995. Retrospective on U. S. health risk assessment. How others can benefit. *Risk: Health, Safety & Environment*, 6: 283–332, see: http://www.fplc.edu/risk/vol6/fall/pausten.htm.

Presidential/Congressional Commission. 1977. Presidential/Congressional Commission on Risk Assessment and Risk Management. Final report, Volume 1. Posted on *RiskWorld* website, January 29, 1997, see: http://www.riskworld.com/nreports/1996/risk_rpt/RR6ME001.HTM.

Rasmussen, N. C., et al. 1975. *Reactor safety study: An assessment of accident risks in U.S. commercial nuclear power plants*. Nuclear Regulatory Commission, NUREG-75/014 (WASH-1400) Washington, DC.

Rechard, R. P. 1999. Historical relationship between performance assessment for radioactive waste disposal and other types of risk assessment. *Risk Analysis*, 19(5): 763–807.

Ruckleshaus, W. D. 1983. Science, risk, and public policy. *Science*, 221(4615): 1026–8.

Metropolis, N., and Ulam, S. M. 1949. The Monte Carlo method. *Journal of American Statistical Association*, 44(247): 335–41.

Starr, C. 1969. Social benefit versus technological risk. What is our society willing to pay for safety? *Science*, 165. 1232–8.

Turner, M. E., Jr. 1975. Some classes of hit-theory models. *Mathematical Biosciences*, 23: 219–35.

U.S. Environmental Protection Agency (EPA). 1976. Health risk and economic impact assessments of suspected carcinogens: Interim procedures and guidelines. *Federal Register*, 41(102), May 25: 21402–5.

U.S. National Academy of Sciences. 1983. *Risk Assessment in the Federal Government: Managing the Process*. National Academy Press, Washington, DC.

U.S. National Research Council (NRC). 1989. *Improving Risk Communication*. NRC Committee on Risk Perception and Communication, National Academy Press, Washington, DC.

U.S. National Research Council. 1996, *Understanding Risk: Informing Decisions in a Democratic Society*. National Academy Press, Washington, DC.

von Neumann, J., and Morgenstern, O. 1944. *Theory of Games and Economic Behavior*. Princeton University Press, Princeton, NJ.

PART ONE

FUNDAMENTAL CHARACTER OF RISK

2

The Conception of Variability in Risk Analyses

Developments Since 1980

Dale Hattis

1. INTRODUCTION

This chapter offers a philosophical and historical perspective on the development of the concept of "variability" in the last few decades. The goal is not to provide a treatment of modern mathematical techniques for the analysis of data indicating variability. The recent literature contains excellent works that document analytical tools and are eminently usable by risk analysis practitioners (Cullen and Frey, 1999; Thompson, 1999). Rather, the object is to reflect on the significance and prospects for quantitative assessment of "variability" as an intellectual innovation that has emerged in part from the interdisciplinary fusion of ideas, techniques, and social needs for information for decision making that is the discipline of risk analysis.

Briefly, the nub of the innovation is distinguishing real variation among things from measurement errors, other sources of uncertainties, and stochastic fluctuations (such as the numbers of cosmic rays arriving at a detector in a specified interval). Where uncertainties reflect the imperfections in available information about the world (and are often seen as an annoying fog that obscures investigators' ability to demonstrate differences among groups, but with no real consequence or interest in itself; Hattis, 1996), real quantitative variation among things/people has real implications for differential behavior among the "things" being studied. Such differences can, for example, take the forms of (a) the relative risks to those things/people and (b) the relative desirability of devoting resources to some things rather than to others (if, for example, the things are categories such as industries that could be the subjects of safety inspections or other resource-consuming activities) (Hattis and Goble, 1994).

Thanks are due to Professors Vicki Bier, Mitch Small, and Rob Goble for insightful suggestions on earlier versions of this manuscript.

Variability exists across virtually all elements of the natural and engineered environment. This variability may be manifest over time, in spatial location, or from item to item (e.g., from one manufactured item to another, one power plant to another, one river basin to another, or one person to another). Variability from item to item often results from spatial or temporal differences/changes in the processes affecting the characteristics of each item; more often than not, these variations are assumed to be purely random. Random variations in items, without regard to order, time or location, are described by the *random variable* properties of the item. A full description of a random variable is provided by the probability distribution function, defined by either the cumulative distribution function (CDF), the probability density function (PDF) for a continuous random variable, or the probability mass function (PMF) for a discrete random variable. Ordered random variables, typically ordered in time or space, are referred to as *random processes*. A random process is characterized by its random variable properties *and* measures of the degree of relationship and correlation along its ordering dimension(s).

Probability models for random variables and processes have been developed to describe variability for a wide range of geological, ecological, and environmental processes (a number of interesting examples are presented in Walden and Guttorp, 1992, and Ott, 1995). The examples in this chapter are focused principally on human interindividual variability in the factors that determine environmental exposure and risk. These include physiological differences among subpopulations and individuals that determine susceptibility to risk, as well as differences in past, current, or projected future exposures that result from different location, occupation, or time-activity patterns.

Section 2 addresses a series of "What?" questions. First, what is the history of the recognition that variability is separate from uncertainty, has different implications for risk management, and requires different techniques for estimation? Next, what is the state of current practice in assessing variability, how is this different from the state of practice in 1980 and earlier, and how is it different again from current and foreseeable future "best" practice? These comparisons are addressed in summary form in Tables 2.1 and 2.2.

Section 3 then surveys some plausible drivers of change in both the social "demand" for information relevant to decision making and the capability of the technical community to begin to "supply" variability information. Briefly, the supply-side drivers include the vast expansion of computerized measurement, data storage, and processing capabilities that have, for example, consigned the slide rule (familiar as a computing device for those of us who were making calculations in the 1960s) to approximately the same status as the abacus. Demand-side drivers include both pressures for greater accountability in the use of resources in the name of health protection and, from the other side of the political spectrum, pressures to recognize the presence of people with unusual susceptibility of various types in the setting of policy. A recent manifestation of this is the Food Quality Protection Act, which mandates consideration of the special sensitivity of developing fetuses and children (although the act, by its device of an added safety/uncertainty factor still reflects what will be argued is an older qualitative/categorical response to the presence of variability).

Finally, Section 4 briefly assesses prospects for further development and practical use of quantitative concepts of variability in environmental health decision making.

Before proceeding it is well first to mention briefly some deeper issues of philosophy of science that either tend to promote/legitimize or pose impediments to the recognition and improvement of variability analyses for different sets of professional practitioners. Major impediments are presented by the lingering influences of positivism (Gale, 1984) in many sectors of the scientific community – in particular the discouragement of theoretical modeling of quantities that cannot be directly observed, and the related notions (a) that all the relevant information for an analysis can be found within the numbers provided in a specific data set and (b) that the chief role of mathematical analysis is to "fit" the data or summarize the data in a more compact form (Suppe, 1977). By contrast, Bayesian traditions of decision analysis do tend to legitimize reasoned integration of information of multiple types from multiple sources to draw inferences. For example, recent discussions of good practice for probabilistic analyses emphasize the role of mechanistic considerations in the choice of distributional forms for application to data (Hattis and Burmaster, 1994; Thompson, 1999). Another example is the need to adjust the spread of a set of quantitative observations to remove the effects of measurement errors from estimates of "real" underlying variability (Hattis and Silver, 1994). That such adjustments are almost never made in current practice even by sophisticated analysts (e.g., see Griffin et al., 1999), and directions for such adjustments are absent even from some of the most advanced recent discussions of probabilistic risk assessment methodology (Cullen and Frey, 1999) are testimony to the potential for improvement in the practical application of the concept of variability to inform risk analyses.

2. THEN VERSUS NOW – COMPARISONS AND CONTRASTS

2.1. Historical Overview

There is room for argument about the degree to which the notion of variability as distinct from uncertainty should really be considered a recent innovation. The term *variability* is in fact used in its modern sense (real differences among things) in the foundational paper by Kaplan and Garrick in the very first issue of *Risk Analysis* in 1981. In that paper, it is suggested that one input to estimates of the uncertainty distribution for the occurrence of a specific failure (a turbine trip) at a particular plant should be the distribution of observations of the frequency of similar failures that have occurred at each of several similar plants in past years. Kaplan and Garrick recommend that such data can be used

as evidence to generate a population variability curve. This curve expresses the fact that different plants in the population have different frequencies of occurrence of this event. The population variability curve is then used as the prior in a second application of Bayes' theorem with information [on the frequency of turbine trips at the particular plant under study] as evidence. Details of this two-stage process will be presented in a subsequent paper. (p. 23)

Kaplan published the subsequent paper in 1983. In its use of a Bayesian framework in the first of the two stages of analysis to estimate a variability distribution, it reflects a very sophisticated understanding of the need to make estimates of real underlying variability from the available observations. The distribution of true underlying failure rates is narrower than the distribution of observed rates because the observed rates must include some spreading owing to random statistical fluctuations in limited observation times (and possibly also measurement/reporting errors[1]). Kaplan and Garrick clearly had the concept of variability as a set of real underlying differences among similar things (power plants, in this case), but their relatively brief mention of it in the 1981 work is as a useful intermediate in the assessment of an uncertainty distribution for the unknown failure rate of a particular facility. Kaplan's subsequent paper (1983), while containing much useful development of estimation methodology for the first-stage (variability) component of the procedure, still treats the variability as of subsidiary interest to the main project of estimating the uncertainty distribution for a specific facility's failure rate.

The earliest clear articulation of the distinction between variability and uncertainty that treats variability as of separate and coequal interest for risk assessment with distinctive implications for risk management is a 1986 Ph.D. thesis by Ken Bogen, and a later book (Bogen, 1990), whose main results are most easily accessible in the form of a paper in *Risk Analysis* by Bogen and Spear in 1987:

We use the term "uncertainty" to refer specifically to a lack of knowledge concerning some risk-related characteristic, either in a statistical sense (regarding some predictable form of inaccuracy or imprecision in parameter estimates) or in a scientific sense (regarding missing or ambiguous information or gaps in scientific theory required to make predictions on the basis of causal inferences). In contrast, the term "variability" is used to refer specifically to interindividual heterogeneity with respect to some risk-related characteristic. (pp. 427–428)

Bogen and Spear go on to make the critical distinction between individual risk and population risk:

"Individual Risks" shall refer to a set of independent risks faced by one or more individuals in an exposed population, whereas 'population risk' shall refer to the probability distribution of the actual health impact (i.e., the total number of additional cases) that may be generated by the imposition of a given set of individual risks. (p. 428)

The same paper also offers extensive mathematical nomenclature and calculation techniques to facilitate quantitative treatment of variability in the context of two-dimensional uncertainty-variability analysis.

By coincidence, the paper that immediately precedes the Bogen and Spear paper in *Risk Analysis* is the author's own first published empirical treatment of human interindividual variability in pharmacokinetic parameters (Hattis, Erdreich, and Ballew, 1987). This was conceived as a start toward evaluating the degree of protection provided by the traditional tenfold "safety" or "uncertainty" factor for differences among humans that was (and still is) almost universally used in estimating "acceptable daily intake" levels, "reference doses," and similar regulatory and advisory criteria for traditional noncancer toxicants worldwide (Lehman and Fitzhugh, 1954; Dourson and Stara, 1983). A year earlier, another paper in *Risk Analysis* gave extensive treatment to the idea of using biomarker measurements to shed light on the extent of human interindividual variability in parameters relevant to various toxic and carcinogenic processes (Hattis, 1986). These papers do not mention the distinction with uncertainty explicitly but do use the term *variability* in the modern sense as real differences among intrinsic characteristics of the human subjects studied. Bier, in a landmark paper in 1989, appears to be the first to have brought together in the same work the engineering ideas of dependencies of failure rates (e.g., common cause failures, cascade failures, and intersystem dependencies) – many of which can be seen as forms of variability – and the idea of person-to-person variability in susceptibility to chemically mediated health effects. Thus, this basic conception of variability was clearly in the intellectual air in the 1980s for the community of researchers who were assessing risks and publishing in *Risk Analysis*.

Despite the appearance of these early papers, it is fair to say that quantitative analysis of variability separate from uncertainty did not quickly spread to the great bulk of practitioners of risk assessment and allied disciplines, and even today in some of the most receptive application areas (such as exposure assessment) one might argue that the distinction is more often honored in rhetoric than fully implemented in practical analyses that inform social decision making. The next two subsections, together with Tables 2.1 and 2.2, discuss the gradual processes by which the risk assessment community has been working out the conceptual and technical implications of the idea of variability as separate from uncertainty and stochastic fluctuations.

2.2. Evolution of Variability Analysis

Tables 2.1 and 2.2 summarize the main threads explored here. The initial discussion will focus on evidence for the "obsolete" label applied to the older practices listed and the intellectual needs to move to more recent concepts. Then the emergence of more modern practices will be described.

2.2.1. Older Practices and the Categorical Conception of Variability

SALIENCE OF QUANTITATIVE VARIABILITY FOR EMPIRICAL STUDY AND CONCEPTION OF THE GOALS/TOOLS FOR STATISTICAL ANALYSIS. The idea of the "salience" of quantitative variability for researchers involves asking some very general questions about research paradigms in fields that contribute to risk analysis, in particular experimental biology (including toxicology), epidemiology, and the engineering and allied disciplines related to the assessment of exposure (e.g., industrial hygiene; fate and transport measurement and modeling). How did typical researchers in these fields conceive the objectives of their research efforts, and to what extent would quantitative assessments of intrinsic variability be helpful in achieving those objectives? The basic idea is that quantitative variability information was probably not a major focus of research prior to the 1980s, at least in most of the risk-related fields derived from the biological sciences because it was not generally seen as a central issue of interest in older research programmes.

Starting with basic biology, it must be noted that biologists have sought to describe intrinsic differences, at least in qualitative/categorical terms, from the very start of the field in the modern (post-Renaissance) era. For example, one essential landmark in the development of modern biology was the articulation of the Linnaean system for categorizing living organisms into groups (Koerner, 2000). Since the middle to late nineteenth-century work of Darwin, these groups are conceived as having historical evolutionary relationships.

Moving a little forward in time, the notion of intrinsic categorical variability was also implicit in Gregor Mendel's famous work on the inheritance of different characteristics through generations of pea plants. Mendel and his later intellectual descendants in the field of genetics in the twentieth century did not hesitate to use mathematics – indeed they extensively developed the implications of combinatorial theories for the ratios of different genotypes and phenotypes among offspring. However, the quantitative analyses are used within a framework of atomistic determinants (genes) of specific characteristics rather than to elucidate the causal factors for quantitative properties of some conceivable interest to agriculture (e.g., yield,

TABLE 2.1. *Conceptual Evolution of Variability Analysis*

State of Practice	Concept of the Salience of Variability for Empirical Study	Mathematical Representation of Variability	Goals and Tools for Statistical Analysis	Implications for the Information Used to Set Policy
Older/obsolete practice	Within-group differences are an annoying detail to be avoided if at all possible because they tend to obscure between-group differences that are the prime objects of study. No distinction between random or systematic measurement errors and real underlying differences that are now defined as *variability*.	Categorical treatment only (e.g., "sensitive subgroups") with no analysis of within-group variability. Sometimes this takes the form of drawing an arbitrary quantitative criterion level in a set of continuous unimodal observations (e.g., hypertension at a blood pressure of 90 mm Hg diastolic; low birth weight at 2500 g).	Standard "testing" of the null hypothesis that two groups are the same at predefined confidence levels.	Dismissal of the relevance of responses of "hypersensitive" people for policy (e.g., ACGIH Threshold Limit Values).
Current usual practice	Some characterization of the spread of observations is desirable, but usually no separate quantitative treatment of variability and uncertainty is attempted.	Simple application of an assumed distributional form without assessment of fit (e.g., characterization of observed chemical concentration levels from environmental or occupational samples with means and standard deviations, even where the standard deviations are similar to the means and the distribution), if examined, is obviously skewed to the right.	(a) Multiple regression analysis to "explain" the amount of the variance of dependent variables that are associated with various independent variables. (b) Simple method-of-moments calculations of standard deviations, skewness, kurtosis. (c) Simple probability plotting.	Allowance for human susceptible subgroups with a defined tenfold "safety" or "uncertainty" factor. "New" recognition of subgroups that were arguably not specifically covered in earlier risk management policy is accommodated by addition of other factors of 10 or 3.
Current/foreseeable future best practice	Intrinsic variability separate from uncertainties. Variability expressed in dimensions of time, space, and other causal factors as appropriate, and residual variability is modeled as a topic of interest in its own right.	Distributional forms chosen based on fundamental mechanistic considerations. Mixed distributions (e.g., two or more normal or lognormal distributions) applied where indicated from both putative mechanisms producing the variability and observation of the data fits (Roeder, 1994; Burmaster and Wilson, 2000).	(a) Maximum likelihood/ likelihood distribution analyses (Cullen and Frey, 1999). (b) Animated displays for exploratory data analysis (Burmaster and Thompson, 1999).	Quantitative analysis of how many people and which groups of people are at how much risk from various policies (Hattis and Anderson, 1999), (ideally with some statement of associated confidence/uncertainty levels).

TABLE 2.2. *Evolution of Practice in Some Technical Aspects of Variability Measurement/Analysis*

State of Practice	Sample Design	Treatment of Measurement Errors	Treatment of Censored Data (i.e., nondetects)
Older/ obsolete practice	Deliberate restriction of the sample to limit variability (e.g., include only young healthy adult males). Alternatively, make measurements in ways that are expected to be biased relative to a random sample (e.g., take samples from a hazardous waste site only in places where a "hot spot" of contamination is suspected).	Only one measurement made per case; no quantitative assessment of error.	Arbitrary assignment of zero for nondetects.
Current usual practice	"Haphazard" or "convenience" samples – include all readily available subjects, without attention to factors that could affect the primary parameter under study.	Some replicate measurements leading to a quantitative assessment of measurement error, but no attempt to subtract the variability attributable to measurement errors from the total observed variance to arrive at an estimate of the "real" intrinsic variability.	Arbitrary assignment of one half of the detection limit for the nondetects.
Current/ foreseeable future best practice	Stratified random sampling with strata constructed to represent groups expected to be different in the studied parameter, with oversampling of relatively rare subpopulations of special interest (National Center for Health Statistics, Data Dissemination Branch, 2003, http://www.col.gov/nchs/about/major/nhis/hissample.htm).	Full modeling of intrinsic variability after adjustment for the spread of the observations caused by measurement errors.	Probability plots and imputation techniques (e.g., estimation of a central tendency value for the nondetects from a distribution fitted to the rest of the data).

sweetness, shelf life). A similar qualitative conception of causal pathways can be seen in the traditional biological research objective of connecting specific structures with specific biological functions.

A final example of the development of categorical descriptions of variability to support qualitative causal pathway conclusions is the medical classification of states of ill-health into a diagnostic system of "diseases." These categorizations are ideally intended to explain originating causes ("etiology"), predict future health states ("prognosis"), and provide a framework for assessing options for appropriate treatment.

These fundamental categorical conceptions of variation and the qualitative ideas of causal pathways necessarily have had an important role in shaping biomedical scientists' paradigms of useful and relevant research. Supporting the articulation of a particular categorical distinction often involves showing that a series of measurements are reliably different between members assigned to one group rather than another. This leads naturally to one basic technique in biostatistics – the idea that it is important for available quantitative data to reject a null hypothesis that members of two putatively distinguished groups are really members of a single population. Important examples of this for risk analysis include the qualitative categorizations that are traditionally included in "hazard identification" (e.g., is this chemical a "carcinogen" or a "neurotoxin")?

In the context of this kind of research effort, within-group differences can easily be seen as an "annoying detail," as has been discussed at greater length earlier (Hattis, 1996). Important practices in toxicology and pharmacology (such as restricting experiments to young healthy adults, often of a single gender) can be understood as efforts to limit within-group experimental variability as much as possible to reveal between-group differences as efficiently and unambiguously as is feasible. Unfortunately, these very practices can often deprive risk analysts of potentially valuable information about the possible diversity in sensitivity and response that might be present in groups whose differences were not deliberately limited in this way, such as the general human population of a city.

MATHEMATICAL REPRESENTATIONS OF VARIABILITY. There is another, more subtle and pernicious effect of the predominant historical conception of variation as categorical in nature. This is that medical and epidemiological researchers have often used arbitrary cut points in continuous distributions to define conditions or "diseases" where other support for making a qualitative distinction is not readily apparent. Thus, we have extensive research into differences between "hypertensives" and "normotensives," even though plots of blood pressures in populations show no obvious departure from a continuous unimodal lognormal distributions, and models of the effect of cardiovascular risk factors on the incidence of adverse effects (e.g., heart attacks and strokes) generally reflect continuous monotonic functions rather than sharp changes at defined cut-points (Hattis and

Richardson, 1980). It is not unreasonable, of course, for medical practitioners to make quantitative cut-points in deciding issues such as which people might benefit from treatment to lower their blood pressure, although even here it is open to question whether the cut-point should be constructed on the basis of blood pressure alone versus some overall calculation of cardiovascular risk that combines the effects of multiple risk factors. But this should not lead professionals to reify the cut-point distinctions as bright line indicators of pathology unless other evidence can be adduced to support the proposition that there is something intrinsically different about, for example, the system of feedback controls that regulates particular continuous parameters.

As another example, epidemiological researchers make counts of babies with "low birth weight" rather than describing relationships between the underlying continuous variable (birth weight) and either causal or outcome parameters of interest. Any dichotomization of the measurements of a continuous variable inherently discards a large amount of the information encoded in those measurements that can be used for group comparisons, or more sensitive multivariate modeling of relevant influences of environmental and other factors on the continuous parameter and associated quantal outcomes such as mortality.

This practice of dichotomization is particularly unfortunate when it tends to obscure the ability of both medical practitioners and researchers to appreciate the dimensions of public health problems. Figure 2.1 shows the relationship between birth weight and infant mortality together with the conventional cutoff used to define low-birth-weight babies. It is certainly correct that babies in the conventional low-birth-weight region are at much greater individual risk than heavier babies, but the relationship is continuous in the region of the conventional cut-point. Moreover, a histogram of total infant deaths by weight indicates that there are two modes in the distribution (Figure 2.2). One of these consists of relatively rare low-birth-weight babies, but another arises from much more frequent babies in the normal-weight region who are at lower individual risk, but in population aggregate terms, amount to a not-inconsiderable fraction of the overall public health problem of infant mortality.

In an earlier work (Hattis, 1998), the author calculated the hypothetical effect of a toxicant or other influence that might marginally change population birth weight distributions on the assumption that the relationship between birth weight and infant mortality might reflect some underlying causal processes, such as the state of development of the baby in preparing to guard itself from the dangers in the outside world. The results of such calculations indicate that a hypothetical toxicant that, like cigarette smoke, changes birth weight distributions, would be expected to cause about equal numbers of excess infant deaths in the low-birth-weight and normal-birth-weight regions if the birth weight change does in fact reflect

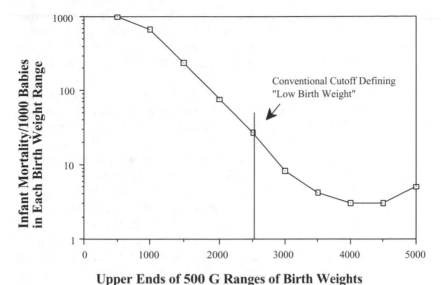

FIGURE 2.1. Relationship between weight at birth (in 500-g increments) and infant mortality. *Source*: Hattis (1998), based on data of Hogue et al. (1987).

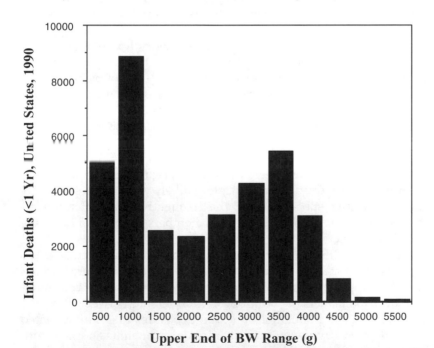

FIGURE 2.2. Distribution of birth weights of babies who died before 1 year of age. (Data from the National Center for Health Statistics (1990).

influences on the underlying causal processes that increase infant mortality (Figure 2.3). In this way, it can be seen that exclusive use of categorical representations of variability can lead to an incomplete view of the nature of a public health problem and the potential rewards from protective interventions.

IMPLICATIONS FOR THE INFORMATION USED TO SET POLICY. As indicated in part earlier, drawing observations mainly or exclusively from normal healthy adults tends to deprive risk assessors of direct observational information on the possible sensitivities of those who might be considered "special" in various ways. Whether this occurs as a result of conventional past practices in animal toxicology, human pharmaceutical testing on normal healthy adults, or occupational health studies limited to putatively healthy human workers, the result is that the gap between the size and diversity of exposed populations and the limited subgroups studied must be bridged with assumptions. In some rare case, such as some workplace standards articulated by traditional industrial hygiene groups (ACGIH, 1998), there is a deliberate exclusion of consideration of possible responses in a minority of "hypersusceptible" workers. More usually, a generic rule-of-thumb "safety," "uncertainty," or "adjustment" factor (usually tenfold) is used to allow for possible differences in sensitivity between the populations studied and the more diverse and numerous groups that are potentially exposed.

However, in the traditional practice of regulatory toxicology, difficulties arise from the categorical conception of variability even in cases where some data are available on responses for a putative "susceptible subgroup" (e.g., sufferers from angina pectoris in the case of air standards for carbon monoxide; offspring of mothers who were exposed to relatively large amounts of methyl mercury in fish). Where even a modest set of such data is available, there is a tendency for regulatory toxicologists to drop the tenfold adjustment factor entirely, thinking that "the" interindividual variability is effectively captured in direct observations of typical members of the subgroup. This implicitly neglects likely variation in sensitivity within the "sensitive subgroup" and the frequently great difference between the number of sensitive subgroup members in the study and the numbers in the general exposed population.

A very similar difficulty is manifest for exposure variability as it has come to be represented in recent times by "Margin Of Exposure" (MOE) analyses. This approach has been strongly promoted as a way of communicating to decision makers the multiplicative difference between exposure levels seen in human populations and the exposures that are associated with "no observed adverse effects" in animal or sometimes human studies. The general problem is that summarizing exposures of a defined group with a single number inherently obscures the distributional nature of the exposures – both the real variation in exposures from person to person

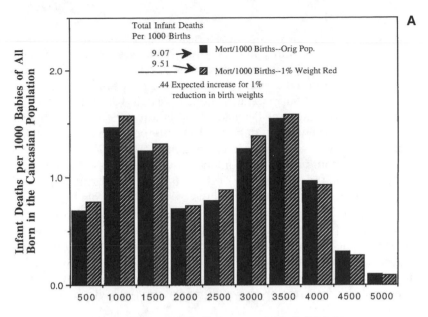

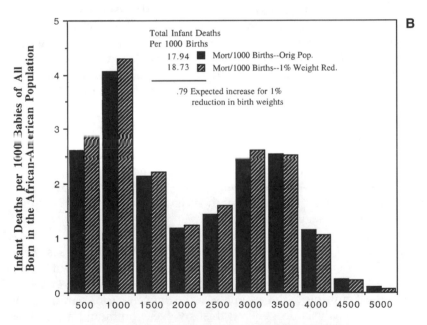

FIGURE 2.3. Expected effect of a 1% reduction in birth weights on the distribution of overall infant deaths per 1000 babies born in (A) Caucasian population; (B) African-American population. *Source*: Hattis (1998).

(and by other causally relevant dimensions such as time, space, and temperatures) and the uncertainty in exposures that arises from imprecision and imperfections in the available information.

The author encountered a recent practical example of this in reviewing a draft risk assessment support document on methyl bromide done by the California Department of Pesticide Regulation (1999). Variability in exposures was represented by defining 160 different scenarios in which different types of workers, community residents, or families returning to fumigated homes could come into contact with methyl bromide. Acute (1-day) exposure levels for calculation of the 160 Margin Of Exposure MOE estimates were drawn from measurements of a single sample in 59 cases; for 43 groups the higher of only two measured values was used; and in the remaining 58 cases the authors chose to use the highest of the more than two available measured values. There is no apparent recognition that the highest of N measured values (where N varies from 1 to about 10) does not represent a consistent percentile of an exposure distribution (even assuming that the data are appropriately representative, which is open to serious question in this case).

That such an analysis could be offered as serious guidance for risk management decision making is testimony to the power of the categorical representation of variability (among groups) to obscure the issue of quantitative variation within groups. It also has the effect of helping decision makers avoid facing hard quantitative policy questions of the form, "In setting standards for (compound W), which percentile (Y) of the variable distribution of true exposures should be kept below X (specific value of exposure level) with a confidence of Z (specific degree of confidence/uncertainty)?" It must be admitted that there has as yet been no overwhelmingly audible clamor from the risk management/decision-making community to face such questions (Hattis and Anderson, 1999). However, sometimes the role of responsible technical professionals must extend to helping the general community (including decision makers) face the realties that the best technical vision of the world can reveal, even if they do not comfortably fit pious fictions of "safety" that may have been embedded in popular culture, wishful thinking, or legislative mandates.

2.2.2. Current and Foreseeable Future "Best" Practices in the Assessment of Variability. The categorical representation of variability is part of a complex of obsolete (but still very widely used) practices that together might be called the "multiple point estimate/adjustment factor" paradigm of regulatory risk analysis. In general, this involves combining a series of point estimates of relevant quantities representing steps along causal pathways from the release of potentially hazardous agents to the production of biological responses [e.g., model-based point estimates of local air transport and deposition in relation to emissions, a soil ingestion rate, a fish

bioconcentration factor, a rate of fish consumption, a fraction of ingested material absorbed, a "typical" body weight for exposed humans, an observed "No Observable Adverse Effect Level" (NOAEL) from a short-term toxicity experiment in a small group of experimental animals, and usually tenfold adjustment or "uncertainty" factors to allow for (a) expected differences for the NOAELs that would be found in chronic versus short-term experiments, (b) expected differences to correct for the difference in dose between "Lowest Observable Adverse Effect Levels" (LOAELs) and NOAELs, (c) expected differences in the doses producing equivalent effects in experimental animals and people, and (d) expected differences among people in the doses that would produce equivalent risks of effects of equivalent severity]. The art of selecting such combinations of point estimates involves combining some relatively "conservative" or "high end" point estimates (the "high end" is usually defined as somewhere between the 90th percentile and the maximum value expected in some defined population) and some "mid range" point estimates to achieve an end result that is appropriately "conservative" for the particular type of decisions that are to be made based on the risk analysis. Unfortunately, human beings, including human experts in various substantive fields related to risk analysis, have notorious difficulties in achieving accurate calibration of subjective probabilities even when dealing with only one type of uncertainty, uncomplicated by variability (Lichtenstein and Fischoff, 1977; Alpert and Raiffa, 1982). Demonstrable biases are particularly serious for the tails of distributions (e.g., the 95–99% extremes), just where the need for information relevant to regulatory decision making is greatest. In light of this, there seems good reason to doubt that the wide Earth contains anyone who can, without formal analytical tools, achieve good subjective calibration of the combined distributional results of a risk or exposure assessment based on as many as three materially contributing distributions.

Gradually there is a movement to replace these point estimates with distributions describing variability (NRC, 1994; EPA, 1995; OEHHA, 1996; California Environmental Protection Agency, 1993; Baird et al., 1996; Hattis, Banati, and Goble, 1999a; Hattis et al., 1999b; Roseberry and Burmaster, 1992) and/or uncertainty and to propagate these in probabilistic risk analyses (EPA, 1997; Baird et al., 1996; Hamey and Harris, 1999; Hattis and Barlow, 1996; Hattis and Froines, 1992; Simon, 1999; Bogen, 1995; Bogen, Conrado, and Robison, 1997) – although, as of yet, the primary use of such distributional analyses has been as a means of assessing and discussing the degree of "conservatism" built into existing formulae based on the point estimates. Progress has been uneven in part because, although the lead U.S. agency in this field, the Environmental Protection Agency, has been receptive to the distributional treatment of exposures [e.g., see the extensive treatment of exposure variability done in support of decision making for recent regulations on radon in drinking water (EPA,

1999; Barry and Brattin, 1998)], to date the EPA has by policy not generally attempted analyses of either quantitative variability or uncertainty related to its estimates of potency for either cancer or noncancer effects (EPA, 1997; EC/R Incorporated, 2000). [A rare exception to this could be cited in the case of some criteria for pollutant risk analyses based on human experimental and epidemiological data – see the ozone example reproduced in (Hattis and Anderson (1999).]

The dominant approach used in the current era to help build the distributional counterparts to the single-point estimates of traditional analyses is to gather together readily available data with some relation to the parameter of interest, and then to summarize them mathematically – either as a series of values at various percentiles, or as a fit to a particular parametric distribution (often lognormal). The orientation toward describing or summarizing data in these ways is relatively unproblematic for a wide range of statistically literate professionals (at least, as a first step). But by itself simple data description or summarization approaches often lead to difficulties when the resulting descriptions are applied in risk analyses:

- The readily available data are generally not from stratified random samples that are plausibly representative of the populations for which the assessor wishes to estimate risks. More usually, the data come from "convenience" or "haphazard" samples assembled for purposes very different from those of the risk analyst. Moreover, the primary data available to quantify variability typically do not contain associated information on other characteristics (e.g. age and gender) with which the variable parameters of interest may have dependencies. (In cases where such information is present, and the dependencies can be defined quantitatively, it may be possible to at least estimate the distribution that would be found in the population of interest.)
- The data usually are collected on a different time scale than the variability of interest for the risk analysis. For example, the researchers may have made measurements of the consumption of a particular food over 1- or 3-day periods, when the toxic effect being modeled is chronic – depending on the buildup of the toxicant or accumulated hazard-related events (e.g., DNA mutations) that occur over time scales of months or years. To produce estimates of variability for relevant time scales, the analyst needs to assess of the degree of within-person correlation of the values of the parameter that would be observed in different time periods. Techniques for this are not yet well worked out, and future practice in the field would greatly benefit from a series of practical examples.
- The data often are not accompanied by the results of systematic efforts to assess measurement errors. This limits the ability of the analyst to separate the degree of spread in the data attributable to this source of uncertainty from true underlying variability.

- The primary data themselves are often not easily available in detailed form but are published in the form of summary statistics (e.g., means and standard deviations). The customs of different fields on the release of primary data differ – with experimental biologists, for example, being generally willing to release primary data but epidemiologists, by and large, are not willing to release primary data, with the notable exception of epidemiologists working directly in governmental agencies.

These difficulties lead naturally to a set of prescriptions for the near-term (next 10–20 years) development of variability analyses, as indicated in the third rows of Tables 2.1 and 2.2.

- *Salience* – Defining variability (and uncertainty) distributions have to be seen by both analysts and managers as an integral part of risk analyses, woven into the routine questions asked at each quantitative step in the study. They cannot be, as is frequently the case at present, an add-on undertaken when 80% of the effort to do an assessment has already been expended. Rather than being seen by hard-pressed technical specialists and managers as "extra work," fairly assessing variability (and uncertainty in the estimated variability) must be understood as an essential part of clear thinking about the technical facts that are relevant for risk management decision making.
- *Mathematical Representation and Analysis* – Because risk projections frequently need to be made beyond the range of available direct observations, it is vital that analysts leverage the predictive power of the available data by evaluating the data in the context of reasonably plausible mechanistic theories about the processes underlying observable variability and causal dependencies. The mechanistic bases of different distributional choices has been discussed at some length in earlier works. A further step that needs to be taken is to join biologists' interest in categorical distinctions (e.g., between sexes, among different alleles at specific genetic loci, and among people with different kinds of pathologies) with mathematical tools capable of representing both these types of dependencies and residual individual variability separate from measurement errors. This will often result in mixed distributions (e.g., two or more lognormal distributions describing different states of some important categorical possibilities). Where information on the numbers of individuals in the tails of distributions is critical to risk management decision making, it is particularly critical that analysis not artificially truncate distributions unless there are strong mechanistic reasons to do so (e.g., solubility limits for concentrations).
- *Technical Aspects (Sample Design and Data Collection)* – Ultimately, for key variables that drive important numbers or types of risk assessments, there is no substitute for deliberate programs of new data collection. These should use modern techniques of sample design including

stratification by suspected interacting variables and oversampling of subpopulations of particular interest for risk assessment. Some such programs are already under way in the areas of exposure assessment (Clayton et al., 1999) and some basic biomedical parameters (National Center for Health Statistics, 1988–94). The cited exposure assessment study (Clayton et al., 1999) seems to be taking the important step of collecting longitudinal data on the same individuals. This necessarily reduces the raw number of people that can be studied, but repeated sampling of the same individuals is vital so that information can be obtained on within-individual correlations in parameter values over time scales that are relevant for the assessment of subchronic and chronic exposures and effects. Similar repeated-sampling studies need to be undertaken for ingestion and other exposures and for biomedical parameters of interest. Studies are also needed of variability in exposures over various periods shorter than a day to assess acute toxic risks (Spicer et al., 1996).

- *Implications for Information Inputs to Risk Management Decision Making* – Increasingly cases such as the effects of airborne particles (EPA, 1996b; Samet, Zeger, and Berhane, 1995; Schwartz, 1994; Schwartz and Dockery, 1992; Dockery, Schwartz, and Spengler, 1992; Dockery et al., 1993; Pope et al., 1993) ozone (EPA, Office of Air Quality Planning and Standards 1996a), and lead (Schwartz, 1996) are forcing the recognition that even for noncancer effects, some finite rates of adverse effects are likely to persist after implementation of reasonably feasible measures to control different exposures. Societal reverence for life and health means "doing the very best we can" (Hattis and Anderson, 1999) with available resources to reduce these effects. And just as choices in the use of finite resources need to be made in health care, responsible risk managers will wish to be as fully informed as possible about the likely consequences of alternative protective policies. Moreover, the wider public constituencies of risk managers will wish to see to it that the managers face the choices squarely and transparently, and with the best technical insights into those consequences that can reasonably be produced. This means that information on quantitative distributions of variability and uncertainty – estimates of how many people are likely to experience what degree of risk (for effects of specific degrees of severity) with what degree of confidence – will need to be effectively produced and considered in cases where highly resource-intensive protective measures are among the policy options. (Decisions in cases where less resource-intensive measures seem likely to be effective, of course, may be responsibly made based on more limited information, but uncertainty and variability considerations should still play a role commensurate with their potential to improve decision making without unduly usurping the resources that would otherwise be available for primary preventive efforts).

3. DRIVERS OF CHANGE

In retrospect, it is perhaps not surprising that the intellectual descendants of Newton (the physical scientists and engineers who are primarily involved in exposure assessment) should have made more rapid progress in the adoption of continuous probabilistic understanding of variability than the intellectual descendants of Linnaeus and Mendel in the experimental biological sciences. This section and the next will briefly survey some of the social forces that seem likely to be involved in bringing about both the recent past and near-term future development of more sophisticated analyses of quantitative variability.

3.1. The Revolution in Information Processing and Storage Capability

It is hard to exaggerate the changes that the last 50 years have seen in computing or the implications of these changes for the practical ability of researchers of all kinds to handle mathematical analyses of variability. This story is not just ever-increasingly capable hardware and software but the somewhat lagged development of human "wetware" – the know-how to use the new tools.

In 1954, when Lehman and Fitzhugh proposed their "100-fold Safety Factor" for arriving at "acceptable daily intakes" from animal toxicology experiments, a computer was an assemblage of tons of vacuum tubes that required at least its own room for housing. The "probit" analysis procedure that was devised somewhat earlier to analyze dose response data, incorporating the idea of a lognormal distribution of thresholds for toxic responses (Finney, 1971), was reportedly designed to avoid negative numbers (by adding 5 to what we would call today a "z-score") because the mechanical calculators of the day could not handle negative numbers. As late as the early 1970s, when the author was completing graduate study in genetics, it was not at all unusual to obtain a Ph.D. in one of the more mathematically oriented biological sciences without ever having a direct personal interaction with a computer (although programming was offered as an option to substitute for the foreign language requirement). Today, of course, it is not at all unusual for reasonably bright second graders to have a working knowledge of the internet – building on practical familiarity with computers accumulated since they were 3. It seems likely that the coming generations of biological researchers will contain a larger proportion of people who are accustomed to thinking of work with large computerized databases as among the things they like to do and are particularly good at.

As it happens, large computerized databases of various kinds are exactly what seems likely to characterize an important part of the coming age of biological research. Microscopes and test tubes are not going to go out of use any time soon, but extracting biologically (and commercially) interesting insights from massive assemblages of data such as those from the Human Genome Project are an increasingly visible part of the action. Knowledge of gene sequences is clearly just a start. Understanding the system will require biologists to first elucidate the functions of the various genes. Then the technical community must build understanding of how these elements of the system are expressed, interact and regulate each other to both maintain homeostatic equilibria, and give the organism the capability to respond appropriately to the opportunities and dangers posed by a changing environment. "Gene chip" technology, which allows simultaneous measurements to be made of the state of expression of thousands of genes in specific tissues will undoubtedly be an important tool for sorting things out, but only with the aid of major computerized data analysis efforts. In this context, important parts of toxicology may come to resemble astronomy or particle physics, in which a few days or weeks of data collection are followed by months or years of analysis, data mining, integrative theoretical model building, and juxtaposition of theory and experimental results.

At a slightly less cosmic level, the revolution in information processing has made possible new kinds of biological observations and data gathering that could hardly have been imagined 20 or 30 years ago. An important example of this is the recent practice of performing 24-hour monitoring of the electrical activity of the heart as human patients go about their normal daily routine. It turns out that the variability in heart rates observed in this way is an important predictor of cardiovascular mortality (Huikuri, 1999). It is a particularly good predictor, apparently, in elderly people (Huikuri, 1998) for whom the predictive power of traditional risk factors declines relative to observations in people under 65. And of great biological interest is the observation that relatively *low* heart rate variability seems to be predictive of increased cardiovascular mortality, not high variability. One possible explanation is that in people with underlying deterioration in function, some physiological parameters are driven closer to their limits – reducing the effective range of parameter values that are available to the organism as reserve capacity to adapt to fluctuations in both the external and internal environment. In this way, measurements of variability could provide clues to the functioning of homeostatic control systems. With simultaneous measurements of other types of parameters, analysts can stop to work out in quantitative detail how the various parts of the system regulate each other, and in fundamental mechanistic terms how environmental stressors, by changing the system in specific ways, change the frequency of rare events (infarctions, life-threatening arrythmias) that lead to serious morbidity and mortality. This line of research is already being pursued in the context of research into the effects of airborne particles and other pollutants (Pope et al., 1999; Peters et al., 2000).

3.2. Contrasting Pressures From Risk Management Considerations

Understanding the real variability in risks is vital for risk management evaluations of both economic efficiency and the equity (fairness) of different policy options. These can be conveniently characterized as demands from the political "right" and "left," respectively. Then some commentary is in order on the "arms race" effect toward advancing the field that arises from the action of U.S. legal processes.

3.2.1. Demands from the "Right" – Accountability for Public Health Results Reasonably Commensurate with the Use of Available Resources.
The field of quantitative risk assessment in many ways traces its origin to the mid-1970s reaction to the environmental and occupational health protection laws passed in the early 1970s.[2] President Ford, as part of his Whip Inflation Now (WIN) program, issued an executive order (No. 11821) that major regulatory actions would need to be accompanied by an Inflation Impact Statement. In parallel with the 1970 National Environmental Policy Act requirement for Environmental Impact Statements, this called for an assessment of the likely economic costs of proposed regulations and an evaluation of whether these costs were justifiable in light of the amounts of expected health and environmental benefits. This was reinforced by a Supreme Court decision in the benzene case (Industrial Union Departments v. American Petroleum Institute, 1980), and similar requirements have been imposed by all subsequent presidents.

Now on its face, this was a fair question to be posed to the health/environmental science community. What do you expect to achieve for the societal efforts you wish to be expended in various ways? The difficulty was that the health/environmental scientists, both within regulatory agencies and without, had no generally applicable tools to produce even approximate quantitative responses that could be juxtaposed with the "quantitative" estimates of compliance costs and other economic consequences that the engineers and economists could produce.

Through the late 1970s and early 1980s federal agencies developed standardized procedures for making some relatively conservative point estimates of carcinogenic risks based on the multistage mechanistic theory of carcinogenesis (EPA, 1986; Crump, Hoel, and Peto, 1976; Albert, 1985), although recent EPA proposals (EPA, Science Advisory Board, 1999) depart somewhat from this orientation toward primary use of the multistage model as the basis for quantification. When combined with conservative procedures for estimating exposures, these approaches form a useful screening tool for relatively rapid assessment of whether specific exposures are likely to be of appreciable concern under a "significant risk/de minimis risk" framework for risk management policy making (Travis et al., 1987; Rosenthal, Gray, and Graham, 1992).

However, it has been noticed by economists and other advocates of cost-benefit considerations in regulatory decision making that this is not exactly what they had in mind for purposes of regulatory impact analysis (OMB, 1990). The intentionally "conservative" procedures designed to screen for potentially significant risks have been labeled as "biased" – and the suggestion has been made that massive misallocation of societal resources has resulted.

The underlying controversy is not really over numerical calculation techniques but over the risk management criteria that the calculations are intended to help implement (Hattis and Minkowitz, 1994, 1996; Hattis and Anderson, 1999). Most of the environmental health and safety legislation was passed primarily to address equity/fairness concerns, such as, "It is unfair for some people, by emitting pollutants into the air, water, land, workplace, or food, to surreptitiously and without informed consent impose risks on other people." This is not the same as the welfare economics concern for the reduction in overall societal allocative efficiency traceable to "externalities." Either by voluntary executive action or mandatory legislation, it appears that estimation of expected societal aggregate health benefits will become more important in regulatory decision making over the next few years.

What does this have to do with interindividual variability? It happens that in most cases both variability and uncertainty distributions are likely to be skewed – with a long tail to higher values (Hattis and Goble, 1991; Hattis and Burmaster, 1994; Hattis and Barlow, 1996). For example, the cancer risks for different individual people may well depend on the product of (a) the concentration of a substance in a specific environmental medium, (b) individual rates of intake of that medium (e.g., a particular food, drinking water, or air), (c) individual rates of metabolic activation of the substance to DNA reactive forms, and (d) the inverse of the rates of metabolic detoxification, systemic excretion, and/or DNA repair.

For a skewed distribution such as lognormal, the appropriate summary statistic for input into a cost-benefit analysis is the arithmetic mean or "expected value" – neither the peak (mode) of the distribution or the 50th percentile (Arrow and Lind, 1970). And in order to calculate the mean, it is vital to know the extent of the skewness (e.g., the geometric standard deviation for a lognormal distribution). The counterintuitive result of this is that if the existing estimates of the extent of variability and uncertainty in cancer potencies are correct, then mean "expected value" estimates of carcinogenic potency for genetically acting, metabolically activated carcinogens may not turn out to be very different in general from conventionally assessed EPA 95% upper confidence limit estimates (Hattis and Barlow, 1996). Of course, in addition to this use in quantifying mean risks, any equity-based analysis of the fairness of a risk distribution for an exposed population means facing a three-dimensional risk characterization – X level of risk, for the Yth percentile of the population (where Y is the variability dimension), with Z degree of confidence (where Z is the uncertainty dimension)(Hattis and Burmaster, 1994; Hattis and Minkowitz, 1994).

Interindividual variability is, of course, even more important for making quantitative estimates of noncancer risks. For traditional toxic effects mediated by individual threshold processes, the human population distribution of those thresholds becomes a central determinant of the numbers of people likely to be affected by any given exposure (Hattis and Silver, 1994). Recently published work (Hattis et al., 1999a) includes efforts to estimate mean "expected values" for such risks in the face of the major uncertainties in the extent of individual variability for particular hazards of this type.

3.2.2. Demands from the "Left" – Demands for Protection on Behalf of Especially Exposed or Especially At-Risk Individuals and Groups.

Where the "right" is seeking to increase the accountability of regulators to the goal of maximizing aggregate economic welfare, there are also voices for increasingly detailed consideration of individual risks from people who think they may be in more jeopardy than usual. Concerns for sensitive subgroups are particularly evident in the 1990 version of the Clean Air Act. The existence of a large amount of interindividual variability for a particular risk can directly raise equity/fairness concerns. This is particularly the case when there is an association between relatively high exposures (and/or risks) and membership in otherwise socially disadvantaged groups (Bullard and Wright, 1993). For example,

- Inner-city African-American children have been observed to have much higher blood lead levels, and possible associated risks of developmental impairment, than the general population (Agency for Toxic Substances and Disease Registry, 1993).
- Some ethnic groups (such as members of some Native American tribes) have substantial numbers of people who live by subsistence fishing and consume much larger amounts of fish than is assumed in standard risk assessment formulae for the general population (Peterson et al., 1994; Columbia River Inter-Tribal Fish Commission, 1994).
- There is concern that unusual dietary patterns and special susceptibility of children has been inadequately reflected in current risk assessment and management procedures (NRC, 1993).
- In the drug field, there is increasing recognition that historical exclusion of females and elderly people from early phases of drug testing (Lai, Fleck, and Caplan, 1988) may cause ultimate recommendations for use to be inadequately adapted to the needs of a general population.

Regardless of how the numbers may come out in specific cases, addressing these kinds of specific concerns may be helpful in recognizing that all these "special" groups have a valued place in a diverse society.

3.2.3. The "Arms Race" Effect of U.S. Legal Processes.

While the health protective mandates of many statutes strongly imply that agencies must consider variability in exposure and human susceptibility, there are no specific requirements for particular forms of quantitative variability analysis in U.S. law. What is generally required [e.g., by the Administrative Procedure Act (Public Law No. 79-404), for Federal regulatory decision makers] is that risk managers and judicial reviewers consider all reasonable alternative views of an issue that are presented to them and apply scientific concepts and analytical tools that are supported by "substantial evidence" in reaching their judgments. This does not mean that decision makers can usually be compelled to draw their analyses from the frontiers of scientific/technical understanding or even that they must adopt a relatively new view of the evidence that would be supported by a majority of practitioners in a relevant technical field. The law recognizes the value of some stability in legal/regulatory judgments. However, when a technical field can be said to have arrived at a clear collective judgment on a concept or a matter of fact, then decision makers use older concepts/approaches at some peril. And giving fair consideration of evidence presented necessitates more than a cursory review and rejection of novel ideas. In general, agencies will need to show that they have applied a level of technical sophistication in their review that is at least reasonably competitive in the marketplace of ideas with the groups offering other views. In the light of the history of ideas reviewed here, it seems reasonable to conclude that understanding and acceptance of the basic concepts of variability and its analysis have advanced to the point that major environmental/health protection agencies undertaking regulatory rule making need to have the capability to understand, analyze, and render risk management judgments in the light of quantitative variability information.

4. LIKELY FUTURE PROSPECTS

Despite the drivers of change reviewed in the previous section, there is also appreciable inertia in both the regulatory system and the scientific disciplines that provide the regulators with technical advice and support. The inertia (particularly at the working level in regional offices of regulatory agencies) resisting the adoption of probabilistic techniques in general has been discouraging for at least some important innovators in the field (D. Burmaster, personal communication). One of the difficulties in the existing system of using multiple "default" point estimates (some "conservative" high-end assumptions and some mid-range assumptions) for routine analyses is that both the technical and the risk management reasoning underlying the default choices is seldom transparent and documented for field workers. This makes it very difficult for working-level assessors to know how to evaluate and utilize the results of probabilistic analyses while keeping faith with overall agency policies to protect human health. The most important impediments to the adoption of more modern techniques for representing variability and uncertainty are that risk managers (in the executive, legislative, and judicial branches of government, in the private sector, and in the general public) have not been challenged with enough quantitative examples for them to see the need to set quantitatively evaluatable goals for risk management policies under different statutes.

The advance of science and technology for understanding and controlling the world changes the context within which both legal and moral judgments must be made. In 1950, if a child developed polio, it would be a tragic circumstance, but no one would usually have been held to be at fault, because at that time there was no ready way to prevent polio infection. Today, however, a new case of polio would rightly be the start of extensive remorse, and perhaps other consequences, because vaccines are readily available to prevent practically all new polio infections. In 1954, the Lehman and Fitzhugh paper proposing 100-fold safety factors was undoubtedly a considerable advance in systematizing existing subjective judgments of appropriately protective exposure levels of potentially hazardous agents. Risk managers should be ashamed, however, if they are still using this kind of mid-twentieth century risk assessment technology 20 years from now.

Notes

1. On the other hand, biased reporting or compilation of data can sometimes lead to the opposite type of error. For example, deletion of "outliers" can lead to estimated variability distributions with *less* variance than the true underlying distribution, if the deleted data points are not really errors in measurement or recording but true cases that happen to lie beyond some arbitrary statistical criterion used to define data points that are accepted as valid.
2. This subsection and the next are adapted from work that has been previously published (Hattis, 1996).

References

American Conference of Governmental Industrial Hygienists (ACGIH). 1998. *Threshold Limit Values for Chemical Substances and Physical Agents and Biological Exposure Indices.* American Conference of Governmental Industrial Hygienists, Cincinnati, OH.

Albert, R. E. 1985. "U.S. Environmental Protection Agency Revised Interim Guideline for the Health Assessment of Suspect Carcinogens" In D. G. Hoel, R. A. Merrill, and F. P. Perera, eds., *Risk Quantitation and Regulatory Policy*, Banbury Report No. 19, pp. 307–29. Cold Spring Harbor Laboratory.

Alpert, M., and Raiffa, H., 1982. A progress report on the training of probability assessors. In D. Kahneman, P. Slovic, and A. Tversky, eds., *Judgment Under*

Uncertainty, Heuristics and Biases, pp. 294–305. Cambridge University Press, New York.

Agency for Toxic Substances and Disease Registry. 1993. *The Nature and Extent of Lead Poisoning in Children in the United States: A Report to Congress*. U.S. Department of Health and Human Services, Atlanta.

Arrow, K. J., and Lind, R. C. 1970. Uncertainty and the evaluation of public investment decisions. *American Economic Review*, 60: 364–78.

Baird, S. J. S., Cohen, J. T., Graham, J. D., Shlyakhter, A. I., and Evans, J. S. 1996. Noncancer risk assessment: A probabilistic alternative to current practice. *Health and Ecological Risk Assessment*, 2: 79–102.

Barry, T. M., and Brattin, W. J. 1998. Distribution of radon-222 in community groundwater systems: Analysis of type I left-censored data with single censoring point, *Human and Ecological Risk Assessment*, 4: 579–603.

Bier, V. M. 1989. On the treatment of dependence in making decisions about risk. *Transactions of the 10th International Conference on Structural Mechanics in Reactor Technology*, Volume P (Probabilistic Safety Assessment), pp. 63–8. American Association for Structural Mechanics in Reactor Technology, Anaheim, CA, August 14–18.

Bogen, K. T. 1986. *Uncertainty in Environmental Health Risk Assessment: A Framework for Analysis and an Application to a Chronic Exposure Situation Involving a Chemical Carcinogen*. Doctoral Dissertation, University of California at Berkeley, School of Public Health, Berkeley, CA.

Bogen, K. T. 1990. *Uncertainty in Environmental Risk Assessment*. Garland Publishing, New York.

Bogen, K. T. 1995. Methods to approximate joint uncertainty and variability in risk. *Risk Analysis*, 15: 411–19.

Bogen, K. T., and Spear, R. C. 1987. Integrating uncertainty and interindividual variability in environmental risk assessment, *Risk Analysis*, 7: 427–36.

Bogen, C. L., Conrado, K. T., and W. L. Robison, W. L. 1997. Uncertainty and variability in updated estimates of potential dose and risk at a U.S. nuclear test site – Bikini Atoll, *Health Physics*, 73: 115–26.

Bullard, R. D., and Wright, B. H. 1993. Environmental justice for all: Community perspectives on health and research needs. *Toxicology and Industrial Health*, 9: 821–41.

Burmaster, D. E., and Thompson, K. M. 1999. Using animated probability plots to explore the suitability of mixture models with two component distributions, *Risk Analysis*, 19: 1185–92.

Burmaster, D. E., and Wilson, A. M. 2000. Fitting second-order finite mixture models to data with many censored values using maximum likelihood estimation, *Risk Analysis*, 20: 261–71.

California Environmental Protection Agency, Department of Pesticide Regulation. 1999. *Methyl Bromide – Risk Characterization Document for Inhalation Exposure*. Draft, Medical Toxicology, Worker Health and Safety, and Environmental Monitoring and Pest Management Branches, Department of Pesticide Regulation, California Environmental Protection Agency, Sacramento, California, October 15.

California Environmental Protection Agency. 1993. *CALTOX – A Multimedia Total Exposure Model for Hazardous-Waste Sites Technical Reports*. Part I, Executive

Summary, pp. 1–31, and Part II, The Dynamic Multimedia Transport and Transformation Model, pp. 1–95. Draft Final, December. Office of Scientific Affairs, Department of Toxic Substances Control, Sacramento, California (Later versions available on the web – http://www.cwo.com/~heroi/caltox.htm.)

Clayton, C. A., Pellizzari, E. D., Whitmore, R. W., Perritt, R. L., and Quackenboss, J. J. 1999. National human exposure assessment survey (NHEXAS): Distributions and associations of lead, arsenic and volatile organic compounds in EPA Region 5. *J Exposure Analysis and Environmental Epidemiology*, 9: 381–92.

Columbia River Inter-Tribal Fish Commission. 1994. *A Fish Consumption Survey of the Umatilla, Nez Perce, Yakama, and Warm Springs Tribes of the Columbia River Basin*, Technical Report 94-3, Columbia River Inter-Tribal Fish Commission, Portland, Oregon, October.

Crump, K., D. Hoel, D., and Peto, R. 1976. Fundamental carcinogenic processes and their implications for low dose risk assessment. *Cancer Research*, 36: 2973–9.

Cullen, A. C., and Frey, H. C. 1999. *Probabilistic Techniques in Exposure Assessment – A Handbook for Dealing with Variability and Uncertainty in Models and Inputs*. Plenum Press, New York.

Dockery, D.,W., Pope III, C. A., Xu, X., Spengler, J. D., Ware, J. H., Fay, M. E., Ferris Jr., B. G., and Speizer, F. E. 1993. An association between air pollution and mortality in six U.S. cities. *New England Journal of Medicine*, 329: 1753–9.

Dockery, D. W., Schwartz, J., and Spengler, J. D. 1992. Air pollution and daily mortality: Associations with particulates and acid aerosols. *Environmental Research*, 59: 362–73.

Dourson, M. L., and Stara, J. F. 1983. Regulatory history and experimental support of uncertainty (safety) factors. *Regulatory Toxicology and Pharmacology*, 3: 224–38.

EC/R Incorporated. 2000. *A Case Study Residual Risk Assessment for EPA's Science Advisory Board Review Secondary Lead Smelter Source Category, Volume I. Risk Characterization*. Report under EPA Contract No. 68-D6–0065, Prepared for the U.S. Environmental Protection Agency. Office of Air Quality Planning and Standards, Research Triangle Park, NC.

Finney, D. J. 1971. *Probit Analysis*. Cambridge University Press, Cambridge.

Gale, G. 1984. Science and the philosophers. *Nature*, 312: 491–5.

Griffin, S., Marcus, A., Schulz, T., and Walker, S. 1999. Calculating the interindividual geometric standard deviation for use in the integrated exposure uptake biokinetic model for fead in children. *Environmental Health Perspectives*, 107: 481–7.

Hamey, P. Y., and Harris, C. A. 1999. The variation of pesticide residues in fruits and vegetables and the associated assessment of risk. *Regulatory Toxicology and Pharmacology*, 30: S34–S41.

Hattis, D. 1986. The promise of molecular epidemiology for quantitative risk assessment. *Risk Analysis*, 6: 181–93.

Hattis, D. 1996. Human interindividual variability in susceptibility to toxic effects – From annoying detail to a central determinant of risk. *Toxicology*, 111: 5–14.

Hattis, D. 1998. Strategies for assessing human variability in susceptibility, and using variability to infer human risks. In D. A. Neumann and C. A. Kimmel, eds., *Human Variability in Response to Chemical Exposure: Measures, Modeling, and Risk Assessment*, pp. 27–57. CRC Press, Boca Raton, FL.

Hattis, D., and Anderson, E. 1999. What should be the implications of uncertainty, variability, and inherent "biases" / "conservatism" for risk management decision making?" *Risk Analysis*, 19: 95–107.

Hattis, D., and Barlow, K. 1996. Human interindividual variability in cancer risks – Technical and management challenges. *Health and Ecological Risk Assessment*, 2: 194–220.

Hattis, D., and Burmaster, D. E. 1994. Assessment of variability and uncertainty distributions for practical risk analyses. *Risk Analysis*, 14: 713–30.

Hattis, D., and Froines, J. 1992. Uncertainties in risk assessment. In Harvey J. Clewell, III, ed., *Conference on Chemical Risk Assessment in the DoD: Science, Policy, and Practice*, pp. 69–78. American Conference of Governmental Industrial Hygienists, Cincinnati.

Hattis, D., and Goble, R. 1991. Expected values for projected cancer risks from putative genetically-acting agents. *Risk Analysis*, 11: 359–63.

Hattis, D., and Goble, R. L. 1994. Current priority-setting methodology: Too little rationality or too much? In A. M. Finkel and D. Golding, eds., *Worst Things First? The Debate over Risk-Based National Environmental Priorities*, Chapter 7, pp. 107–31. Resources for the Future, Washington, DC.

Hattis, D., and Minkowitz, W. W. 1996. Risk evaluation: Criteria arising from legal traditions and experience with quantitative risk assessment in the United States. *Environmental Toxicology and Pharmacology*, 2: 103–9.

Hattis, D., and Minkowitz, B. 1997. Risk evaluation: Legal requirements, conceptual foundations, and practical experiences in the United States. Originally presented at the *International Workshop on Risk Evaluation* (June 20–21, 1994), Schloss Haegerloch, Germany. Discussion paper No. 93, ISBN 3-932013-16-6, Center of Technology Assessment in Baden-Wuerttemberg.

Hattis, D., and Richardson, B. 1980. *Noise, General Stress Response, and Cardiovascular Disease Processes: Review and Reassessment of Hypothesized Relationships*. MIT Center for Policy Alternatives CPA-80-02, June. Massachusetts Institute of Technology, Cambridge, MA.

Hattis, D., and Silver, K. 1994. Human interindividual variability – A major source of uncertainty in assessing risks for non-cancer health effects. *Risk Analysis*, 14: 421–31.

Hattis, D., Banati, P., and Goble, R. 1999a. Distributions of individual susceptibility among humans for toxic effects – For what fraction of which kinds of chemicals and effects does the traditional 10-fold factor provide how much protection? *Annals of the New York Academy of Sciences*, 895: 286–316.

Hattis, D., Banati, P., Goble, R., and Burmaster, D. 1999b. Human interindividual variability in parameters related to health risks. *Risk Analysis*, 19: 705–20.

Hattis, D., Erdreich, L., and Ballew, M. 1987. Human variability in susceptibility to toxic chemicals – A preliminary analysis of pharmacokinetic data from normal volunteers. *Risk Analysis*, 7: 415–26.

Hogue, C. J. R., Buehler, J. W., Strauss, M. A., and Smith, J. C. 1987. Overview of the national infant mortality surveillance (NIMS) project – Design, methods, results. *Public Health Reports*, 102: 126–38.

Huikuri, H. V., Makikallio, T., Airaksinen, J., Mitrani, R., Castellanos, A., and Myerburg, R. J. 1999. Measurement of heart rate variability: A clinical tool or a research toy? *Journal of the American College of Cardiology*, 34: 1878–83.

Huikuri, H. V., Makikallio, T. H., Airaksinen, K. E. J., Seppanen, T., Puukka, P., Haiha, I. J., and Sourander, L. B. 1998. Power-law relationship of heart rate variability as a predictor of mortality in the elderly. *Circulation*, 97: 2031–6.

Industrial Union Department v. American Petroleum Institute, 448 U.S. 607 (1980).

Kaplan, S. 1983. On a two-stage Bayesian procedure for determining failure rates from experimental data. *IEEE Transactions on Power Apparatus and Systems*, PAS-102: 195–202.

Kaplan, S., and Garrick, B. J. 1981. On the quantitative definition of risk. *Risk Analysis*, 1: 11–27.

Koerner, L. 2000. *Linnaeus – Nature and Nation*. Harvard University Press, Cambridge, MA. Reviewed by Schiebinger, L. 1999. Dreams of Arctic tea plantations. *Science*, 287: 1761.

Lai, A. A., Fleck, R. J., and Caplan, N. B. 1998. Clinical pharmacokinetics and the pharmacist's role in drug development and evaluation. In A. E. Cato, ed., *Clinical Drug Trials and Tribulations*, Chapter 5, pp. 79–98. Marcel Dekker, New York and Basel.

Lehman, A. J., and Fitzhugh, O. G. 1954. 100-fold margin of safety. *Association of Food and Drug Officials of the United States, Quarterly Bulletin*, 18: 33–5.

Lichtenstein, S., and Fischoff, B. 1977. Do those who know more also know more about how much they know? *Organizational Behavior and Human Performance*, 20: 159–83.

National Center for Health Statistics, Data Dissemination Branch, Third National Health and Nutrition Examination Survey (NHANES III), 1988–94. CD-ROM available from the National Technical Information Service, Computer Products Office, Springfield, VA. Also see http://www.cdc.gov/nchs.

National Center for Health Statistics. 1990. *Linked Birth and Death Records*. CD-ROM.

National Research Council (NRC). 1993. *Pesticides in the Diets of Infants and Children*. National Academy Press, Washington, DC.

National Research Council (NRC). 1994. *Science and Judgment in Risk Assessment*. National Academy Press, Washington, DC.

Office of Environmental Health Hazard Assessment (OEHHA). 1996. *Technical Support Document for Exposure Assessment and Stochastic Analysis*. Technical Review Draft, December. California Environmental Protection Agency, Berkeley, CA, Available on the web at www.calepa.cahwnet.gov/oehha.

Office of Management and Budget (OMB). 1990. Current regulatory issues in risk assessment and risk management. In *Regulatory Program of the United States Government, April 1, 1990–March 31, 1991*, pp. 13–26. Office of Management and Budget, Executive Office of the President of the United States, Washington, DC.

Ott, W. R. 1995. *Environmental Statistics and Data Analysis*. Lewis Publishers, Boca Raton, FL.

Peters, A., Liu, E., Verrier, R. L., Schwartz, J., Gold, D. R., Mittleman, M., Baliff, J., Oh, J. A., Allen, G., Monahan, K., and Dockery, D. W. 2000. Air pollution and incidence of cardiac arrhythmia. *Epidemiology*, 11: 11–17.

Peterson, D. E., Kanarek, M. S., Kuykendall, M. A., Diedrich, J. M., Anderson, H. A., Remington, P. L., and Sheffy, T. B. 1994. Fish consumption patterns and blood mercury levels in Wisconsin Chippewa Indians. *Archives of Environmental Health*, 49: 53–8.

Pope III, C. A., Thun, M. J., Namboodiri, M. M., Dockery, D. W., Evans, J. S., Speizer, F. E., and Heath Jr. C. W., 1995. Particulate air pollution as a predictor of mortality in a prospective study of U.S. adults. *American Journal of Respiratory and Critical Care Medicine*, 151: 669–74.

Pope, C. A., 3rd, Verrier, R. L., Lovett, E. G., Larson, A. C., Raizenne, M. E., Kanner, R. E., Schwartz, J., Villegas, G. M., Gold, D. R., and Dockery, D. W. 1999. Heart rate variability associated with particulate air pollution. *American Heart Journal*, 138: 890–9.

Public Law No. 79–404, Stat 237 (1946) codified as amended at 5 USC section 551–706 (1982).

Roeder, K. 1994. A graphical technique for detemining the number of components in a mixture of normals. *Journal of the American Statistical Association*, 89: 487–95.

Roseberry, A. M., and Burmaster, D. E. 1992. Lognormal distributions for water intake by children and adults. *Risk Analysis*, 12: 99–104.

Rosenthal, A., Gray, G. M., and Graham, J. D. 1992. Legislating acceptable cancer risk from exposure to toxic chemicals. *Ecology Law Quarterly*, 19: 269–362.

Samet, J. M., Zeger, S. L., and Berhane, K. 1995. The association of mortality and particulate air pollution. *Particulate Air Pollution and daily Mortality – Replication and Validation of Selected Studies – The Phase I Report of the Particle Epidemiology Evaluation Project*, August. Health Effects Institute, Cambridge, MA.

Schwartz, J. Air pollution and daily mortality. 1994. A review and meta-analysis. *Environmental Research*, 64: 36–52.

Schwartz, J. 1993. Beyond LOEL's, p values and vote counting: Methods for looking at the shapes and strengths of associations. *NeuroToxicology*, 14(2–3): 237–46.

Schwartz, J., and Dockery, D. W. 1992. Increased mortality in Philadelphia associated with daily air pollution concentrations. *American Review of Respiratory Diseases*, 145: 600–4.

Simon, T. W. 1999. Two-dimensional Monte Carlo simulation and beyond: A comparison of several probabilistic risk assessment methods applied to a Superfund site. *Human and Ecological Risk Assessment*, 5: 823–43.

Spicer, C. W., Buxton, B., Holdren, M. W., Smith, D. L., Kelly, T. J., Rust, S. W., Pate, A. D., Sverdrup, G. M., and Chuang, J. C. 1996. Variability of hazardous air pollutants in an urban area. *Atmospheric Environment*, 30: 3443–56.

Suppe, F. 1977. The search for philosophic understanding of scientific theories. In: F. Suppe, ed., *The Structure of Scientific Theories*, 2nd ed. University of Illinois Press, Champaign.

Thompson, K. M. 1999. Developing univariate distributions from data for risk analysis. *Human and Ecological Risk Assessment*, 5: 755–83.

Travis, C. C., Richter, S. A., Crouch, A. C., Wilson, R., and Klema, E. D. 1987. Cancer risk management: A review of 132 federal regulatory decisions. *Environment Science and Technology*, 21: 415–20.

U.S. Environmental Protection Agency (EPA). 1986. Guidelines for carcinogen risk assessment. *Federal Register*, 51: 33992.

U.S. Environmental Protection Agency (EPA). 1995. *Exposure Factors Handbook.* External Review Draft, Exposure Assessment Group, EPA/600/P-96/002B, Washington, DC. More recent versions available on the web at www.epa.gov.

U.S. Environmental Protection Agency (EPA). Office of Air Quality Planning and Standards. 1996a. *Review of National Ambient Air Quality Standards for Ozone: Assessment of Scientific and Technical Information.* OAQPS Staff Paper, U.S. EPA Technology Transfer Network Electronic Bulletin Board, EPA-452\R-96-007, June.

U.S. Environmental Protection Agency (EPA). 1996b. *Review of the National Ambient Air Quality Standards for Particulate Matter: Policy Assessment of Scientific and Technical Information, Office of Air Quality Planning and Standards.* EPA-452-R-96-013. U.S. EPA, Washington, DC.

U.S. Environmental Protection Agency (EPA), Risk Assessment Forum. 1997. *Guiding Principles for Monte Carlo Analysis.* EPA/630/R-97/001, March. This is available in HTML format on the EPA website–http://epa.gov/osp/spc/probpol. htm.

U.S. Environmental Protection Agency (EPA). 1999. Radon in drinking water, health risk reduction and cost analysis; Notice. *Federal Register,* 64: 9560–80.

U.S. Environmental Protection Agency (EPA), Science Advisory Board. 1999. *Review of Selected Sections of the Cancer Risk Assessment Guidelines by the SAB/SP Subcommittee on Cancer Guidelines.* SAB Review Draft of a report to EPA Administrator Carol Browner, 5/17/99.

Walden, A. T., and Guttorp, P., Eds. 1992. *Statistics in the Environmental & Earth Sciences.* Edward Arnold Publishers, London.

3

Mechanistic Considerations in the Harmonization of Dose-Response Methodology

The Role of Redundancy at Different Levels of Biological Organization

Lorenz R. Rhomberg

1. INTRODUCTION

There is a great deal of interest in developing more biologically insightful approaches to dose-response analysis, and also in achieving harmonization of the rationale and methods for analysis of cancer and noncancer endpoints (Beck et al., 1993; Conolly, 1995; Barton, Andersen, and Clewell, 1998; Conolly, Beck, and Goodman, 1999). These goals require one to think about how events at underlying levels of molecular and physiological activity are modulated by the action of a toxic agent, and how such low-level effects then propagate to produce the overt toxic effects seen in the living organism. In essence, we need to delve into how and why it is that varying levels of a toxicant lead to varying levels of response, dissecting the chains of cause-and-effect relationships involved. This paper explores some of these issues.

2. BASIC DOSE-RESPONSE APPROACHES

There are two broad approaches to analysis of dose-response relationships, and they correspond to two alternative rationales for why responses are observed to vary as a function of dose (Rees and Hattis, 1994). The first is the idea that a dose-response curve describes a tolerance distribution (Figure 3.1). Under this view, the population at risk consists of individuals who vary in the amount of an agent they can experience without ill effect (Krewski et al., 1999). In essence, each individual is assumed to have an exposure threshold, but the value of this threshold differs from one individual to the next (Eaton and Klaassen, 1996). This tolerance variation may be attributed to metabolic differences as well as to differences in the

This paper was produced under funding from the National Center for Environmental Assessment, U.S. Environmental Protection Agency, order number 0W-2612-NASX.

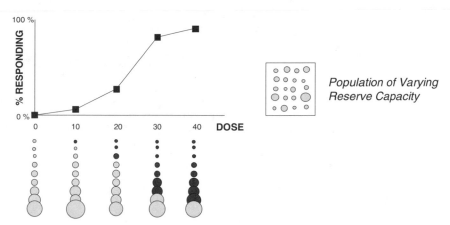

FIGURE 3.1. Tolerance distribution dose-response curve. The diameter of each circle indicates the tolerance of an individual. Higher exposures represent greater challenges to this tolerance, and as dose increases, an increasing fraction of the challenged population have their individual tolerances exceeded and so manifest the effect.

reserve capacity held by different individuals, an excess of whatever it may be that is diminished by the exposure, the exhaustion of which leads to adverse health effects (Rees and Hattis, 1994). At low rates of exposure, the tolerance of a relatively few individuals is exceeded, and so only in those few is a toxic response seen. At higher levels of exposure, the personal ability to tolerate the agent has been exceeded for a higher proportion of the population, and so more responses are seen. A sufficiently high dose exceeds the tolerance of all members of the population. Under this view, the dose response relationship exists due to preexisting underlying variation in reserve capacity or chemical processing efficiency that different individuals bring to ward off the toxic agent's effects.

Historically, noncancer risk assessment has used the tolerance distribution idea as its presumption about the underlying reason that higher doses yield higher response rates (Eaton and Klaassen, 1996). By making an assumption about the shape of the statistical distribution of tolerances, one can describe the dose-response curve as tracing out the associated cumulative distribution (Rees and Hattis, 1994; Holland and Sielken, 1993). For example, if the distribution is assumed to be lognormal, then the dose-response is described by a log probit curve. Since it is assumed that there is some dose that constitutes a population threshold (i.e., a dose below the individual tolerances of essentially all of the population), noncancer risk assessment has taken as its key goal the identification of a dose level believed to be near this population threshold. In practice, one finds either a No Observed Adverse Effect Level (NOAEL), or a dose corresponding to a

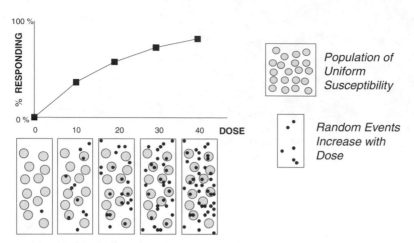

FIGURE 3.2. Stochastic event dose-response curve. The individuals are of inherently equal susceptibility in that they are all subject to random events that can cause transformation to a responding state. The frequency of such events increases with exposure, and so an increasing fraction of the population is expected to experience the chance events.

low percentile of sensitivity, the Benchmark Dose (BMD; Rees and Hattis, 1994; Crump, 1984), and then divides this by a series of uncertainty factors (Dourson and Stara, 1983).

The second major approach to dose-response analysis is the stochastic event rationale (Figure 3.2; Rees and Hattis, 1994; Krewski et al., 1999; Holland and Sielken, 1993; Olin et al., 1995). Under this view, the population at risk is essentially uniform, but each individual is subject to transformation from the healthy to the diseased state if one or more uncommon events happen. These events are assumed to be discrete changes that are engendered by the toxic substance at a rate that increases with exposure. At low levels of exposure, the transforming events are rare, and few individuals are so unlucky as to experience them. At higher exposure levels, the events become more common, and by chance a higher fraction of the population experiences the requisite random occurrences, leading to their response. At sufficiently high doses, essentially no individuals escape the transformation events because they have become too probable. Under this view, the reason a dose-response relationship exists is not because of variation in tolerance, but because the events happen stochastically, and the random chance of escaping them diminishes as they become more frequent. An individual who succumbs while his fellows do not does so owing to "bad luck" rather than a particular preexisting vulnerability.

Historically, this approach has been applied to the dose-response assessment of carcinogens on the grounds that normal cells are thought to be transformed to the malignant state by a series of rare somatic mutations or mutation-like events that bestow on the cells that bear them, and on their cellular descendents, the capacity to evade normal controls on cell multiplication, differentiation, and limitation by surrounding tissue (Rees and Hattis, 1994; Moolgavkar, Krewski, and Schwartz, 1999). The mathematical form of the dose-response curve depends on the particular equation chosen to represent the increasing probability of the requisite stochastic events as a function of dose (Rees and Hattis, 1994). In practice, most models in use make the assumption that the probability of transforming events may become low at low exposure levels, but it is never entirely extinguished, and any exposure, no matter how low, has some (perhaps exceedingly small) probability of effecting transformation.

The last 20 years of risk assessment practice has seen some refinement and elaboration of these two approaches, but their basic essences remain. The difference in their fundamental underlying rationales and assumptions has divided carcinogen risk assessment from analysis of noncancer endpoints (Rees and Hattis, 1994). In recent years, there has been considerable interest in harmonizing cancer and noncancer approaches, breaking down the barriers between the methods used in the two realms of quantitative risk assessment.

This interest has several motivations. First, there is an increasing understanding that carcinogenesis may sometimes arise as a secondary consequence of noncancer toxicity (Cohen and Ellwein, 1990; Butterworth et al., 1992; Butterworth, Conolly, and Morgan, 1995). If classical noncancer toxicity leads to a milieu within a tissue (because of rapid regenerative cell division following cytotoxicity, high oxidative stress, or other such effects) that induces transformation of cells that would not occur in the absence of such toxicity, then the dose-response curve for carcinogenesis should be largely governed by the dose-response for the noncancer toxicity. There are also agents that act directly on mitogens or that may affect the relative propensity of normal and spontaneously initiated cells to divide. In all these cases, the toxic agent does not act by directly affecting the mutation rates of cell transformation but rather by creating other toxicity or physiological perturbations that in turn alter the chances of transformation. Thus, models that assume a direct impact of the agent on transformation, and the lack of a threshold for such effects, may be inappropriate. Moreover, it would seem appropriate that the methods used to analyze noncancer toxicity in its own right be logically compatible with those used when the same toxicity is considered as a precursor to carcinogenesis. This line of thinking does not deny (or at least it should not deny) that malignancy results from discrete transforming events happening to individual cells, but it emphasizes that such events are only indirectly influenced by the agent as a consequence of its other actions.

So-called biologically based dose-response models for carcinogens attempt to incorporate such effects explicitly (Moolgavkar et al., 1999). In

form, they are elaborated stochastic event models in which the probabilities of transforming events are explicitly framed in terms of rates of
mutation and effects of the agent on clonal growth of intermediate cells
(those with some, but not all, of the events necessary for full transformation to malignancy). As currently formulated, these models make no al-
lowance for interindividual variability. They do, however, try to describe
the overall dose-response in terms of the consequences of specified cellular events that can be investigated biologically. They focus attention on
the dose-response relationship for these intermediate endpoints occurring
at the cellular level. For instance, the rates of mutation and the rates of
clonal expansion of intermediate cells must be specified as functions of
dose. In practice, this is done with equations that are as simple or simpler
than those traditionally used to describe overt whole-organism toxicity,
but they open the door to the use of insights into *underlying dose-response
relationships at lower levels of organization*, describing the generation of frank
toxic effects as quantitatively analyzed consequences of such precursors.
A second impetus toward harmonization of methods is to better describe
the "noncancer" parts of such biologically based models of carcinogenesis
and to make better use of empirical data on the variation of intermediate
endpoints with dose.

Indeed, there is increasing interest in using intermediate endpoints as
markers of the dose-response relationship. Such endpoints are often more
observable than the final toxicity of concern because the precursor effects
may be more common, may appear earlier, may be more easily diagnosed in
a larger population of subjects, and may be observable at lower dose levels.
Making use of such information requires being able to tie varying levels of
the precursor to the risks of the frankly toxic final effect and to extrapolate
this relationship to dose levels where the precursor can be observed but the
frank effect cannot (Swenberg et al., 1987). This endeavor would be aided
by more insight into how events at lower levels of organization propagate
their influences, ultimately causing overt toxicity.

The U.S. EPA's recent proposal to revise its carcinogen assessment guidelines (1996) includes a method for carcinogen assessment that bears many
similarities to that used for noncancer assessment. Rather than extrapolate
stochastic event models to low doses, they propose to use empirical curve-
fitting in the range of observation of tumors to describe the part of the
dose-response relationship that can be established empirically. A point on
this curve from the low end of the observable range is chosen as a point of
departure, conceptually similar to a benchmark dose or a NOAEL. A linear
extrapolation from this point can be undertaken, but if evidence suggests
that the low-dose behavior of the relationship should be expected to be
markedly nonlinear, then doses sufficiently below the point of departure
can be considered to be essentially without added cancer risk. (A difficulty
is defining how far below the point of departure one must go for confidence

that the risk is minimal.) Since this choice between linear extrapolation and a "margin of exposure" approach invokes the role of toxicity or of a necessary precursor endpoint that is thought to have threshold behavior, the choice of such a point of departure should be done in view of the dose-response relations of the intermediate endpoints thought to be critical. This process would be aided by the harmonization of methodology and a focus on understanding how intermediate or precursor endpoints relate to dose on the one hand and to the generation of cancer risk on the other.

All the issues discussed previously have in common that they seek to reconcile differing approaches to dose-response analysis through reliance on the notion that *intermediate* endpoints provide information about the nature of the toxic effects and the dose-response shape to be expected. They explicitly invoke the notion that dose-induced physiological effects at lower levels of biological organization can have consequences on overt toxicity at the whole organism level. To understand such toxicity, one wishes to understand both how the intermediate precursor response varies as a function of dose and how the overt toxicity varies as a function of the precursor. Harmonization approaches seek to explain cancer and noncancer responses as consequences of similar kinds of underlying biological events, constructing a single conceptual model that consistently and logically accommodates the variety of modes of action that may apply.

These questions demand an understanding of how toxic effects operate at various levels of biological organization and how effects at one level come to have consequences for other levels. The balance of this chapter attempts to raise some issues and to investigate some underlying principles. It is based on an invited presentation I gave at the 2000 Society of Toxicology meeting in Philadelphia (Rhomberg, 2000).

I begin with a discussion of levels of biological organization. I then consider how the direct effects of a toxic chemical act at low levels of organization (i.e., through direct molecular interactions). Then, I discuss the concept of redundancy at the molecular level, arguing that the differentiation between the fate of individual molecules and of the biochemical function they carry out arises because molecular functions are carried out by many independently acting units. I offer this notion as a useful idea in understanding how events at low levels of organization come to affect (or not affect) events at higher levels. Next, I discuss how the concept of redundancy can be applied more generally to other levels of biological organization, noting the critical role of lack of redundancy at the genetic level. Finally, I attempt to draw some conclusions about how dose-response patterns at different levels of organization influence how the consequences of low-level events are propagated through chains of causation to yield the overall dose-response relationship for overt toxic effects at the whole-organism level.

3. LEVELS OF BIOLOGICAL ORGANIZATION

I will consider toxicity to be the causation of ill health by exposure to a chemical substance. Health, and hence ill health, are properties of the whole living organism, and so toxicity manifests itself at this highest level of biological organization. Moreover, health consists of maintenance of the body in a well-functioning and controlled state; consequently, ill health is characterized by loss of normal function and loss of the ability of the body to maintain its state within normal bounds.

Toxicity manifests itself at the whole-organism level, but it is initiated at the molecular level. Fundamentally, the sole *direct* effect of a toxicant on its target tissues is through interaction of its molecules with particular molecules of the affected tissues (Gregus and Klaassen, 1996). The propagation of the consequences of these interactions through higher levels of organization eventually results in the manifestation of ill health at the whole-organism level. Strictly speaking, the higher-level responses are not directly caused by the toxic chemical itself but rather as a consequence of certain effects the chemical has on a lower level. The effects at lower levels become the causes at higher ones, and this connection by chains of causation is how effects propagate from the immediate molecular interactions to the whole organism. In this propagation, there is a series of dose-response relationships at different levels, each intermediate outcome becoming the causative factor for responses at another level, ultimately leading to the whole-organism response that we observe in toxicity testing.

Thus, understanding and characterizing the overall dose-response relationship through characterization of underlying mechanisms (or modes) of action entails tracing the ways in which effects propagate through the levels of organization, from the primary interactions with target molecules through the impacts on cells and tissues, until they manifest themselves as frank toxic effects. One must address not only the causal linkages but also the propagation of the *dose-dependence* of effects through the levels of the hierarchy.

Table 3.1 lists the usually recognized levels of biological organization along with the main features of function and control that operate primarily at that level. Beginning at the bottom, the genetic level provides coded instructions but, by itself, no means to carry them out. The molecular level comprises the components of the immediate biochemical machinery. The components of control processes include receptors, their ligands, and the machinery of gene expression. The material basis of epigenetic expression occurs at this level. Organelles spatially organize and separate the molecular components to achieve coordinated functions of biochemical action. Cells comprise the main self-sustaining units of organization, coordinating internal maintenance and control functions while providing a variety of specialized effector functions for the organism as a whole, including

TABLE 3.1. *Levels of Biological Organization and Their Control Functions*

Level	Key Contribution
Whole organism	Maintenance of healthy state
Organs / organ systems	Coordination, control, physiological response
Tissues / functional units	Physiological function
Cells	Cellular behavior
Organelles	Biochemical action
Molecules	Machinery, ligands / receptors / gene expression
Genes	Code

secretion, metabolism, transport, and structure. Cells are organized into tissues, but more importantly from our point of view, this organization is often in the form of repeated microscopic structures, each carrying out its small part of the collective function of the organ in which it is found. This is the level where such basic physiological functions as gas exchange and filtering of wastes are provided for the organism. Organ systems coordinate these functions to provide organized physiological systems and responses. At the top is the whole-organism level, which constitutes the ultimate level of coordination and control among parts, and at which loss of this coordination and control becomes manifest.

To reiterate, the direct effects of a chemical are at the molecular level, but the effects of ultimate interest (i.e., toxicity) manifest themselves at the whole-organism level. Accordingly, it is useful to consider some general rules and principles that apply to the dose-response properties of (1) the primary effect – the interaction of molecules of the toxicant with molecules of the organism – and (2) the propagation of dose-dependence of effects at the molecular level to higher levels, ultimately producing toxicity on the whole organism level.

4. MOLECULAR LEVEL

At the molecular level, it is important to distinguish among (1) events at the level of individual molecules and individual encounters between molecules, (2) collective properties of all molecules of a given type in a cell or a tissue, and (3) the suite of interacting molecular processes, with multiple steps, feedbacks, and controls, that go on within a cell or organelle. All three of these might legitimately be called "the molecular level," but each sublevel has its own properties that bear some consideration.

4.1. Individual Molecules and Their Interactions

With some exceptions, no matter how toxic a substance may be, the mere presence per se of its molecules in a tissue is of no consequence;[1] it is the interaction of these molecules with certain molecules belonging to the organism (the "target" molecules), and the consequences of those interactions, that lead to toxicity.[2]

Depending on the case, the interaction may be a chemical reaction (perhaps leading to covalent binding of the toxicant to the target) or a noncovalent, reversible binding (Gregus and Klaassen, 1996). In either case, at the level of individual molecules, the interactions are pairwise – one molecule of toxicant with one target molecule.[3] The interaction may be one of a reactive agent with a macromolecular target (e.g., a free radical with a lipid molecule or formation of a DNA adduct), an enzyme and its substrate (e.g., activation of benzo(a)pyrene to an epoxide by aryl hydrocarbon hydroxylase), or a receptor and a ligand (e.g., binding of nonylphenol to the estrogen receptor).

For an individual pair of molecules encountering one another, the interaction is all-or-none – they either bind/react or they do not. The outcome is quantal in the sense that only a few defined states are possible: the target molecule either (1) remains as it was before, (2) is effectively destroyed, or (3) becomes a new molecular entity (a bound complex or a reaction product) with a new set of fixed properties regarding its biological function (a different catalytic activity, a different affinity for a receptor, etc.). That is, the change in a single reacting target molecule is not graded as a function of the "strength" of the interaction, nor does it depend on the concentration of the reactants. The outcome of each pairwise encounter is independent of the fates of all the other molecules (reactants and targets). In this sense, the effect is unconnected to the concentration of the toxicant, although the frequency of encounters between target and toxicant molecules depends on their concentrations.

The changes in molecular state of the target molecules, through the propagation of their consequences for the organism's physiological control and functioning, ultimately lead to toxicity. In one important sense, the toxicant's *direct* effect on the organism begins and ends with this molecular outcome. The consequences for the organism's health are caused by those altered molecules (and the resulting chain of events that alter the organism's physiology) and not directly by the molecules of the toxic chemical per se.

4.2. Collective Properties of a Set of Molecules

This level of *individual* molecules needs to be contrasted to that of the *collective* properties of all of the molecules (of a given type) in a cell or a tissue. As noted, the frequency of pairwise interactions between molecules, and hence the rate of reaction, increases with their concentrations. The reaction rate is a collection of independent quantal transformation events with a per-encounter probability of happening, so the overall rate will follow a law of mass action. At low concentrations, it rises linearly with the concentrations of the reactants. At higher concentrations, the reaction rate begins to "saturate" (increasing less than proportionally to reactant concentrations because already-reacted molecules and those in the process of interacting are effectively withdrawn from the pool of available reactants). Essentially the same phenomenon underlies Michaelis-Menten kinetics of an enzyme catalyzing transformation of its substrate and the kinetics of receptors binding their ligands.

The important general principle is that, at this level of direct interaction of toxicant molecules with their molecular targets, no threshold is expected. Indeed, based on very general principles of reaction kinetics, at sufficiently low concentrations, the rate of reaction of toxicant and target is expected to be linear.

4.3. The Suite of Interacting Molecular Processes

Other factors, such as feedback inhibition, enzyme induction, and cascades of linked reactions, can induce nonlinearity, but these are properties of the interactions among a number of reactions, each of which obeys the simple principles just outlined. Such suites of interacting molecular processes can better be considered as part of the control and homeostatic processes occurring at the cellular level, as discussed later in this chapter.

To recapitulate, at the level of individual molecules of a toxic compound, the effect is all-or-none on individual target molecules and is independent of concentration. At the level of all of the molecules of a given type in a cell or tissue, the "dose-response relationship" consists of a change in the fraction of molecules in different distinct states, a change that should be proportional to concentration of the toxic agent in the tissue, acting without a threshold. Any grading of response consists of differences in the number of molecules in discrete active versus inactive or normal versus altered states.

The problem then becomes to understand how responses of this kind at the molecular level go on to become the causes of further processes at higher levels of organization, ultimately being manifested as overt toxicity. We need to consider not just the qualitative cause-and-effect links, but also the quantitative relationship through which the nonthreshold events at the molecular level become mapped into dose-response relationships of different types and shapes.

The key consideration is whether the impact of the molecular effects on the rest of the body is mediated through the changes in individual molecules or through the biochemical or physiological function that all the molecules of a given type collectively provide for the cell or tissue. In most cases, it is the latter – the functional role of the attacked molecules (be it catalytic, synthetic, signaling, or even structural) – that is critical to the maintenance of control and a healthy state at a higher level of organization.

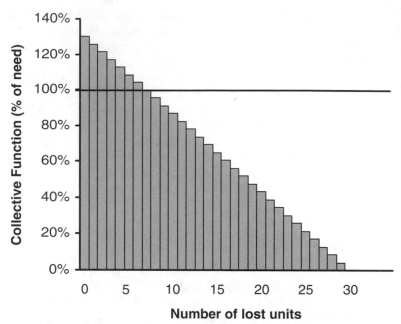

FIGURE 3.3. Effect of loss of units on collective function in a redundant system.

Each molecule contributes some small marginal amount to this function, but the fate of individual molecules is not important, only the sufficiency of the collective functionality of the whole complement of molecules present in the cell or tissue matters to the rest of the organism.

For illustration, consider an organophosphorus pesticide that covalently binds to acetylcholinesterase, resulting in complete inactivation of bound molecules. At the level of individual acetylcholinesterase molecules, each one is either bound and inactivated or unbound and fully enzymatically active. On the level of a neuronal synapse, however, the total amount of acetylcholinesterase activity is graded depending on how many active molecules remain. When the total number of molecules is large (and the marginal contribution of each individual molecule to the total enzymatic activity is small), the response is effectively continuous. What matters to the normal functioning of the synapse is the total amount of cholinesterase activity that can be mounted. If this is "enough" or "more than enough," the neurotransmitter is cleared from the synaptic junction quickly, but if the cholinesterase activity is inadequate to accomplish this clearance sufficiently rapidly, some degree of impairment of the normal patterns of impulse transmission occurs, and the impairment becomes worse to the degree that the cholinesterase activity falls farther below what is required.

Figure 3.3 diagrams this concept. For simplicity, a function fulfilled by thirty molecules is imagined, although in real cases, the number of molecules of a given type in a cell will usually be much greater. For each successive molecule that is lost via reaction with (and consequent inactivation by) the toxic chemical, the total amount of activity is decreased by one-thirtieth, being extinguished entirely once all thirty molecules are lost. If, when the full complement of thirty molecules is present, the total activity is 130% of the minimum needed, then (in this illustrative example) the loss of one or two molecules is hardly noticeable, the loss of seven is just tolerable, and the loss of more than seven leads to increasing degrees of inadequacy of the total activity, presumably leading to some degree of failure of the cell in which these molecules are contained to maintain itself or function properly (which failure has further consequences at still higher levels of organization, perhaps eventually leading to overt toxicity in the organism).

It is important to recognize that because the function is accomplished by many identical units (the individual molecules) operating identically, independently, and in parallel, it is possible for the state of the overall function (the catalytic activity) to be considered separately from the fate of individual units. Since each molecule contributes a small independent bit to the overall catalytic activity, the loss of only a few of them (even the complete loss as envisaged here) does not necessarily adversely affect the total. When the number of repeated functional units becomes very large (as it often is), the collective function becomes effectively continuous, and we may forget (as we often do) that additions or losses to the total function actually come in discrete amounts as a result of all-or-none addition or loss of the contributing functional units.

It is worthwhile to consider a contrasting hypothetical system with a low level of redundancy (i.e., with a very few repeated functional units comprising the whole source of physiological activity). Such a system is diagrammed in Figure 3.1. In this case, there are but three units (molecules) of the given type in the cell. Loss of even one of them leads to a one-third drop in total catalytic activity, leaving insufficient total activity to serve the needs of the cell, which presumably suffers toxicity as a consequence. In such a case, the fate of the function the molecules are carrying out remains tied to the fate of each molecule. This concept of the redundancy of a system is useful for understanding processes not only at the molecular level, but also at higher levels of organization. It is discussed more fully next.

So far we have taken a rather static view of the questions at hand, but the living system is dynamic and it changes over time, partially in response to changes imposed upon it. In particular, the loss of molecules through reaction with molecules of the toxic substance does not go without response.

The extent of true molecular repair – in the sense of restoration of individual molecules to their prereaction state – is actually rather small for most

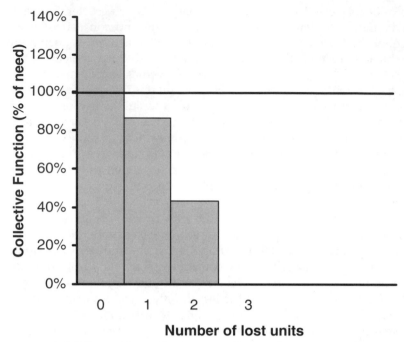

FIGURE 3.4. Effect of loss of units on collective function in a low-redundancy system.

systems. The major exception is DNA repair; there are elaborate, extensive, and efficient systems for undoing covalent reactions with the genetic material. There are also less extensive systems for repair of lipid molecules and a few cases of specialized repair systems for particular protein adducts, especially on hemoglobin. By and large, however, nature has answered the age-old question, "repair or replace?" with the second option (Gregus and Klaassen, 1996). Most of what we conventionally think of as repair is actually replacement of damaged units via de novo synthesis. Once again, the two concepts of redundancy and levels of biological organization are useful. "Repair" at one level usually consists of replacement of redundant functional units at a lower level of organization. Thus, a cell repairs itself by replacing losses to its complement of damaged molecules or organelles, and a tissue repairs itself by replacing lost cells by mitosis.

If losses can be replaced, then there are means to keep the total number of units at a sufficient level, for instance, to keep the number of molecules in the system illustrated in Figure 3.3 near thirty, even in the face of some rate of loss. This illustrates that what is key is not only the rate of generation of new damage (on the molecular level, of new reactions with target molecules) but also the rate of resynthesis of the lost units. In fact, the balance of damage and repair, or more properly, of destruction and replacement, dictates whether a sufficient collective functional capacity is maintained.

This has some important implications for thinking about dose-response analysis and its use in risk assessment. First, if damage *rate* and repair *rate* are so critical, it is evident that the element of time plays too little a role in much of our analysis of toxicological data. We dodge the issue by considering artificial exposure regimes in animal experiments construed so that the rate issue does not arise (e.g., single doses, doses spaced at regular intervals, but rarely dose-rate as an experimentally manipulated variable) by artificially dividing toxicity questions into "acute," "subchronic," and "chronic" realms, within which rate issues can be played down (Rozman, 2000) and by considering human exposures as a hypothetical continuous daily exposure, even if this is not the actual pattern. Such measures may be adequate for cases where exposures of interest really are without rate fluctuation or where repair is negligible and toxicity is a function of cumulative damage integrated over time (Hattis, 1986). But these artificial cases are assumed to apply much more widely than they truly do.

A second impact of the balance of damage and repair is on cross-species scaling. Much of our efforts to measure exposures and to allow for differences in humans and experimental animals in the uptake, distribution, and metabolism of compounds has been aimed at arriving at supposedly equal concentrations of the active agent in target tissues. That is, it is mostly oriented toward defining exposures that will lead to given damage rates, implying that if such damage rates were equalized, toxic response would be as well. In view of the fact that larger species (such as humans) have systematically slower rates of physiological processes, one would expect that the rates of resynthesis and replacement of damaged components would proceed at a slower rate. This point of view suggests that scaling of doses should not aim at equalizing tissue concentrations but rather at equalizing the balance of damage and repair rates. The prima facie basis for doing so is to scale dose rate in proportion to the general pace of physiological processes in differently sized organisms. Such a goal is a tenet of so-called weight-to-the-$\frac{3}{4}$-power scaling of doses. The logic of such scaling, usually applied to scaling of carcinogens, would seem to apply a fortiori to noncancer toxicity and threshold endpoints, where the maintenance of an adequate cushion of functional capacity in the face of its attack and replenishment is the central reason that certain dose levels are thought to be without adverse effect.

In the preceding discussion, I have spoken of "damage" and "repair" as the opposing processes, but these descriptions should not be taken too literally. It is also possible that the effect on target molecules is one of activation and the opposing process, one of restoration to an unactivated state.

5. PROPAGATION OF MOLECULAR-LEVEL EFFECTS TO HIGHER LEVELS

The outcomes at the molecular level have their effect on higher levels of organization through their impact on the functional activity that the target molecules carry out. That is, the outcome (response) at the lower level becomes the input or cause (in a sense, the "dose") for an effect at a higher level. In the example discussed earlier, the level of acetylcholinesterase activity in a synaptic junction is simultaneously the "response" to a dose-response relationship at the molecular level and a "dose" to a dose-response relationship at the synapse level. One could draw a dose-response relationship between acetylcholinesterase activity on the dose axis, and some measure of synapse function on the response axis. In turn, some measure of nervous system function could be thought of as a response to various degrees of synaptic impairment.

The point of this discussion is not to call for the complete description of organism-level dose-response relationships in terms of its component underlying processes – such descriptions will usually be beyond what can be achieved with readily obtainable information. Rather, the aim is to emphasize that the cascade of effects that follow from the presence of molecules of a toxic substance tend to comprise a chain of cause-and-effect connections in which failures of units or functional impairments at lower levels of organization come to cause changes at higher levels. These connections are quantitative and describable in terms of the relationship between changing levels of the outcome variable as a function of changing levels of the causal variable. Outcomes in turn become causes of further effects. Ultimately, effects on the health of the whole animal are recognized as toxic responses.

As at the molecular level, it is easiest to imagine loss of a critical function as the means of impact on the living system, but the argument really applies more generally to any quantitative function that is altered beyond the range of values compatible with normal operation, including functions in which a harmful *increase* is induced. It can also apply to alternations in responsiveness to signals or alteration of the integrity of homeostatic feedback mechanisms.

Recognizing the importance of certain key intermediate events and understanding how to relate measures of them to the ultimate degree of health impact are central to the strategy of finding commonality between cancer and noncancer risk assessment as well as for the process of using such intermediate events in illuminating the properties of the overall dose-response relationship for frankly toxic effects.

5.1. Redundancy

The concept of redundancy is very useful in thinking about how the events at one level of organization relate to effects on a higher level of organization. The central idea is that many systems are made up of a large number of essentially identical functional units or modules, each acting independently and in parallel. The total functionality of such a system comprises the collective contributions of the many modules, each one of which makes up a minor part of the total.

This concept was introduced earlier in the context of the many copies of a particular kind of molecule contained in a cell, but it can be generalized to other levels of organization. A cell's organelles comprise repeated modular units that carry out certain functions. A tissue contains many cells of a given type, each one independently carrying out some of the tissue's functions; the functioning of the tissue as a whole is not generally dependent on any specific individual cells but rather on the collective functioning of all of the cells of a given type. Many organs are comprised of repeated functional units, such as the alveoli of the lung, the acini of the liver, and the nephrons of the kidney.

To constitute such a redundant system, the modules must each be able to carry out the function (although the quantitative contribution of each to the total is marginal), and they must act in parallel. For example, the ribs can be thought of as constituting a moderately redundant system. The several ribs are similar although not identical, and they work in parallel (literally as well as figuratively) to provide certain structural properties to the thorax. A broken rib may diminish the rib cage's effectiveness, but the collective function of the remaining ribs still retains most of its function. In contrast, the vertebrae act in *series* (again, literally as well as figuratively). Even though they constitute a set of more or less similar units that operate in similar fashion, the failure of one leads to a broken spine for all.

Redundancy allows the collective action to be different from the individual fates of the components. The units can fail, but the function continues (Hayflick, 2000). Even if failure of units is all-or-none (inactivation of molecules, death of cells, atresia of oocytes), the function as a whole can be effectively continuous and graded rather than changing in large steps.

The maintenance of a certain level of overall function, even in the face of losses or reduced contributions from the individual units, depends on the balance of damage and repair. As noted in the discussion of the molecular level, "repair" is usually, in fact, replacement of units or modules, and the damage at a higher level of organization (impaired functionality) is repaired by the replenishment of the pool of functioning modules. Cells repair themselves by replacing damaged molecules or organelles, tissues repair themselves by replacing damaged cells, organs repair themselves by replacing damaged functional units such as acini or nephrons.

UNIVERSITY OF HERTFORDSHIRE LRC

Some kinds of damage are not readily repaired in this way, and toxic endpoints that involve such damage can often be thought of as a cumulative loss in function (Hattis, 1996). There are two chief reasons for lack of repair by replacement. First, the machinery for replacement of modular units may not be present. For example, there are no means for replacing damaged alveoli. Second, nonredundant aspects of the system that cannot be made up for simply by replacing modules may be damaged. For example, a crushed bone cannot be restored to its proper shape and structure through the reworking of its cellular and crystalline structure alone. In such a case, the damage is to a feature (the morphology) that is inherent to the higher level of organization, and not merely an emergent property of the collection of units from which it is made.

The issue of disrepair must also be kept in mind. Some toxic effects emerge through faulty or incomplete repair of damage, rather than from the damage itself (Gregus and Klaassen, 1996). Emphysema in lung and cirrhosis in liver result in part from scarring as a result of disrepair of acute damage, and mutations in proliferating cells attempting to repair cytotoxic damage to tissues may be a cause of carcinogenesis.

5.2. Redundancy at Different Levels of Organization

The level of redundancy varies from system to system, but for many levels of organization, it is quite high (Table 3.2). For most risk assessment applications, the individual is the unit of concern, and there is no redundancy on the whole-organism level.[4] At the level of organs and organ systems, there is rather low redundancy; most organs occur singly, and a few are in pairs. A loss, failure, or shortfall in function at this level will result in effects to the whole organism since the functional contribution of the organ cannot meet the whole organism's demand.

At the level of tissues and functional units within organs, the level of redundancy varies widely among systems. It should be clear that the issue here is not the number of distinct tissues within an organ (i.e., multiplicity,

TABLE 3.2. *Redundancy at Different Levels of Biological Organization*

Level of Biological Organization	Redundancy (No. of Modules)
Whole organism	1
Organs / organ systems	1–2
Tissues / functional units	1 to 10^8 per organ
Cells	100 to millions per unit
Organelles	a few to 100s per cell
Molecules	1000s to many thousands per cell
Genes	1 or 2 (+)

UNIVERSITY OF HERTFORDSHIRE LRC

TABLE 3.3. *Examples of Redundancy in Functional Units*

Functional Unit	Redundancy	Reference
Certain blood vessels; nerve fibers	1 to ~10	Guyton and Hall, 1996 Berne and Levy, 1988
Bronchi	~100 to 1000	Guyton and Hall, 1996 Rhoades and Tanner, 1995
Liver lobules; oocytes	~10,000 to 10^5	Guyton and Hall, 1996
Nephrons; alveoli	~10^6 to 10^8	Berne and Levy, 1988
Spermatozoa; erythrocytes	~10^9 to 10^{12}	Berne and Levy, 1988

or number of types) but redundancy in the sense of numbers of repeated independently functioning units or modules. At this level, the relevant focus is on structurally repeated functional units such as acini or nephrons. Some organs are not really subdivided into modular components, and so they have essentially no redundancy at this level.

Some of the diversity at this level is presented in Table 3.3. Such high-redundancy structures as alveoli and liver acini have been discussed. In cases where individual cells act as the functional units, the redundancy can be very high, with spermatozoa and erthyrocytes serving as examples. On the other hand, systems that are arrayed in the form of branching networks have rather low levels of redundancy despite having many similar repeated units. This is because the different branches serve different parts of the applicable organ, and failure of one branch can cut off an entire part of that organ. For example, the reason that blockage of small blood vessels serving the brain are capable of causing a stroke is that particular areas of the brain are served by only one or a very few vessels. Similarly, a particular muscle may be served by a nerve fiber consisting of rather few axons. The fact that there are other blood vessels or other nerve fibers elsewhere in the body provides no help for the local area that is deprived of its own branch. This kind of limit on the redundancy of branching networks becomes of most consequence when the tissues served have low redundancy themselves. Blockage of a minor airway may have little effect because the alveoli that are served are highly redundant (although all that are served by a given blocked airway will be cut off), but loss of a small area of the brain by cutting off its blood flow may destroy unique functional parts of the brain that have no backup.

Redundancy is quite high at the level of cells within a tissue or organ. Many secretory, metabolic, and biochemical functions are carried out at the cellular level, and the function for the organ (and indeed for the whole organism) is the collective activity of all the cells similarly engaged. The situations where redundancy is more moderate are where the organ is highly modularized into functional units, each of which is rather small. A good example is the kidney, where the individual nephrons – or more to the point, key structural and functional parts of nephrons – are small enough that they are made up of only moderately large numbers of cells. Remembering that the reason for examining redundancy is to assess the vulnerability of the whole resulting from failure of some of its component units, the question becomes whether there are cases where loss of a relatively few key individual cells causes the whole unit of which they are a part to fail as well. From this point of view, it is useful to think of the nephron as a fairly complex structure with subparts acting in series. Failure of any component (be it the glomerulus or the proximal convoluted tubule) will lead to failure of the nephron as a whole. So the question becomes how much cellular redundancy is in individual functional parts of a nephron, which begins to become a fairly small number.

At the level of organelles within cells, most cells have a few copies of their key organelles, and some cells may have hundreds of copies. The high degree of redundancy at the level of molecules within cells has already been discussed. Clearly, the number of molecules of a given type in a cell varies from one type to another, and the numbers of some, such as receptors, may be quite modest.

The lowest in the hierarchy of levels of biological organization considered here is that of the gene, and at this level the amount of redundancy becomes very low. While it is estimated that each cell has copies of several tens of thousands of genes, and while each cell has its own set, the number of copies of a given gene in a single cell is typically two (for autosomal genes), but often one (for X-linked genes and those that are maternally imprinted to inactivate one copy). A relatively few genes are duplicated, so that two or more essentially identical loci have two copies each. Some cells are polyploid and may have more copies, such as tetraploid or hexaploid hepatocytes. Even in the case of autosomal genes, the two copies may represent different alleles in some individuals, one of which may be nonfunctional or of limited function. Thus, on average, the redundancy at the level of particular genes within a cell is something less than two.

Aside from transcribed genes there is the issue of promoter and initiator regions of DNA to which molecular signals must bind to turn on transcription of downstream genes. For some signals, there may be several such elements in the genome, but they serve as the basis for allowing one signal molecule to affect transcription of several genetic loci, and the redundancy of such sites for control of transcription of particular genes is also low, usually one per gene copy.

It is this very low redundancy at the genetic level (and the lack of a still lower level from which to draw replacements) that accounts for the extraordinary efforts the living system undertakes to repair DNA and to avoid mutation to its precious few copies of each gene. Clearly, mutagenic activation of protooncogenes and genetic changes that can effect malignant transformation of cells is a major issue, but every genetic locus is susceptible to loss of its function through loss or inappropriate activation of its few copies. Many such changes will have little or no effect on a cell's viability and ongoing function, but some may kill the cell owing to loss of a critical maintenance or metabolic mechanism. It is because the cells themselves are highly redundant that moderate rates of such losses do not cause effects at higher levels of organization, although heavy losses can have effects.

The major toxicological concern is for that small fraction of genetic changes that leads to carcinogenic transformation of the cell. Because such a change does not induce loss of the cell but rather uncontrolled clonal expansion, a change at the level of the genes of a single cell (at a point of low redundancy) can be amplified to affect the health of the whole organism. In a sense, carcinogenesis reverses the protection afforded by redundancy at the level of cells because many cells make for many opportunities for rare events in some single one among them to occur – events that happen at a level not protected by redundancy. Because a transformed cell is not affected by the collective properties of its companion normal cells or by normal functions at higher levels of organization, the phenomena of propagation of causes and effects to higher levels (and the protection afforded by excess capacity and redundancy of units along the way) is essentially bypassed.

5.3. Propagation of Effects

I have expressed the propagation of consequences as though it were predominantly a linear chain of cause and effect working its way up the hierarchy of levels of biological organization. Clearly, this is a simplification that, untempered, leads to an inappropriately reductionist view of toxicity as merely the whole-organism consequences of molecular-level interactions.

First, it is clear that consequences can propagate up *and down* through levels of organization. As I write this sentence on my computer, my whole-organism desire to press the appropriate keys leads to effects on a host of nerve and muscle cells that act together to carry out the manipulation, and these cells act by polarization changes in membranes, emptying of neurotransmitter vesicles into synapses, contraction of myofibrils, and other subcellular events. These in turn are effected by opening and closing ion channels, pumping potassium and sodium ions, binding of calcium troponin and tropomysin, and so on. Somewhere in my body there is an individual ATP molecule in an individual muscle cell whose hydrolysis by a particular ATPase site on an individual myosin molecule (i.e., a pairwise interaction of two particular individual molecules) is caused by my wish to type the period at the end of this sentence. (Moreover, my communication of this idea to you, the reader, leads to propagation of effects down to the level of individual molecules in *your* body.) Whether the molecular events ultimately cause the typing or the typing the molecular events is a matter of philosophy.

The propagation of effects is actually a web of causative interactions (although it is frequently possible to recognize the key events as forming chains). Nonetheless, each link in the web can be understood as a functional relationship, and the questions remain to understand how redundancy and sufficiency of functionality influence the shape of the effect curve as a consequence of different magnitudes of the cause, and how those effects go on to become causes of other toxicologically relevant processes.

Looking at the system globally, under the stress imposed by a toxicant, the key to the overall organismal response will be the "weak" links in the web of interactions. The factors that tend to make a link "weak" are:

- low redundancy, leading to the potential for large consequences if some modular units are lost or severely compromised in function;
- low excess capacity, leading to vulnerability of falling below the required level of collective functioning when stressed;
- lack of protection from propagation of effects that act as inputs (i.e., no insulation from stresses because earlier processes in the chain have abundant excess capacity);
- critical consequences that, if passed on, lead subsequent processes to fail.

Death, the ultimate endpoint, illustrates the notion of weak links and low redundancy. Although we think of death as all-or-none and having a definite time of occurrence, these properties really only accrue to a very few low-redundancy, high-level life processes: heart and brain function. The proximate cause of death (whatever the ultimate cause) is often heart failure due to ischemia of the myocardium (Nuland, 1994). We define death in terms of cessation of heartbeat because we have but one heart, because its cessation of function is discrete and detectable, and because the rest of the body is highly dependent on its function, with the consequences of its failure proliferating rapidly and extensively. The lack of redundancy of the heart means that, once it fails, all the things it functionally supports (i.e., perfusion of all the body's tissues) fail as well. In fact, however, the deaths of the various tissues happen much more gradually, in an order depending on their sensitivity to lack of oxygen (an effect upon which organ transplantation depends). Within each tissue the point of death is ill-defined; the redundancy among the many cells leads to gradual loss of tissue function as the numbers of viable cells decline.

6. CONCLUSIONS AND PRINCIPLES

It is now time to try to draw some conclusions and identify some useful principles that arise from the twin concepts of propagation of effects across levels of organization and redundancy. The following are conclusions that I think follow from the arguments I set out in this chapter:

1. The only direct action of a toxic substance is at the molecular level. Molecular events have consequences, however, and it is through the propagation of these consequences rising to higher levels of organization that overt toxicity arises at the level of the whole organism. This propagation can be thought of as a series of dose-response relations, with the effects at one level serving as the causes for phenomena at a higher level.

2. At the level of rates of interaction among molecules, thresholds are not expected. Indeed, linearity is expected at low to moderate exposure levels. A central question, then, is whether the effects at higher levels depend on the state of individual molecules or on the collective function of the set of all molecules of a given type. Typically, it is the latter, but situations exist (such as mutation) where the state of individual molecules matters.

3. The high level of redundancy at the molecular level allows the collective function to continue in the face of loss of individual molecules. Similarly, it is the high level of redundancy at the cellular level that allows tissues to function in the face of loss of some of their cells (and the molecules in those cells). If the collective function displays some excess capacity, then losses can be tolerated in the short run, and even in the long run if replacement of lost units keeps pace with the rate of impairment.

4. The concept of redundancy also applies to higher levels of organization. Whenever the needs from a higher level of organization consist of the collective function of the redundant units, and when there is some excess capacity, loss of some units can be tolerated.

5. Repair consists primarily of replacement of lost modules at a lower level of organization. (True repair is rare except in the case of DNA.) Accordingly, sustainable rates of ongoing damage do not get ahead of the process by which damaged modules (be they molecules or cells) are replaced. This fact should be informative for assessment of cross-species equivalence of chronic dose rates and for the assessment of doses that vary in delivery rate over time.

6. To get a threshold effect at a certain level of organization, there must be true excess capacity of function provided by a lower level. A true excess means that the function is completely sufficient to the need, that higher levels provide no benefit, and that lower levels (though still above the threshold) lead to no impairment. (If capacity

in excess of need still provides some added benefit, there might be nonlinearity, but no strict threshold. Similarly, if partial deficits lead to decreased function that is still within tolerable or normal range, then the effects of this decreased function might be felt as the consequence of decreased input to another process, which might not be as tolerant of the loss.)

7. Thresholds do *not* come from the lack of effects at low levels of organization when exposures are low. Generalized notions of "defenses" or "repair" do not lead to thresholds because they operate at the molecular level where thresholds are not expected. Repair contributes to thresholds only when it serves to maintain excess capacity in the face of losses, but the excess capacity itself is key.

8. A no-threshold effect at a lower level (e.g., molecular interactions) can lead to a threshold effect at a higher level if the lower level consists of a redundant system with excess capacity. On the other hand, a threshold effect at a lower level can prompt a no-threshold effect at a higher level if the many redundant units have different individual thresholds, and if exposures are great enough that some of them are exceeded. That is, the tolerance distribution model applies at the level of varying redundant units, and no effect is propagated to the higher level only when the population threshold among units at the lower level is not exceeded.

9. An effect at a lower level of organization is "adverse" only in the sense that it leads to a whole-organism adverse effect. Adversity is best judged at the level of overall health of the organism. Lower-level effects are not adverse in and of themselves; their consequences may be adverse.

10. In addition to the effect of a toxic agent, other agents, background processes, and spontaneous effects may act to tax excess capacity. The propagation of an effect depends only on the total functional capacity contributed by a redundant system, and not on the source of its impairment, so it is expected that these other effects will interact with the dose of an agent through their mutual impact on functional capacity. The question of whether two agents (or an agent and a spontaneous background process) act via a "common mechanism" can be regarded as a question about whether this kind of joint impact on a key function at an intermediate level of organization is at issue.

11. While many aspects of the living system have high redundancy, there are some key aspects that have moderate, or even low, redundancy. In particular, the genetic level has very low redundancy, and oxygenation of tissues is very dependent on the functioning of the single heart.

12. The low genetic redundancy makes cells vulnerable to mutations, since one or two molecular changes can affect all aspects of a cell's

life that stem from the affected genes. Most such changes will lead to loss of the cell through disruption of some vital maintenance function. This typically has little consequence owing to the redundancy at the level of cells. For mutations involved in malignant transformation, however, the effects on individual genes can cause the cells that bear them to grow into a clone that affects the whole organism. This constitutes the most prominent instance of effects on individual molecules having consequences for higher levels of organization. It arises because malignancy is not a collective effect of many cells, but a change of state in a single cell that is caused by changes in a system of low redundancy.

13. When carcinogenesis is secondary to some noncancer toxicity or physiological effect, the generation of a malignant cell still arises because of rare, molecular-level events, but the impact on induction of such events by the compound is only through the toxic effect, not directly. If the toxic effect has a threshold, so presumably does the induction of increases in the risk of cancer. To invoke this mechanism, then, one would want to investigate the dose-response of the causative toxicity and establish doses at which the effect does not occur, and hence has no secondary consequences on carcinogenesis.

14. Other toxicities aside from cancer may be vulnerable to low-redundancy systems. These include certain cardiovascular effects where single blood vessels serve critical tissues (such as cardiac tissue) that themselves have low redundancy. Neurotoxicity may also involve loss of a relatively few of a small number of nerve fibers serving specific tissues.

15. Rather than distinguishing cancer from noncancer risk assessment, it might be wiser to distinguish systems where the key events happen at low redundancy versus high-redundancy levels. In the former case, the possibility of stochastic effects looms larger, since loss of but a few component units can prompt a failure of the whole.

Implications for Assessment of Mixtures

Under the view espoused here, whether a low-level molecular or physiological effect propagates to affect whole-organism health depends on whether the functional contribution of the affected process (on which the organism as a whole depends) is eroded beyond the extent of its excess capacity. If several agents affect the same functional process, each contributing to some loss, then the loss of functionality could be cumulative. The capacity for mixtures to have toxic effects different from those predicted by their individual components can be gauged by examining their potential for having impacts on the same underlying functional process. Even if they erode function by different mechanisms, it is at the level of erosion of function that their interaction can become manifest. This being said, whether the effects on function will propagate to affect the health of the organism will still depend on the nature of the function affected, its degree of redundancy, its balance of damage and repair rates, and whether there is reserve functional capacity that insulates processes that are functionally and causally downstream from impacts. It is only by characterizing the interactions of compounds at the levels of biological organization where they jointly come to bear that one can establish the expected joint effect.

Implications for Risk Assessment and the Research Supporting It

In practice, actually untangling the chains of causation from the molecular level through the various cellular, physiological, and organismal consequences to the manifestation of frank toxicity will be difficult. The point of calling attention to the way these underlying processes interact is not to call for their explicit description as a means to conduct risk assessment (although this can be an ideal goal). Rather, the aim is to provide a framework for reasoning about the impacts to be expected from various observed effects at lower levels of biological organization when such effects are put forth as a possible basis for assessing the potential of exposures to chemicals to cause toxic effects.

For instance, the arguments presented previously make clear that one should not expect thresholds at the level of molecular interactions, but that the absence of such thresholds does not preclude the existence of thresholds at higher levels of organization. Moreover, if one is using an "intermediate endpoint" or "precursor event" as a basis for understanding dose-response relationships, then it is important to establish how the dose-response for the observed effect is related to the genesis of the frank toxic effect at the whole-organism level. Whether the impact is on a system with high or low redundancy is informative, and the consequences of failures of modules on the overall functional contribution of the system to the organism's health needs to be factored in.

The arguments presented here do suggest, however, that one needs to better understand modes of action of toxicants in order to define sensible approaches to dose-response characterization and in order to discern the dose-response properties (shapes, existence of thresholds, role of interindividual variability) that it is reasonable to expect a toxicant to display. The former division into carcinogens and noncarcinogens (or of carcinogens into genotoxic and nongenotoxic agents) is seen to be overly simple. There are some toxicities that, owing to the involvement of critical points with low redundancy, one might expect to evince stochastic influence even when genetic mutations are not involved. On the other hand, taking a threshold approach to carcinogens necessitates identifying the way in which an excess capacity of a system with adequate redundancy is involved in a critical way such that the low-redundancy genetic effects are effectively protected from impact.

Notes

1. Notable exceptions include simple asphyxiants that dilute the oxygen in inhaled air, detergents and similar agents that disrupt the integrity of cell membranes via solvent effects, and crystals that exert damage via physical injury or irritation.
2. We are concerned primarily with "pharmacodynamic" interactions – reactions of the proximate toxicant with its target molecules as the first event in toxicity – but the principles also apply to molecular interactions in pharmacokinetics, such as metabolic biotransformations and binding to serum proteins.
3. When several reactants are involved, the molecular mechanism proceeds in a series of pairwise steps as when, for example, a newly formed receptor-ligand complex itself binds with another such complex to form a dimer. Enzymes with multiple active sites or the cooperative binding of several oxygen molecules by hemoglobin, while involving more than two molecular actors, can also be seen as a series of pairwise interactions, with the complex formed in one step being one of the pair of reagents in the subsequent step.
4. For ecological risk assessment, however, one may focus on higher, supraindividual levels of organization, and the individual may be thought of as but one unit of a kinship group or a local population. Even a whole population may be thought of as but one unit in a whole guild of species fulfilling some ecological role in an ecosystem, and loss of a few such populations may be deemed tolerable as long as the ecological function, and hence the ecosystem, persists. The redundancy concept provides an avenue for harmonization not only of cancer and noncancer health assessment but also for these and ecological risk assessment.

References

Barton, H. A., Andersen, M. E., and Clewell III, H. J. 1998. Harmonization: Developing consistent guidelines for applying mode of action and dosimetry information to cancer and noncancer risk assessment. *Human and Ecological Risk Assessment*, 4: 75–115.

Beck, B. D., Conolly, R. B., Dourson, M. L., Guth, D., Hattis, D., Kimmel, C., and Lewis, S. C. 1993. Improvements in quantitative noncancer risk assessment. *Fundamental and Applied Toxicology*, 20: 1–14.

Berne, R. M., and Levy, M. N., eds. 1988. *Physiology*, 2nd ed. C. V. Mosby St. Louis.

Butterworth, B. E., Conolly, R. B., and Morgan, K. T. 1995. A strategy for establishing mode of action of chemical carcinogens as a guide for approaches to risk assessments. *Cancer Letters* 93: 129–46.

Butterworth, B. E., Popp, J. A., Conolly, R. B., and Goldsworthy, T. L. 1992. *Chemically Induced Cell Proliferation in Carcinogenesis*. IARC Scientific Publications No. 116, pp. 279–305. International Agency for Research on Cancer, Lyon, France.

Cohen, S. M., and Ellwein, L. B. 1990. Cell proliferation in carcinogenesis. *Science*, 249: 1007–11.

Conolly, R. B. 1995. Cancer, and non-cancer risk assessment: Not so different if you consider mechanisms. *Toxicology*, 102: 179–88.

Conolly, R. B., Beck, B. D., and Goodman, J. I. 1999. Stimulating research to improve the scientific basis of risk assessment. *Toxicological Sciences*, 49: 1–4.

Crump, K. S. 1984. A new method for determining allowable daily intakes. *Fundamentals of Applied Toxicology*, 4: 854–71.

Dourson, M., and Stara, J. F. 1983. Regulatory history and experimental support of uncertainty (safety) factors. *Regulatory Toxicology and Pharmacology*, 3: 224–38.

Eaton, D. L., and Klaassen, C. D. 1996. Principles of toxicology, In C. D. Klaassen, ed., *Casarett and Doull's toxicology: The basic science of poisons*, 5th ed., pp. 13–33. McGraw-Hill, New York.

Gregus, Z., and Klaassen, C. D. 1996. Mechanisms of toxicity. In C. D. Klaassen, ed., *Casarett and Doull's toxicology: The basic science of poisons*, 5th ed., pp. 35–74. McGraw-Hill, New York.

Guyton, A. C., and Hall, J. E., eds. 1996. *Textbook of Medical Physiology*, 9th ed. W. B. Saunders, Philadelphia.

Hattis, D. 1986. The promise of molecular epidemiology for quantitative risk assessment. *Risk Analysis*, 6: 181–93.

Hayflick, L. 2000. Hormesis, aging, and longevity determination. *BELLE Newsletter*, 9(3): 8–9.

Holland, C. D., and Sielken, Jr., R. L. 1993. *Quantitative Cancer Modeling and Risk Assessment*. PTR Prentice-Hall, Englewood Cliffs, NJ.

Krewski, D., Cardis, E., Zeise, L., and Feron, V. 1999. Empirical approaches to risk estimation and prediction. In S. Moolgavkar, D. Krewski, L. Zeise, E. Cardis, and H. Møller, eds., *Quantitative Estimation and Prediction of Human Cancer Risks*, IARC Scientific Publications No. 131, pp. 131–78. International Agency for Research on Cancer, Lyon, France.

Moolgavkar, S., Krewski, D., and Schwartz, M. 1999. Mechanisms of carcinogenesis and biologically-based models for estimation and prediction of risk. In S. Moolgavkar, D. Krewski, L. Zeise, E. Cardis, and H. Møller, eds., *Quantitative Estimation and Prediction of Human Cancer Risks*, IARC Scientific Publications No. 131, pp. 179–237. International Agency for Research on Cancer, Lyon, France.

Nuland, S. B. 1994. *How We Die: Reflections on Life's Final Chapter*. Knopf, New York.

Olin, S., Farland, W., Park, C., Rhomberg, L., Scheuplein, R., Starr, T., and Wilson, J. 1995. *Low-Dose Extrapolation of Cancer Risks: Issues and Perspectives*. ILSI Press, Washington, DC.

Rees, D. C., and Hattis, D. 1994. Developing quantitative strategies for animal to human extrapolation. In A. W. Hayes, ed. *Principles and Methods of Toxicology*, 3rd ed., pp. 275–315. Raven Press, New York.

Rhoades, R. A., and Tanner, G. A. 1995. *Medical Physiology*. Little Brown and Co., Boston.

Rhomberg, L. R. 2000. Mode of action as a guide to quantitative analytical approaches. *The Toxicologist*, 54: 194.

Rozman, K. K. 2000. The role of time in toxicology or Haber's c x t product. *Toxicology*, 149: 35–42.

Swenberg, J. A., Richardson, F. C., Boucheron, J. A., Deal, F. H., Belinsky, S. A., Charbonneau, M., and Short, B. G. 1987. High to low dose extrapolation: Critical determinants involved in the dose-response of carcinogenic substances. *Environmental Health Perspectives*, 76: 57–63.

U.S. Environmental Protection Agency (EPA). 1996. *Proposed Guidelines for Carcinogen Risk Assessment*. EPA/600/P-92/003C. U.S. EPA, Office of Research and Development, Washington, DC.

4

Risk of Extreme and Rare Events

Lessons from a Selection of Approaches

Vicki M. Bier, Scott Ferson, Yacov Y. Haimes,
James H. Lambert, and Mitchell J. Small

> GUILDENSTERN: We have been spinning coins together since I don't know
> when, and in all that time . . . I don't suppose either of us was more than
> a couple of gold pieces up or down. I hope that doesn't sound surprising
> because its very unsurprisingness is something I am trying to keep hold
> of. The equanimity of your average tosser of coins depends upon a law, or
> rather a tendency, or let us say a probability, or at any rate a mathematically
> calculable chance, which ensures that he will not upset himself by losing too
> much nor upset his opponent by winning too often. This made for a kind of
> harmony and a kind of confidence. It related the fortuitous and the ordained
> into a reassuring union which we recognized as nature. The sun came up
> about as often as it went down, in the long run, and a coin showed heads
> about as often as it showed tails. Then a messenger arrived. We had been
> sent for . . . Ninety-two coins spun consecutively have come down heads
> ninety-two consecutive times.
>
> Tom Stoppard
> *Rosencrantz and Guildenstern Are Dead*

1. INTRODUCTION

Extreme and rare events have captured our imagination. They have inspired fear, introspection, art, literature, religion, law, science, and engineering. Are they acts of God or acts of man; destined or random; to be
expected, designed for, and perhaps controlled, or rather ignored, left off

The authors are grateful to the members of the Society for Risk Analysis for discussions
leading to the current paper. Selected portions of this paper are based upon work supported
in part by the U.S. Army Research Laboratory and the U.S. Army Research Office under grant
number DAAD19-01-1-0502 to the University of Virginia. The authors are also grateful to the
Virginia Transportation Research Council and the Virginia Department of Transportation for
partial support.

of our "worry budgets," and responded to only if they occur? Few elements of risk assessment have as deep an implication for our perception and response to the world around us as do extreme events. If they yield a positive outcome, a joining of "the fortuitous and the ordained into a reassuring union" (Stoppard, 1967), we might call them natural (the blessings of nature), or perhaps even miraculous. If the results are negative, of low probability and high consequence, we might ascribe the event to a freak of nature, simple bad luck, or a result of forces beyond our control; or we may determine that the event was foreseeable and avoidable, and initiate a search for negligence and blame.

With so much of the practice of risk science now addressing, assessing, protecting against, and designing for extreme and rare events, a clear definition of their nature and impact is needed (Kaplan and Garrick, 1981). Similarly, a delineation of the tools available for better understanding and evaluating implications is sought. That is the purpose of this chapter. In particular, we provide an overview of current methods for characterizing and considering extreme and rare events and identify those areas in need of further research and methods development. In particular, we hope to help those working on the mathematical underpinnings and empirical demands and limits of real risk problems to appreciate the more fundamental elements of their work.

Several methods for defining extreme events have been proposed. One proposal is to define extreme events in terms of their low *frequency*. However, the concept of frequency is often relative to a specific context, frame of reference, or problem framing. Consider, for example, a fair lottery where one billion tickets have been sold and you are the owner of one of these tickets.[1] Your chances of winning are 1 in 10^9. If you do win, this is indeed a rare and fortuitous event. Now consider three separate and independent fair lotteries. For each, 1,000 tickets have been sold and you are the owner of three tickets, one for each lottery. Your chances of winning each lottery are 1 in 10^3 and your chances of winning *all three lotteries* are 1 in 10^9, the same as your chances in the single, larger lottery.

Is winning *all three* of the smaller lotteries a rare and extreme event, just like the winning of the single larger lottery? From your perspective, perhaps it is. They have the same probability, and conceivably even the same implications for your future wealth (assuming that each involves the same price per ticket and the same percentage payoff). However, they are fundamentally different, especially from the perspective of state lottery officials, and perhaps the crime and racketeering unit of the police as well. For you, the first event is *extraordinary*, but for the lottery officials (and the police), it is quite *ordinary*. One ticket must win, and your ticket is no better or worse than any other. The probability that someone will win the lottery is 1. However, the second event is (assuming fair and independent lotteries) extraordinary from all perspectives. The probability that someone will win the first of the smaller lotteries is 1; the probability that the *same* individual will win the next two lotteries as well (assuming that they own one ticket for each) is 1 in 10^6.

Moreover, while the probability of your winning a fair billion-ticket lottery (or three fair thousand-ticket lotteries) is low, it is not as low as the (zero) probability of a continuous random variable taking on an exact value. For example, an outdoor temperature of exactly 22.0°C (= 71.6°F) in a temperate location during the summer might sound rather common, but so long as this number is exact (i.e., 22.00000000 . . . °C), its probability of occurrence is zero. The probability of a continuous random variable, such as temperature, taking on a specific value is defined only in terms of its probability *density*, with probabilities being defined only over *an interval or range* of the continuous variable (computed by integrating the probability density over this interval). The probability of taking on a value within a small range, plus or minus ε around 22.0°C, can still be made arbitrarily low as ε decreases toward zero. Nonetheless, occurrences within this range, while technically "rare," may still be considered ordinary, and not extraordinary in the sense that we intend for an extreme and rare event.[2]

Those rare events that are truly extreme are generally found at the *tails* of probability distributions; for example, extremely high or low temperatures. Thus, in an area where the mean daily summer temperature is 22°C and the standard deviation is 7°C (12.6°F), a temperature of 50°C (122°F) would be considered very hot, while a temperature of –6°C (21.2°F) would be considered quite cold. Both values are four standard deviations away from the mean and, assuming a normal distribution, daily average temperatures that low or lower (or that high or higher) in the summer are each expected to occur with probability 3.17×10^{-5}. These low probabilities are computed from the *cumulative distribution function* (CDF) of the variable, $F_X(x) = P[X \leq x]$. So long as this probability is very close to zero (at the lower, "cold" tail) or 1.0 (at the upper, "hot" tail), the event is a candidate for being both rare and extreme.

Are low probabilities obtained at the tails of distributions sufficient for classification of an event as rare and extreme? Not unless the event is *also* the source of some notable impact (usually negative) – for example, on somebody's health or welfare. The temperatures noted earlier might fulfill this requirement for an extreme event based on the *severity* of the outcome, depending on the location; the availability of heating or air conditioning to protect individuals against the extreme temperatures; the extent to which people are accustomed to coping with occasional, large temperature excursions;[3] the presence of plants or crops that are sensitive to freezing or heat damage; and so on. An event that is considered extreme in one system or context may be viewed as a normal occurrence, causing little impact or notice, in another. Furthermore, peoples' evaluations of the severity of a risk are affected by a number of attributes of the risk that go beyond its direct effects (deaths, damage, etc.), including how well the risk is understood and whether it is controllable, voluntary, and distributed equitably across affected individuals or groups (Fischhoff et al., 1978; Slovic, 1987).

For many variables, severe outcomes occur at both tails of the distribution. Ambient temperatures, long-term precipitation, and stream-flow values are examples – floods and droughts are both severe occurrences. For other variables, severe outcomes might occur at one end of the distribution, but not at the other. In the domain of exposure to anthropogenic toxins, doses at the high end of the population distribution are often associated with serious health effects, but extremely low doses are usually of no concern.[4] In contrast, unusually high strengths in the structural components of a bridge are not expected to result in any harm; only extremes at the low end of the distribution of strength resistance are cause for concern. In fact, given the link to severe consequences as a key criterion, an outcome well within the central range of a distribution could emerge as a candidate for consideration as an extreme event. Thus, while an ambient temperature at or near the mean summer value may be of little interest, an election result at or near a 50 : 50 split between the two leading candidates could have serious consequences for a representative democracy, as occurred in the Florida balloting of the United States (Bush/Gore) presidential election of 2000.

Probabilities of occurrence are also sensitive to the applicable scales of aggregation. The probability that *at least one* of many rare outcomes will occur increases dramatically when an experiment is repeated over many trials or repetitions. For example, a 100-year storm is a rare event for a single drainage basin. However, if there are 100 independent drainage basins in a given country, we can expect on average one of these floods per year somewhere in that country.[5] With news coverage now more global and intensive, it may seem like rare natural disasters and similar events are occurring more frequently. This is not necessarily the case – we may simply be more aware of events at the full range of locations over which floods, droughts, storms, and other disasters take place. In a similar fashion, a large meteorite or asteroid hitting the Earth and causing great damage and mass extinction is perhaps the classic exemplar of extreme, disastrous events, even with its very small probability of occurrence over many human lifetimes (e.g., Remo, 1997; Gerrard and Barber, 1997; Gerrard, 2000; Peebles, 2000).[6] However, when one considers a large set of candidate planets throughout many solar systems, it is likely that meteorites hit planets on a fairly regular basis. Perhaps at some time in the distant future an enterprise will promote the sale of asteroid protection for planets by noting that somewhere in the universe a planet is struck by a damaging meteorite every two minutes (just as is now done in the promotion of automobile antitheft devices).

Still, with a focus on a particular outcome in a particular location during a particular time period, very low probabilities can be computed (or assumed), rendering many different types of natural, technological, or social events as candidates for treatment as extreme and rare events. How low a probability is low enough for the corresponding event to be considered rare? In contrast, how high must the probability be before the event is worthy of concern and attention? The criterion of expected loss avoidance dictates that as an event becomes more severe, the threshold probability for concern becomes correspondingly lower. This raises the question of how confident we can be of probability estimates that get *really* low. Consider the following exchange in Isaac Asimov's novel, *The Robots of Dawn* (1983). A millennium into the future, Detective Elijah Baley questions master roboticist Fastolfe on the likelihood that the mental freeze-out of an advanced, humaniform robot is the result of internal "positronic drift" versus an unplanned sequence of human commands. Fastolfe responds:

"I imagine that a mental freeze-out through positronic drift might have a probability of 1 in 10^{12}; that by accidental pattern-building 1 in 10^{100}. That is just an estimate, but a reasonable one. The difference is greater than that between a single electron and the entire Universe – and it is in favor of the positronic drift."

In our current millennium, most of us consider a probability of 1 in 10^{12} small enough to be essentially zero. Distinguishing this small probability from one that is even smaller seems like an exercise in irrelevance (or, in this case, entertaining literary fancy). As noted in the sections that follow, our ability to compute or estimate such low probabilities is notoriously poor and subject to numerous sources of error. We dealt with some of these, such as the failure to recognize that an experiment is repeated many times, previously. Others, such as the failure to recognize common-cause failure events and a *lack* of independence, are addressed in further detail later. For now, we shall assume that our criteria for defining extreme events (rarity of occurrence and severity of impact) are clear enough for readers to appreciate their significance, despite the many difficulties involved in characterizing these attributes.

2. PROBABILITY MODELS FOR EXTREME EVENTS

Probability models for extreme events can be developed to characterize the probability that an event exceeding a given size (or severity) will occur over a given time period and/or spatial domain. These models often begin with a fundamental characterization of the random variation of the targeted variable over time and space (or from trial to trial in a sequence of repeated experiments) to derive the probability distribution for the event magnitude over a given time period and area of aggregation.

While many different types of probability distributions have been used to study extreme events, particular classes of distributions tend to arise when considering mathematical properties and behavior at the tails of distributions. This section begins with a review of these "extreme value distributions" and then considers other classes of probability models often used for extreme events, including mixture models, random process models, and modeling/decomposition approaches. Methods for fitting or eliciting the parameters of these models are next addressed. Alternative frameworks to those based upon traditional probability models are also considered.

2.1. Extreme Value Theory and Probability Distributions for Extreme Events

A convenient mathematical approach is to conceptualize an extreme event as the maximum (or minimum) of a sequence of underlying events, which may or not be independent. This is the basis for extreme value theory, pioneered by Fisher and Tippett (1927), Gnedenko (1943), and Gumbel (1958). Recent summaries of extreme value theory methods and applications are found in Ang and Tang (1984), Kinnison (1985), Castillo (1988), Bierlant, Teugels, and Vynckier (1996), Reiss and Thomas (1997), and Metcalfe (1997). In its simplest form, extreme value theory provides asymptotic distributions for the minimum or maximum of identically distributed (and typically independent) variables, as the number of such variables becomes arbitrarily large. In particular, when the number of variables is large, the distributions for the maximum and minimum depend only on the tails of the original distribution, and their distributions generally converge to one of three limiting distribution types.

Consider n observations from a random variable X_i: $i = 1, \ldots, n$. For example, X_i might correspond to the daily stream flow at a particular location on a river on day i, out of a period of n days. The annual maximum flow would then be the largest of these values over consecutive sequences with $n = 365$ (assuming no leap years). The probability that the largest observation in a given year X is less than or equal to some given value x is equal to the probability that X_1 is less than or equal to x, *and* X_2 is less than or equal to x, *and* X_3 is less than or equal to $x, \ldots$, *and* X_n is less than or equal to x:

$$P[X_{\max} \leq x] = P[X_1 \leq x \cap X_2 \leq x \cap \cdots \cap X_n \leq x] \tag{1}$$

That is, all n values of X_i must be less than or equal to x. If the X_i are independent,[7] this can be computed as the product of the individual probabilities:

$$\Pr[X_{\max} \leq x] = \Pr[X_1 \leq x] \times \Pr[X_2 \leq x] \times \cdots \times \Pr[X_n \leq x] \tag{2}$$

Each of these factors is simply equivalent to the CDF evaluated at x, so that, if the X_i are identically distributed with CDF $F_X[x]$, the CDF for the maximum is given by[8]

$$F_{X_{max}}[x] = \prod_{i=1}^{n} F_{X_i}[x] = \prod_{i=1}^{n} F_X[x] = \{F_X[x]\}^n \tag{3}$$

The probability density function (PDF), $f_{X max}[x]$, is found by taking the derivative of the CDF with respect to x, yielding

$$f_{X_{max}}[x] = n\{F_X[x]\}^{n-1} f_X[x] \tag{4}$$

The CDF and PDF of X_{max} can thus be derived as simple and direct functions of the CDF and PDF of the parent variable X. Similar results can be derived for the minimum of a sequence of n independent, identically distributed observations:

$$F_{X_{min}}[x] = 1 - \{1 - F_X[x]\}^n \tag{5}$$

$$f_{X_{min}}[x] = n\{1 - F_X[x]\}^{n-1} f_X[x] \tag{6}$$

These exact distributions for X_{min} and X_{max} may be computed given any distribution of X and any value of n. However, for large n, these exact distributions tend to converge to one of three principal classes of distribution, depending on the behavior in the tails of the parent distribution for X:

1. A Type I, or Gumbel, distribution asymptotically arises for the maximum (or minimum) when the upper (or lower) tail of the parent distribution exhibits exponentially decreasing probability as the value of X increases (decreases);
2. A Type II, or Frechet, distribution asymptotically arises for the maximum (or minimum) when the upper (or lower) tail of the parent distribution exhibits polynomially decreasing probability (i.e., $1/X^\alpha$) as the value of X increases (decreases); and
3. A Type III, or Weibull, distribution asymptotically arises for the maximum (or minimum) when the upper (or lower) tail of the parent distribution is limited; that is, when the parent distribution exhibits a finite upper (or lower) bound.

These three types of distributions are thus often observed and applied to describe extremely large (or small) outcomes of a random variable that is sampled many times. Examples include: wind speed (Thom, 1954; Jenkinson, 1955; Delaunay, 1987; Solari, 1996); earthquake strength and seismic risk to structures (Nordquist, 1945; Burton and Makropoulos, 1985; Bolotin, 1993); annual maximum daily precipitation (Stedinger et al., 1993); annual maximum flood flows (Natural Environment Research Council, 1975; Jain and Singh, 1987; Resendiz-Carrillo and Lave, 1987); air pollution concentrations (Roberts, 1979a,b; Chock and Sluchak, 1986; Surman, Bodero, and Simpson,

1987; Kuchenhoff and Thamerus, 1996); civil engineering and dam safety (Haimes, 1998; Lambert and Li, 1994); and sea-level/tidal elevations or wave heights (Muir and El-Shaarawi, 1986; Tawn, 1992; Mathiesen et al., 1994). As an example of the utility of extreme value distributions in analysis and design, their use has allowed the simplification of multiobjective optimization problems involving risks of extreme events (e.g., Haimes et al., 1990; Haimes 1998; Lambert et al., 1994; Pannullo, Li, and Haimes, 1993; Wang, Lambert, and Haimes, 1999a,b; Frohwein and Lambert, 2000; and Frohwein, Haimes, and Lambert, 2000).

A generalized extreme value distribution can also be formulated to allow for transitions from one type of extreme value distribution to another (Jenkinson, 1955; Hosking, Wallis, and Wood, 1985; Johnson, Kotz, and Balakrishnan, 1995). Other distribution forms that are often used for processes with heavy tails (that hence do not converge to one of the extreme value distributions) include: the standard (two-parameter) lognormal distribution (Aitcheson and Brown, 1957); the three-parameter lognormal distribution, in which the minimum value need not equal zero (Sangal and Biswas, 1970; Burges, Lettenmaier, and Bates, 1975; Stedinger, 1980; Small, Sutton, and Milke, 1988); the log-Pearson Type III distribution, in which the logarithm of the variable follows a three-parameter gamma distribution (Bobee, 1975; Interagency Advisory Committee on Water Data, 1982; Tung and Mays, 1981; Phien and Hira, 1983; Wallis and Wood, 1985; Gross and Small, 1998); the Wakeby family of distributions (Landwehr, Matalas, and Wallis, 1979); and the generalized Pareto distribution (Hosking and Wallis, 1987; Anderson and Meerschaert, 1998). Since some extreme events may be defined by rare outcomes in the *co-occurrence* of two or more unusual events or variables (e.g., high winds and high rains, or high temperature and drought), bi- and multivariate distributions are needed in some applications (e.g. Coles and Tawn, 1991, 1994; Metcalfe, 1997; Kotz and Nadarajah, 2000).

2.2. Mixture Models

Mixture models are used to characterize a random variable that is sampled from more than one underlying population. *Finite* mixture models allow for a finite number of parent populations (typically two or more, but usually less than ten) (Aitkin and Wilson, 1980; Everitt and Hand, 1981; Titterington, Smith, and Makov, 1985; Singh, 1986; Burmaster and Wilson, 2000), whereas *continuous* mixture models allow the parameters of the distribution of the random variable to themselves be sampled from (continuous) distributions (Johnson, Kotz, and Balakrishnan, 1995). In the latter case, each observation of the random variable is in fact sampled from a different distribution, since the parameters of the parent distribution change for each sample. A mixture model has a hierarchical structure that can be used to capture variability at different spatial and temporal scales, and as a function of other explanatory variables (Crawford et al., 1992).[9]

Mixture models can be effective for describing variables affected by different conditions at different times, such as, different weather patterns giving rise to 100-year versus 1000-year floods, or normal versus upset operating conditions at an industrial operation (Smith, Bradley, and Baeck, 1994; Bradley and Smith, 1994). This may be especially useful when, as is often the case, extreme events tend to occur under special conditions that typically apply only a fraction of the time. Failure to account for these significantly higher risk rates during unusual or upset conditions (when "the analysis only focuses upon normal operating conditions") can lead to misleading risk estimates (Freudenburg, 1988). Gumbel (1958) and Rossi, Fiorentino, and Versace (1984) illustrate the use of two-component mixtures of extreme value distributions for the evaluation of flood flows. Crawford et al. (1992) demonstrate the application of a mixture model to lake chemistry in north central Wisconsin where two distinctly different types of lakes are present (drainage lakes with contributing watersheds, and seepage lakes where virtually all water inputs are the result of direct precipitation). Mixture models are also commonly used to describe the reliability and failure time distributions of normal components versus weak or defective parts and components (e.g., Arjas, Hansen, and Thyregod, 1991; Block, Mi, and Savits, 1993).

2.3. Random Process Models for Extreme Events

Extreme events often arise as the result of an underlying stochastic process occurring over time and space. Continuous climate, weather, geologic, hydrologic, environmental, economic, and social processes can lead to extreme events that reflect excursions that go beyond usual conditions, but that can nonetheless be occasionally expected to occur given the overall evolution of the random process. Significant research has been done to characterize the nature of "line-crossing" events arising from different classes of underlying random processes (e.g., Rice, 1944; Leadbetter, Lindgren, and Rootzen, 1982; Vanmarcke, 1983; Resnick, 1987), as a basis for characterizing extreme excursions. A common limiting case of these representations leads to a purely random Poisson point process for the excursion events. More complicated statistical and physical processes can lead to deviations from the Poisson model, with alternative point-process models for different classes of events, for event clustering or exclusion, and for space-time interactions (Vere-Jones, 1970, 1992; Kagan and Knopoff, 1977; Smith and Karr, 1983, 1985; Small and Morgan, 1986; Foufoula-Georgiou and Guttorp, 1987; Smith and Shively, 1995). Multidimensional stochastic models of various types can be used to predict the concurrent behavior of physical variables at different times and locations, thus simultaneously modeling both routine outcomes and associated extreme events (Bras and Rodriguez-Iturbe, 1985).

The occurrence of extreme events can be affected by other variables that vary over time or space, and the explicit inclusion of dependencies between these variables and the occurrence of an extreme event can lead to more accurate and effective prediction (Anderson and Nadarajah, 1993). As a first step, occurrence models can be specified with parameters reflecting seasonal variation (e.g., North, 1980) or long-term trends (Horowitz and Barakat, 1979; Leadbetter et al., 1982; Katz and Brown, 1992; Olsen, Lambert, and Haimes, 1998), if they are present. A more sophisticated model relating the occurrence of an extreme event to environmental conditions or other variables is the Cox regression model, in which the occurrence rate is expressed as a function of the explanatory variables (Cox, 1972; Vere-Jones, 1978; Smith and Karr, 1986).

Random process models can also be used to evaluate the magnitude of an excursion event, in addition to its frequency – for example, by characterizing the heaviness of the associated probability distribution tails (Taqqu, 1987; Leadbetter, 1991; Bacchi, Brath, and Kottegoda, 1992; Anderson and Meerschaert, 1998). These models can then be linked with physical models for hydrologic response (Eagleson, 1972; Weiss, 1977; Robinson and Sivapalan, 1997), environmental and/or geo-chemical impact (Di Toro, 1980; Ramaswami and Small, 1994; Adams and Papa, 2000), or material and structural damage (Zhang and Der Kiureghian, 1994) to predict the statistical response of the natural or engineered system that is affected by the excursion.

2.4. Modeling and Decomposition

System decomposition is often used to develop reliability models for the purpose of assessing the frequencies of extreme events, and designing methods to control such events, in both natural and engineered systems. Probabilistic risk analysis (PRA) models based on system simulation, event trees, and/or fault trees are used in a number of applications involving complex engineered systems with many stochastic inputs and interrelated mechanical, electrical, and human components. Examples of this type of model are found in Powers and Tompkins (1974), U.S. Nuclear Regulatory Commission (1975), Powers and Lapp (1976), Pate-Cornell (1984), Harr (1987), Kumamoto and Henley (1996), Kottegoda and Rosso (1997), Youngblood (1998), Bedford and Cooke (2001), and Vaurio (2001). Models have advanced well beyond the early representations used for nuclear and other applications, some of which did not adequately account for the vagaries of human error and common-cause failures. Human error and common-cause failures can reduce the reliability of facilities with supposedly redundant, independent safety systems, since such common-cause failure modes make the various safety systems *not* independent in reality (Fleming et al., 1983, 1986; Heising and Guey, 1984; Guey and Heising, 1986; Heising and Luciani, 1987; Vaurio, 1994, 1995, 1998).

A broader, hierarchical system perspective is now often utilized, allowing a wider set of causative factors and interrelationships to be considered in characterizing the risks associated with extreme events (Sharit, 2000; Lambert et al., 2001). Haimes (1998), Frohwein (1998), and Frohwein and co-workers (1999, 2000) have introduced multiobjective decision trees incorporating the statistics of extremes. Behavioral, institutional, and organizational factors are now included as fundamental components of risk analysis models and risk mitigation studies; see the work of Pate-Cornell and co-workers (1990, 1992, 1993, 1997), Mosleh and co-workers (Apostolakis, Bier, and Mosleh, 1988; Mosleh et al., 1991, 1997; Macwan and Mosleh, 1994; Julius et al., 1995; Hsueh and Mosleh, 1996; Smidts, Shen, and Mosleh, 1997), and Davoudian, Wu, and Apostolakis (1994a,b). Political unrest and security concerns can similarly affect risk analyses (e.g., Munera, Canal, and Munoz, 1997), and fundamental assessments of societal infrastructure and its protection against the threats of war and terrorism have also adopted probabilistic approaches (Haimes et al., 1998; Haimes, 1998, 1999; Longstaff et al., 2000; Haimes and Jiang, 2001).

Recent examples of PRA applied to complex technological human systems include studies of hazardous materials transport (Purdy, 1993; Pet-Armacost, Sepulveda, and Sakude, 1999), structural safety (Moses, 1998), pipeline failure (Cooke and Jager, 1998), ship transportation risk (Merrick et al., 2000; van Dorp et al., 2001; Fowler and Sorgard, 2000), and space mission safety (Frank, 2000). In each of these cases, domain-specific knowledge and experience is combined with quantitative methods to assess system vulnerabilities, failure probabilities, and risk mitigation options. With many complex systems now highly dependent upon computer control systems, code validation procedures to identify and explore the implications of faulty or degenerate cases in program instruction are also an important component of many safety evaluations (Leveson, 1995). Given the complexity of many codes, probabilistic methods are sometimes used here as well, even though such codes are, in principle, deterministic (Hart, Sharir, and Pneuli, 1983; Vardi, 1985; Manna and Pnueli, 1992; Johnson, 1993; Pnueli and Zuck, 1993; Hartonas-Garmhausen et al., 1995; Hartonas-Garmhausen, 1998).

It is clear that refinements in the use of modeling and system decomposition have the potential to improve further the assessment of extreme event probabilities. The basic approach is to model the probabilities of interest in terms of quantities about which we know more. The use of decomposition is widely accepted as a useful technique not only in risk analysis but also in a number of related fields, such as decision analysis, industrial engineering, and long-range forecasting. Armstrong (1985) notes that, although there has been little rigorous empirical work documenting the benefits of decomposition in practice, "the limited evidence suggests that it is a technique with much potential. Research on how and when to use decomposition would be useful." In particular, we believe that further research on which approaches to analysis and decomposition yield the best results (as measured, for example, by accuracy, consistency, efficiency, and robustness) would be especially valuable.

2.5. Parameter Estimation for Extreme Event Models

Regardless of whether event frequency and magnitude are modeled with simple, univariate probability distributions or using a complex model, estimates are required for the parameters and inputs to these models. When adequate data are available, traditional methods of parameter estimation (e.g., classical statistics) can be used. However, because of the nature of extreme events and the conditions that lead to them, the availability of such data is often severely limited. Creative approaches involving a combination of expert judgment, limited data, and theoretical limits and bounds are therefore frequently required.

For extreme value and related distributions, parameter estimation from observed data can sometimes be pursued using classical statistical methods such as (in generally increasing order of complexity and sophistication): the method of moments; probability plot correlation methods; and maximum likelihood methods. Because extremes are often analyzed in the context of *order statistics* (Wilks, 1948, 1962; David, 1981; Galambos, 1987) – that is, by considering the value of the largest (or smallest) of a sequence of *n* observations, the next to the largest (or next to the smallest) of those observations, and so on – parameter estimation methods based on order statistics can also be effective (Hosking, 1990; Hosking and Wallis, 1997; Johnson and Albert, 1999). Probability weighted moments, which use the observed percentile of each sample in an empirical CDF to weight its value in moment calculations, are particularly useful for estimating the parameters of extreme value distributions (Greenwood et al., 1979; Hosking, Wallis, and Wood, 1985).

Another approach to probability estimation is the use of Bayesian methods. In Bayesian estimation, probability estimation is approached by combining a subjective (though hopefully well-informed) *prior* distribution with the *likelihood* of the observed (sometimes limited) set of outcome data to yield a *posterior* probability distribution for the quantity of interest (Press, 1989; Leonard and Hsu, 1999; DeGroot and Schervish, 2002). Both Bayesian and classical methods can be used in situations of abundant data. However, in many cases, little or no available data may be directly relevant to the assessment of rare events. In such situations, classical statistical methods are often of little use. Bayesian methods become both more desirable than classical methods and more controversial when data are sparse. On the one hand, unlike classical statistical procedures, Bayesian analysis can be used in situations of sparse data, since important subjective and other nonstatistical types of evidence (in the form of an informative prior distribution) can be used in the decision or inference process. However, with sparse data, the results of Bayesian analyses become more sensitive to the analyst's interpretation (e.g., in the choice of a judgmental distribution to represent nonstatistical evidence) for both the prior and the likelihood (Small and Fischbeck, 1999). Hence, Bayesian methods can be more subjective and less readily accepted when data are sparse.

The subjective nature of judgmental distributions is particularly problematic when assessing small tail probabilities. In particular, behavioral research shows that people are poor at assessing small probabilities (Lichtenstein et al., 1978), so there may be little reason to expect that people's judgmental distributions will be reliable or reproducible in the extreme tails. Preposterior analysis (Martz and Waller, 1982) is one approach for evaluating the reasonableness of an assessed prior. The idea here is to perform a number of Bayesian updates of the prior distribution with *hypothetical* data sets (including some data sets that would be considered extreme or surprising) before analyzing the actual data. This is particularly useful for ensuring that the tails of an assessed prior distribution are reasonable, something that is difficult to do otherwise.

Particular difficulties arise for probabilistic assessment of rare events when no occurrence has, as yet, been observed. Statistical methods for addressing this case are available when it is clear that a number of "trials" have taken place, with zero failure events (Bailey, 1997). Not surprisingly, classical and Bayesian methods both suggest lower and lower estimates for the failure rate as the number of failure-free trials increases. However, one should be careful about interpreting this as evidence of ongoing improvements in performance (or system "learning"), especially when subsequent trials take place under changing conditions that might not have been experienced or tested in previous experiments. Methods for analyzing data with no failures have in the past been used to build a case that the safety of risky technologies has improved over time (e.g., Duane, 1964; Vaurio, 1984); they are conceptually related to reliability models with "bathtub-shaped failure rates," whereby failure rates are initially high during the system or component "burn-in" period and then decrease until subsequent wear-out and deterioration cause them to increase again in the future (Block and Savits, 1997). Such models must be informed as much by science, engineering, and behavioral insight as by statistics, and can imply a level of confidence not warranted by the realities and demands of ever-changing operations, as demonstrated by the *Challenger* disaster (Vaughan, 1997). See also the comment by Lynn and Singpurwalla, published with the paper by Block and Savits (1997). Recognizing how data might be misleading is just as important as recognizing where it appears to be leading (see, for example, Royall, 2000), although this is usually easier with the perfect 20 : 20 vision of hindsight.[10]

The idea of maximum entropy distributions, based on information theory (Shannon, 1948; Goldman, 1953; Greeno, 1970), is to use whatever partial information is available about the uncertain quantity of interest (e.g., mean, median, or mean and variance), but as little additional information as possible beyond that, to avoid inadvertently assuming more than is actually known. Therefore, the maximum entropy distribution is defined to be the least informative distribution that matches the specified constraints (Jaynes, 1995; Levine and Tribus, 1978). The resulting distribution can then be used either as a prior distribution for Bayesian analysis (if additional data are available) or as a partially informative distribution with no updating.

As a result of the limitations of the various approaches discussed previously, much interest has recently focused on so-called "robust" Bayesian methods and other bounding approaches. Robust Bayesian methods (Berger, 1985) generally involve updating an entire *class* of priors (rather than a single prior distribution) with observed data. Note that for some class specifications, the computational effort involved in considering all distributions in the class will not be substantially greater than for considering a single distribution. If all (or most) priors in a suitably broad class give similar results (e.g., a similar posterior distribution or optimal decision), this can lead to greatly improved confidence in the results of the analysis. (Note, however, that robustness does not necessarily imply correctness. The result might be stable, but still wrong.)

In a similar spirit is probability bounds analysis (Ferson and Long, 1995; Ferson, 1995, 2002), which is a bounding approach used for propagating uncertainties in models (rather than for Bayesian updating). In this approach, the analyst specifies bounds on the CDF of each parameter used as input to a model, rather than a single informative distribution for each input value. These input bounds are then propagated through the model. The uncertainty propagation process, which is computationally quite efficient, yields valid bounds on the CDF for the final result of the model (e.g., an estimated risk level).

Expert Opinion for Extreme Event Probabilities. When Bayesian or related subjective methods are used to assess probability distributions, expert opinions are often sought as a basis for these estimates. For some problems where estimates are needed for conditions outside of any range of past or current conditions, expert opinion is the only basis for assessment. Expert opinion has been widely used in practice (Cooke, 1991; Mosleh, Bier, and Apostolakis, 1987), and a number of guidelines have been suggested for assessing informative judgmental distributions

(Hogarth, 1975; von Winterfeldt and Edwards, 1986; Morgan and Henrion, 1990). Many of the behavioral procedures that are useful in general are likely to be especially valuable when dealing with rare and/or extreme events, where naively generated estimates can be quite unreliable. More specifically, several elicitation methods seem particularly well suited to the analysis of extreme events. For example, Armstrong (1985) has advised that, in long-range forecasting, one should "avoid hiring the most expensive expert.... Instead, spread the budget over a number of less expensive experts." The reason for this advice is that almost by definition, there will be little opportunity to observe and study events that are outside the normal range of experience of the system in question. Therefore, above some minimal level of expertise, the "best" expert may have little if any more knowledge than others in the field.

When multiple experts are used, a number of issues arise regarding methods for comparing or combining their assessments (Genest and Zidek, 1986; Clemen, 1989; Winkler and Clemen, 1992). Among these concerns is the extent to which assessments from multiple experts are actually independent (Winkler, 1981; Batchelor and Dua, 1995). Recognizing that dependency among experts is unavoidable, but often difficult to characterize, methods for promoting explicit interaction among experts have been used as part of group-elicitation procedures. Budnitz et al. (1998) examine methods for integrating the input from multiple experts in the context of a probabilistic assessment of seismic risk. They propose a process where experts first act as independent evaluators, and then, with the help of a technical facilitator/integrator, work together as integrators to represent their overall field of expertise. Guidance is provided on the selection of experts, their interaction, elicitation training, expert elicitation, analysis, and documentation.

Creative methods are also needed to characterize and display differences among experts and "schools of thought" within a given domain. For example, such methods can be used to illustrate differences among experts in what they believe future research may reveal, and how their assessed probabilities could change in response to information of different types (Morgan and Keith, 1995; Stiber, Pantazidou, and Small, 1999).

2.6. Alternatives to Probabilistic Methods

In some cases, it is arguable that the problem of estimating the risk of an extreme event may be too information-sparse to handle in an explicitly probabilistic manner. This is a controversial view, since Bayesian statisticians believe that all problems can be addressed probabilistically through reliance on what little information is available and the concept of "degree of belief" or subjective judgment. Nonetheless, it is worth reviewing other potentially useful approaches for analyzing extreme events that do not have as strict a set of information requirements as do more formal probabilistic techniques.

One set of approaches to be considered might be described as nonprobabilistic or quasi-probabilistic methods, such as fuzzy set theory (Zadeh, 1965; Kaufmann and Gupta, 1991; Zimmermann, 1991), Dempster-Shafer theory (Shafer, 1976; Dong and Wong, 1986), or the theory of interval-valued probabilities (Walley, 1991). Ferson (1993) and Ferson and Ginzburg (1996) have argued that such methods relax some of the axioms of probability theory and, hence, may be applicable when it is not clear that those axioms are reasonable for modeling the problem at hand. Alternatively, methods such as fuzzy set theory and interval analysis can be viewed as giving bounds (albeit sometimes quite loose bounds) on the results of a full probabilistic analysis when precise estimates of numerical probabilities and dependencies are unreliable or unavailable (Zadeh, 1978). Such bounding methods may, therefore, be more appropriate than full probabilistic analyses, which can sometimes imply a false sense of precision.

Some nonprobabilistic but quantitative approaches, such as chaos and catastrophe theories, can provide insights about particular classes of extreme events. Chaos theory is concerned in part with understanding "the complex boundaries between orderly and chaotic behavior" (Gleick, 1987). Since some extreme events may be characterized as transitions from orderly to chaotic behavior (or vice versa), methods that allow explicit characterization of such phase transitions may be desirable. Similarly, catastrophe theory is concerned with "*sudden* changes caused by *smooth* alterations in the situation" (Poston and Stewart, 1978) (emphasis in original). Although not specifically intended for the study of disasters (despite the connotations of its name), this approach may nonetheless be well suited to studying those extreme events that are associated with nonlinear system behavior. Study of such phase transitions may be potentially useful, for example, in the analysis of global climate change, where physical feedback mechanisms and system interactions could lead to unexpected and potentially severe outcomes (Lehman and Keigwin, 1992; Keigwin and Jones, 1994; Bond et al., 1993; Broecker, 1994, 1995; Webb et al., 1997).

Finally, qualitative approaches to the analysis of extreme events can provide useful insights, either in place of quantitative methods or as a supplement to them. For example, some have attempted to develop a theory of "surprise." Even though some of this work is not concerned with extreme events per se, it might well be relevant to defining and assessing the probabilities of extreme events. For example, Shackle (1972) distinguishes between "counter-expected events" and "unexpected events." Here, counter-expected events are events that had been imagined, but consciously rejected. By contrast, unexpected events are events that had never been imagined. Occurrences of either type of event would generally result in surprise, and both counter-expected and unexpected events will often tend to be extreme (either in the sense of severity of consequences, or in the sense of being outside the normal range of experience of the system in question).

Similarly, Fiering and Kindler (1984) developed a typology of surprising events that can affect the operation of water resource systems. These include structural collapse of a system component, errors in system design, changes in the hydrologic regime, new laws or environmental standards, surprises resulting from lack of understanding of how a system will respond to particular conditions, and demographic changes. Sometimes, these possibilities are explicitly considered during the design of the system, and an explicit decision is made that they do not need to be addressed; in this case, they would be considered counter-expected events if and when they do occur. More often, in practice, such events are merely overlooked in the design phase; in this case, should they occur, they would be unexpected events. Such taxonomies, while not directly answering the question of how we can assess the probabilities of extreme events, can be helpful in identifying which types of extreme events we should be considering, thereby minimizing the number of unexpected (as opposed to counter-expected) events.

3. COMMUNICATING EXTREME RISKS

Communicating the risks of rare and extreme events is often a challenge (Bier, 2001), in part because people tend to overweight small probabilities in decision making. See, for example, work by Ali (1977), Bostrom (1997), Griffith (1949), Lichtenstein et al. (1978), Lopez Gomez (1990), Nogee and Lieberman (1960), Preston and Baratta (1948), and Sprowls (1953). The National Research Council (1996, pp. 56–70) discusses the importance of the choice of risk measures in framing the interpretation of risk analysis results.

Studies reported in the literature have compared the effectiveness of verbal, numerical, and graphical methods of communicating probabilistic information. Interestingly, Gonzalez-Vallejo and Wallsten (1992, p. 855) found that verbal presentation of probability information seemed to yield fewer "preference reversals" or inconsistencies on the part of experimental subjects than the use of numerical information. However, past studies on the correspondence between verbal and numerical presentations of probabilistic information have found large variability in how people translate probabilistic phrases (such as "highly unlikely") into numerical values, and these studies have generally not extended to probabilities less than 1%. See for example Beyth-Marom (1982), Budescu and Wallsten (1985), Budescu, Weinberg, and Wallsten (1988), Gonzalez-Vallejo and Wallsten (1992), Hakel (1968), Lichtenstein and Newman (1967), Pepper (1981), Shiloh and Sagi (1989), Simpson (1944, 1963), and Stone and Johnson (1959). Survey articles by Budescu and Wallsten (1995), Clark (1990), and Wallsten and Budescu (1995) provide overviews of this literature. Thus, there does not appear to be any definitive guidance on how to translate numerical probabilities into lay language.

Kaplan, Hammel, and Schimmel (1986) found that graphical representations of extremely small side-effect probabilities increased the probability that experimental subjects would express the willingness to take a vaccine. This suggests that graphical portrayals of small probabilities may reduce the tendency to overemphasize those probabilities in decision making by effectively illustrating how small they really are. However, relatively little work has been done in this area, and the results do not yield definitive recommendations as to the most appropriate graphical format to use. For example, Loomis and Du Vair (1993, p. 297) found that a "risk ladder . . . does a better job providing information on relative risk" than a pie chart, while pie charts were effective "for communicating the absolute level of risk reductions" to be achieved by a particular program. However, Sandman, Weinstein, and Miller (1994) found that changing the scale on risk ladders changed subjects' perceptions of risk even when the same quantitative risk information was presented.

Siegrist (1997) briefly reviews some of the literature (e.g., Halpern, Blackman, and Salzman, 1989; Cosmides and Tooby, 1996, Gigerenzer and Hoffrage, 1995) on alternative numerical formats for risk communication, such as probabilities, odds, frequencies, and return periods, and provides further interesting results. In particular, Siegrist suggests that the frequency format (e.g., "600 in 1,000,000") more effectively communicates the differences in occurrence rates between risks of different magnitudes, while the probability format (e.g., "0.0006") tends to obscure such differences. Again, however, the scope of the analysis was limited, and Siegrist notes the need for more comprehensive studies. Igerenzer (1992) offers advice to nonprofessional audiences on how to avoid confusion in interpreting statements from risk assessments. This advice includes a collection of simple, general suggestions based on behavioral research that can often allow for substantial improvements in understanding.

Further work is clearly needed to explore the implications of different risk communication formats. This is particularly important, since current thinking (e.g., National Research Council, 1996) favors the adoption of an "analytical-deliberative" approach to public decision making regarding extreme risks, in which stakeholder understanding of key risk assessment results is a critical element.

4. IMPLICATIONS FOR DECISION MAKING

In the face of such unforeseen calamities as airplanes crashing, buildings collapsing, bridges falling, dams bursting, and nuclear power plants failing, we are more willing to consider the occurrence of "extreme" events in our decision calculus. Modern decision analysts therefore are working to develop a more robust treatment of extreme events, in both a theoretical and a practical sense. Furthermore, managers and decision makers are often concerned with the risk associated with extreme events, not just with the average outcome that could result from a range of situations. Indeed, people in general are not risk-neutral. For example, they are often more concerned with low-probability, catastrophic events than with more frequently occurring but less severe accidents.

The pitfalls of simple expected-value decision making are well known, but are likely to be even greater when there is the potential for extreme events (e.g., catastrophic accidents or bankruptcy) (Haimes, 1988). Essentially, expected-value decision making treats low-probability, high-consequence (i.e., extreme) events the same as more routine high-probability, low-consequence events, despite the fact that they are unlikely to be viewed in the same way by most decision makers. This shortcoming of expected-value decision making is addressed by the use of expected-utility (as opposed to expected-value) decision making; see for example Keeney and Raiffa (1976), von Winterfeldt and Edwards (1986), Pratt, Raiffa, and Schlaifer (1995), Raiffa and Schlaifer (1961), DeGroot (1970), Clemen (1991), and Clemen and Reilly (2001).

In expected-utility decision making, a numerical utility function is assessed over a set of potential outcomes. This utility function is intended to capture the values and preferences of the decision maker, including any important nonlinearity in those preferences. Thus, for example, expected-utility decision making can capture the fact that a 3-foot flood may cause little damage beyond some soggy backyards, while a 20-foot flood may cause millions of dollars of damage and even some fatalities; similarly, a $10 million loss (if sufficient to cause bankruptcy for a company) may be much more than ten times as bad as a $1 million loss. The shape (e.g., convexity) of the utility function has implications for the level of "risk aversion" (or conversely, "risk proneness") of the decision maker; see for example Keeney and Raiffa (1976) and Clemen and Reilly (2001). If the utility of each possible outcome is quantified, and the distribution of outcomes associated with each decision alternative is characterized, then under straightforward and reasonable assumptions, the preferred alternative is the one that maximizes the expected value of the utility.

Despite the theoretical appeal of expected-utility decision making, some practical issues may need to be resolved, particularly when applying expected-utility decision making to the risks of extreme events. Challenges in applying the theory include situations where the worst-case outcome is abhorrent at even very low probabilities (e.g., a "nuclear winter" or a large asteroid impact involving the extinction of human life). It may be difficult if not impossible to assign finite utilities to such extreme events; worldwide extinction could have an infinite negative utility for many decision makers. Moreover, since we have seen that the probabilities of extreme events are likely to be highly unreliable, the expected utilities computed based on these probabilities will tend to be unreliable as well. If minor differences in a small probability can result in major changes to the expected utility, and hence the optimal decision, expected-utility decision making may be unable to yield definitive recommendations as to the best course of action. Lastly, decision makers and/or stakeholders may not have the time to have their utility functions assessed, or may be unwilling to reveal their utility functions to others, particularly for extreme events with strong public and moral overtones (e.g., Spash, 2000), or when the manner in which the elicited information will be used is not clearly specified (Fischhoff, 2000).

Some other techniques have also been suggested for lending visibility to low-probability, high-consequence events so that they are not camouflaged by expected-value calculations. For example, the use of *conditional* expected values (e.g., conditional on exceeding some threshold value) (Asbeck and Haimes, 1984; Haimes, 1998) can help focus the decision maker's attention on the tail of a distribution. Similarly, approaches such as multiple-criteria decision making (Chankong and Haimes, 1983; Haimes et al., 1990, 1992; Haimes, 1991, 1998; Haimes and Li, 1991; Haimes and Steuer, 2000) can treat the risk of extreme events separate from the overall (i.e., unconditional) expected value. Since they are less well known than expected utility (and do not require the assessment of a decision maker's complete utility function), these approaches are explored in more detail here.

The conditional expected value of the risk of extreme events (among other conditional expected values) generated by the partitioned multiobjective risk method (Asbeck and Haimes, 1984; Haimes, 1998), when used in conjunction with the (unconditional) expected value, avoids treating routine and extreme events as commensurate. Such multiple-objective analysis has been demonstrated as being complementary to benefit-cost analysis in applications (Haimes and Hall, 1974; Haimes, 1998).

The partitioned multiobjective risk method (PMRM) is a risk analysis method developed for solving multiobjective problems of a probabilistic nature (Asbeck and Haimes, 1984; Haimes, 1998). In particular, the PMRM generates a number of conditional expected-value functions, termed *risk functions*, to represent the risk associated with specific ranges of the extreme event probability. A conditional expectation is defined as the expected value of a random variable, given that this value lies within some prespecified probability (or damage) range. Clearly, the values of conditional expectations are dependent on where the probability (or damage) axis is partitioned. The analyst makes the choice of where to partition subjectively, in response to the characteristics of the problem and the preferences of the decision maker. For example, if the analyst or the decision maker is concerned about a one-in-a-million-year catastrophe, the partitioning could be chosen to emphasize this catastrophic risk.

The PMRM generates multiple conditional expected-value functions, each associated with a particular range of exceedance probabilities (or damage severities). The resulting conditional expected-value functions, in conjunction with the traditional expected value, provide a family of risk measures associated with a particular policy. For each of the partitioning ranges, the conditional expected damage (given that the damage is within that particular range) provides a measure of the risk associated with that range. These measures are obtained through the definition of the *conditional expected value*. For example, if the total range of possible outcomes is partitioned into three ranges, the resulting measures of risk are: $f_2(\bullet)$, the expected value in the range of high exceedance probability and low severity; $f_3(\bullet)$, the expected value in the range of medium exceedance probability and moderate severity; and $f_4(\bullet)$, the expected value in the range of low exceedance probability and high severity. These are given by:

Risk Functions:

$$f_2(\bullet) = E[X \mid X \le \beta_1]$$
$$f_3(\bullet) = E[X \mid \beta_1 < X \le \beta_2] \tag{7}$$
$$f_4(\bullet) = E[X \mid X > \beta_2]$$
$$f_5(\bullet) = E[X]$$

(The function $f_1(\bullet)$ is reserved for the cost associated with the management of risk.)

The function $f_4(\bullet)$, for example, is computed as

$$f_4(\bullet) = \frac{\int\limits_{\beta_2}^{\infty} x p(x) dx}{\int\limits_{\beta_2}^{\infty} p(x) dx} \tag{8}$$

Thus, for a particular policy option, this approach yields three measures of risk – $f_2(\bullet)$, $f_3(\bullet)$, and $f_4(\bullet)$ – in addition to the traditional expected value, denoted here by $f_5(\bullet)$. In the PMRM, all (or some subset) of these measures are balanced in a multiobjective formulation. Figure 4.1 illustrates partitioning on the damage axis x. In parallel with the partitioning on the damage axis, Equations (9) are measures of the same conditional expected values with partitioning on the probability axis (see Figure 4.2).Relationships between the statistics of extremes and the PMRM have also been developed (Karlsson and Haimes, 1988a,b; Mitsiopoulos, Haimes, and Li, 1991; Romei, Haimes, and Li, 1991; and Haimes, 1998).

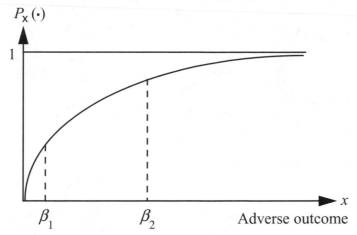

FIGURE 4.1. Partitioning on the damage axis.

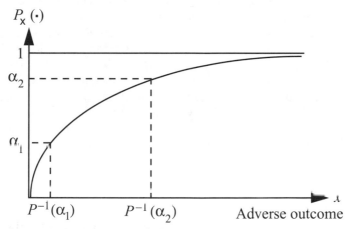

FIGURE 4.2. Partitioning on the probability axis.

Risk Functions:

$$f_2(\bullet) = E[X \mid X \leq P^{-1}(\alpha_1)]$$
$$f_3(\bullet) = E[X \mid P^{-1}(\alpha_1) < X \leq P^{-1}(\alpha_2)] \quad (9)$$
$$f_4(\bullet) = E[X \mid X > P^{-1}(\alpha_2)]$$
$$f_5(\bullet) = E[X]$$

To illustrate the use of the additional information provided by the PMRM, consider Figure 4.3, where the cost of preventing groundwater contamination, f_1, is plotted against: (a) the conditional expected value

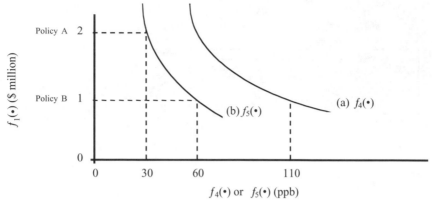

FIGURE 4.3. Cost function versus (a) conditional expected value of contaminant concentration $f_4(\bullet)$; and (b) expected value of contaminant concentration $f_5(\bullet)$.

of contaminant concentrations at the low probability of exceedance/high-concentration range, f_4; and (b) the unconditional expected value of contaminant concentration, f_5 (Haimes, 1998). Note that with Policy A, an investment of \$2 million in preventing groundwater contamination results in an expected-value contaminant concentration of 30 parts per billion (ppb); however, under the more conservative view represented by f_4, the conditional expected value of contaminant concentrations (given that the state of nature will be in a low-probability/high-concentration region) is twice as high (60 ppb). Policy B, with \$1 million of expenditure, reveals similar results – 60 ppb for the unconditional expectation f_5, but 110 ppb for the conditional expectation f_4.

Also, note that the slopes of the noninferior frontiers with policies A and B are not the same. The change in f_5 between Policies A and B is smaller than that in f_4, indicating that a further investment beyond \$1 million would contribute more to a reduction of the extreme-event risk f_4 than it would to the unconditional expectation f_5. The trade-offs described by $\lambda_{1i} = -\Delta f_1 / \Delta f_i$ provide valuable information. More specifically, the decision maker is provided with additional insight into the risk trade-off problem through the function f_4 (and similarly f_2 and f_3). In particular, the analysis presented here reveals that Policy B, while yielding an expected contaminant concentration of 60 ppb, could under extreme conditions yield expected concentrations greater than 100 ppb. (If, for example, the partitioning is made on the probability axis, and in addition a normal probability distribution is assumed, then the likelihood of this outcome could be quantified in terms of a specific number of standard deviations.) However, with an additional expenditure of \$1 million (to adopt Policy A), even an extreme event would yield a conditional expected concentration of only 60 ppb – which might be within the range of acceptable standards.

To further demonstrate the value of the additional information provided by the conditional expected value $f_4(x)$, consider the following results obtained by Petrakian et al. (1989) in a study of modification options for the Shoohawk Dam. Two decision options are considered – raising the dam's height and increasing the dam's spillway capacity.

Table 4.1 and Figure 4.4 present the values of $f_1(x)$ (the cost associated with increasing the dam's height and the spillway capacity), and $f_4(x)$ and $f_5(x)$ (the conditional and unconditional expected values of damages, respectively). These values are listed for each of the selected scenarios. Note that the unconditional expected value of the damage, $f_5(x)$, is in the range of $16.5 million to $161.7 million for all scenarios. On the other hand,

TABLE 4.1. *Cost of Improving Dam Safety and Corresponding Conditional and Unconditional Expected Damage*

Management scenarios[a]	Cost, $f_1(x)$ ($ million)	High-damage, conditional expected value, $f_4(x)$ ($ million)	Unconditional expected value, $f_5(x)$ ($ million)
1	0	1260	161.7
2	20	835	161.6
3	26	746	161.6
4	36	719	161.5
5	46	793	160.5

[a] Involving different combinations of engineering programs to change the dam height and spillway capacity.

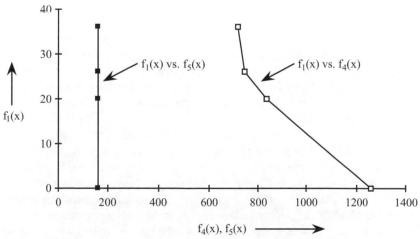

FIGURE 4.4. Pareto optimal frontiers of $f_1(x)$ versus $f_4(x)$ and $f_5(x)$.

the range of the low-frequency, high-damage conditional expected value, $f_4(x)$, varies between \$719 million and \$1,260 million – a marked difference. Thus, while an investment in dam safety at a cost, $f_1(x)$, ranging from \$0 to \$46 million does not appreciably reduce the unconditional expected value of damages, such an investment markedly reduces the conditional expected value of extreme damage from about \$1,260 million to \$720 million. This significant insight into the probable effect of different policy options on the safety of the Shoohawk Dam would have been lost without the consideration of the conditional expected value derived by the PMRM, or similar expected utility calculations that appropriately value extreme damage outcomes.

In sum, metrics to represent the risk of extreme events (such as the PMRM or expected utility) are needed for decision making to supplement and complement the expected value measure of risk, which is only a summary average representing the central tendency of events. There is more work to be done in this area, including extensions of the PMRM. For example, research efforts directed at using the statistics of extremes to represent risks of extreme events in the PMRM are promising.

5. POLICY ISSUES CONCERNING EXTREME RISKS

This section addresses some issues with risk-based policy and regulation of potential extreme events. Risk-informed decision-making (i.e., the explicit use of risk analysis results in making regulatory or risk management decisions regarding extreme risks) has the potential to reduce risk while simultaneously reducing compliance cost; see for example Bier and Jang (1999).

In nuclear power, existing regulations for component testing were largely established before PRA techniques were available, and may thus apply the same requirements to components with very different contributions to plant risk. Because of that, increased testing of the few most risk-critical components can significantly decrease the risk of extreme events for little increase in cost, while decreased testing of less-significant components can substantially reduce compliance cost for a small increase in risk, thus creating a win-win situation for both regulators and regulated parties (Specter and Braun, 1994). Despite these benefits, such risk-informed regulation of nuclear safety has been slow to be adopted in practice. The first nuclear plant PRA was published more than 25 years ago (U.S. Nuclear Regulatory Commission, 1975). However, the U.S. Nuclear Regulatory Commission (NRC) did not issue regulatory guidance for risk-informed regulation until quite recently (U.S. NRC, 1998a–i), after spending nearly two decades attempting to formulate and apply an alternative approach – quantitative safety goals – with little concrete success (see, for example, Bier, 1988, 1994).

By contrast, the design of dams is typically based on a "probable maximum flood" – defined as "the flood that may be expected from the most severe combination of critical meteorological and hydrologic conditions that are reasonably possible" (National Research Council, 1985) – despite the difficulties in defining "reasonably possible." In fact, some even view the explicit recognition that dams have a nonzero probability of failure as "designing for failure," and find risk-based standards ethically unacceptable for that reason (Moses, 1998; Lave and Balvanyos, 1998). Moreover, some agencies are still not comfortable with a probabilistic approach at all.

Griesmeyer and Okrent (1980) note that "there will always be a lack of assurance about the estimates of low frequency, high consequence events," and that "the subjectivity, the subtlety and the novelty [of risk analysis methods] leave analysis open to bias which is unintended, as well as to outright abuse." They conclude: "The variation of both societal values and societal risks, as well as the uncertainties in the estimation of those risks, ensures that there will always be conflict in the management of risk." This is particularly true when there are asymmetries in information between regulators and the parties that they regulate, and conflicts between the interests of the two groups.

Regulators will rarely have either the staff time or the specific knowledge needed "to monitor and enforce a performance-based standard" (Center for Strategic and International Studies, 1998). Moreover, pervasive informational asymmetries may make complete verification difficult or impossible to accomplish even in the absence of budget limitations. Regulators are therefore often dependent on analyses performed by regulated parties. Regulators have a natural incentive (and often a mandate) to seek large safety margins for extreme events (e.g., by ensuring that risks are estimated conservatively; see, for example, Finkel, 1989). By contrast, while regulated parties also have an incentive to ensure the safety of the businesses that they own and operate, in their case this is balanced by a competing desire to minimize costs, which can provide an incentive to underestimate risks in order to minimize regulatory burden. Even if a regulated party does not engage in *deliberate* bias for this purpose, there is typically enough room for judgment in structuring a risk model and estimating key parameters that risk analysts can easily "shade" their analyses favorably without departing from the range of credible assumptions (Applegate, 1991). For example, there is a natural tendency to review the dominant risk contributors identified in a PRA and "sharpen one's pencil" (i.e., perform a refined analysis to see whether the risks from the dominant contributors have been overestimated).

Performing a comprehensive, rigorous, and objective risk analysis can be a costly and time-consuming undertaking that requires detailed site-specific or technology-specific knowledge. The consequences of potential extreme events often cannot be measured directly (Chinander, Kleindorfer, and Kunreuther, 1998). In such cases, predictive models of potential extreme events must replace direct measurement (e.g., of emission levels), and the accuracy of the resulting "measurement" system will depend critically on the extent to which the risk analyses can be verified and validated. If the verification process is to generate high confidence in the results of the risk analysis, the levels of effort and expertise required for verification may be comparable to those required for performing the analysis in the first place.

One solution to the difficulty of developing, verifying, and validating site- and facility-specific risk analysis results is the adoption of officially approved default modeling assumptions and parameter values, which are presumptively acceptable with only minimal review. Moreover, the use of defaults also facilitates the process of performing the PRA in the first place, and can help to ensure more uniform standards of PRA quality, particularly in industries where regulated parties differ widely in their technical expertise. Currently, agencies differ in their approach to default assumptions in the assessment of extreme events, and the implications of these differences are not yet well understood. At the U.S. Nuclear Regulatory Commission, defaults have generally been set at or near the mean of the nuclear power industry, while other agencies are more conservative. A recent report (U.S. NRC, 1997) states, "Generic data sources should be representative of the plant components." Similarly, Regulatory Guide 1.174 (U.S. NRC, 1998f) states, "When actual conditions cannot be monitored or measured, whatever information most closely approximates actual performance data should be used."

Regulators and other risk managers are struggling to find an appropriate balance in their use of risk analysis methods in decision making. The situation is well described as a combination of ongoing advancement in the application of risk-based methods for handling extreme events, coupled with hesitancy or distrust of their use in practice. Such tensions arise in part because of the distinction (Lowrance, 1976) between the quantification of risk (an empirical process) and the determination of safety (a normative process). Despite the best of intentions, analysts' assumptions and beliefs inherently have normative implications, making it impossible to uncouple fully the (ideally objective) task of risk quantification from the value judgments involved in the determination of safety.

The role of the analyst (sometimes hidden, sometimes explicit) is not unique to risk assessment and management, but rather is inherent to the process of modeling and decision making in any field. However, the large uncertainties associated with the assessment of risks from rare or extreme events make the challenges even greater here than in other fields, since empirical data will often be too limited to resolve ongoing disputes.

6. SUMMARY

We have reviewed (1) the conceptualization and definition of extreme events, (2) methods of modeling and estimating the frequencies of extreme events, including extreme value theory, mixture models, random process models, decomposition, parameter estimation, and nonprobabilistic models, (3) communication of the risk of extreme events, (4) implications for decision making about extreme events, and (5) policy issues concerning extreme events. We have observed that assessment of the risks of extreme events is different in kind from many of the problems faced in other fields, as a result of the small amounts of data that are typically available. Moreover, assessing the probabilities of extreme events is a difficult task, and is likely to remain difficult even given the results of research on better approaches for doing so. However, further work specifically focused on extreme events and their implications for risk analysis and the future management of technology remains essential. These advances are needed to improve the technical foundations for assessing the risks of extreme and rare events, as well as the societal experience of using these methods in setting priorities and policies.

Notes

1. The example that follows is taken from Schlesinger (1990).
2. All events and outcomes may, from a sufficiently distant perspective, be viewed as extraordinary, with our existence itself being so improbable as to argue that all of our experience is an extreme and rare event. However, as noted by J. Monod in his essay *Chance and Necessity* (1971): "Among all the events possible in the Universe, the *a priori* probability of any particular one of them occurring is next to zero. Yet the Universe exists. particular events must exist in it, the probability of which before was infinitesimal."
3. For consideration of the role of human experience and adaptation in affecting health outcomes from extreme temperatures, see, for example, Larsen (1990a,b), Langford and Bentham (1995), Patz (1996), Kalkstein and Greene (1997), Chestnut et al. (1998), and Smoyer (1998).
4. Note that if the chemical is also an essential nutrient, which is unlikely for anthropogenic compounds, low doses take on more importance.
5. Because of positive correlation in events occurring between adjacent and nearby basins, *more* than 100 actual basins are required to yield the statistical equivalent of 100 *independent* basins.
6. As noted in Peebles (2000, p. 159), Edmond Halley was the first, in 1694, to raise concern about the impact of a comet on Earth leading to worldwide extinction. Lord Byron was also very concerned about the effects of an asteroid or comet impact and first promoted the exploration of options for planetary defense in 1822.
7. The independence assumption is usually *not* good for daily stream flow and other variables that exhibit a high degree of autocorrelation; however,

if longer averaging times are used, this assumption can be more nearly satisfied.

8. We assume that the X_i are sampled from the same population, independent, and identically distributed (iid).
9. Mixture models are sometimes referred to as "contamination mixture" models, especially when a small subset of the population differs from the general population in a distinct manner.
10. Hora and Iman (1987) present an insightful analysis of Bayesian methods for failure rates when learning is a priori assumed and then combined with evidence from repeated failure-free operation (alternative prior models with forgetting, and outcomes *with* failure events, were also considered for illustrative purposes). Though not presenting space shuttle and related rocket launch data directly (since these data were restricted), the authors based their example on the performance history of such launches and the evidence of learning suggested by them. The authors had the misfortune of submitting their paper for publication in February 1985, with eventual publication in May 1987. In between, on January 28, 1986, the space shuttle *Challenger* exploded shortly after takeoff, resulting in the death of its seven crew members. Thus, in retrospect, such models may teach us more about what we have *not* learned than about what we have.

References

Adams, B. J., and Papa, F. 2000. *Urban Stormwater Management Planning with Analytical Probabilistic Model*. Wiley, New York.

Aitcheson, J., and J. A. C. Brown, 1957. *The Lognormal Distribution*. Cambridge University Press, New York.

Aitkin, M., and Wilson, G. T. 1980. Mixture model, outliers, and the EM algorithm. *Technometrics*, 22 (3): 325–31.

Ali, M. M. 1977. Probability and utility estimates for racetrack bettors. *Journal of Political Economy*, 85: 803–15.

Anderson, C. W., and Nadarajah, S. 1993. Environmental factors affecting reservoir safety. In V. Barnett and K. F. Turkman, eds., *Statistics for the Environment*, pp. 163–82. Wiley, Chichester.

Anderson, P. L., and Meerschaert, M. M. 1998. Modeling river flows with heavy tails. *Water Resources Research*, 34: 2271–80.

Ang, A. H. S., and Tang, W. H. 1984. *Probability Concepts in Engineering Planning and Design*, Vols. 1 and 2. Wiley, New York.

Apostolakis, G. E., Bier, V. M., and Mosleh, A. 1988. A critique of recent models for human error rate assessment. *Reliability Engineering and System Safety*, 22: 201–17.

Applegate, J. S. 1991. The perils of unreasonable risk: Information, regulatory policy, and toxic substances control, *Columbia Law Review*, 91: 261–333.

Arjas, E., Hansen, C. K., and Thyregod, P. 1991. Estimation of mean cumulative number of failures in a repairable system with mixed exponential component lifetimes. *Technometrics*, 33: 1–12.

Armstrong, J. S. 1985. *Long-Range Forecasting: From Crystal Ball to Computer*. Wiley, New York.

Asbeck, E., and Haimes, Y. Y. 1984. The partitioned multiobjective risk method. *Large Scale Systems*, 6: 13–38.

Asimov, I. 1983. *The Robots of Dawn*. Spectra Publishing, New York.

Bacchi, B., Brath, A., and Kottegoda, N. T. 1992. Analysis of the relationship between flood peaks and flood volumes based on crossing properties of river flow processes. *Water Resources Research*, 10: 2773–82.

Bailey, R. T. 1997. Estimation from zero-failure data. *Risk Analysis*, 17 (3): 375–80.

Batchelor, R. A., and Dua, P. 1995. Forecaster diversity and the benefits of combining forecasts. *Management Science*, 41: 68–75.

Bedford, T., and Cooke, R. 2001. *Probabilistic Risk Analysis: Foundations and Methods*. Cambridge University Press, Cambridge.

Berger, J. O. 1985. *Statistical Decision Theory and Bayesian Analysis*. Springer-Verlag, New York.

Beyth-Marom, R. 1982. How probable is probable? Numerical translation of verbal probability expressions. *Journal of Forecasting*, 1: 257–69.

Bier, V. M. 1988. The U.S. Nuclear Regulatory Commission safety goal policy: A critical review. *Risk Analysis*, 8: 559–64.

Bier, V. M. 1994. After 15 years of quantitative safety goals, now what? *American Nuclear Society Executive Conference on Policy Implications of Risk-Based Regulation* (March 13–16, 1994), Washington, DC.

Bier, V. M. 2001. On the state of the art: Risk communication to the public. *Reliability Engineering and System Safety*, 71: 139–50.

Bier, V. M., and Jang, S. C. 1999. Defaults and incentives in risk-informed regulation. *Human and Ecological Risk Assessment*, 5: 635–44.

Bierlant, J., Teugels, J. L., and Vynckier, P. 1996. *Practical Analysis of Extreme Values*. Leuven University Press, Leuven, Belgium.

Block, H. W., and Savits, T. H. 1997. Burn-in (with comment by Lynn and Singpurwalla). *Statistical Science*, 12 (1): 1–19.

Block, H. W., Mi, J., and Savits, T. H. 1993. Burn-in and mixed populations. *Journal of Applied Probability*, 30: 692–702.

Bobee, B. 1975. The log Pearson type 3 distribution and its application in hydrology. *Water Resources Research*, 11: 681–9.

Bolotin, V. V. 1993. Seismic risk assessment for structures with the Monte Carlo simulation. *Probabilistic Engineering Mechanics*, 8: 169–77.

Bond, G., Broecker, W., Johnsen, S., McManus, J., Labeyrie, L., Jouzel, J., and Bonani, G. 1993. Correlations between climate records from North Atlantic sediments and Greenland ice. *Nature*, 365: 143–7.

Bostrom, A. 1997. Vaccine risk communication: Lessons from risk perception, decision making and environmental risk communication research. *RISK: Health, Safety & Environment*, 8: 173–200.

Bradley, A. A., and Smith, J. A. 1994. The hydrometeorological environment of extreme rainstorms in the southern plains of the United States. *Journal of Applied Meteorology*, 33 (12): 1418–32.

Bras, R. L., and Rodriguez-Iturbe, I. 1985. *Random Functions and Hydrology*. Addison-Wesley, Reading, MA.

Broecker, W. S. 1994. Massive iceberg discharges as triggers for global climate change. Nature, 372: 421–4.

Broecker, W. S. 1995. Chaotic climate. *Scientific American*, November, pp. 62–8.

Budescu, D. V., and Wallsten, T. S. 1985. Consistency in interpretation of probabilistic phrases. *Organizational Behavior and Human Decision Processes*, 36: 391–405.

Budescu, D. V., and Wallsten, T. S. 1995. Processing linguistic probabilities: General principles and empirical evidence. In J. Busemeyer, D. L. Medin, and R. Hastie, eds., *Decision Making from a Cognitive Perspective*. Academic Press, San Diego.

Budescu, D. V., Weinberg, S., and Wallsten, T. S. 1988. Decisions based on numerically and verbally expressed uncertainties. *Journal of Experimental Psychology: Human Perception and Performance*, 14: 281–94.

Budnitz, R. J., Apostolakis, G., Boore, D. M., Cluff, L. S., Coppersmith, K. J., Cornell, C. A., and Morris, P. A. 1998. Use of technical expert panels: Applications to probabilistic seismic hazard analysis. *Risk Analysis*, 18: 463–9.

Burges, S. J., Lettenmaier, D. P., and Bates, C. L., 1975. Properties of the three-parameter lognormal distribution. *Water Resources Research*, 11 (2): 229–35.

Burmaster, D. E., and Wilson, A. M. 2000. Fitting second-order finite mixture models to data with many censored values using maximum likelihood estimation. *Risk Analysis*, 20: 261–71.

Burton, P. W., and Makropoulos, K. C. 1985. Seismic risk of circum-Pacific earthquakes. II. Extreme values using Gumbel's third distribution and the relationship with strain energy release. *Pure and Applied Geophysics*, 123: 849–66.

Castillo, E. 1988. *Extreme Value Theory in Engineering*. Academic Press, Boston.

Center for Strategic and International Studies (CSIS). 1998. *The Environmental Protection System in Transition: Toward a More Desirable Future – Final Report of the Enterprise for the Environment*. CSIS Press, Washington, DC.

Chankong, V., and Haimes, Y. Y. 1983. *Multiobjective Decision Making: Theory and Methodology*. Elsevier, New York.

Chestnut, L. G., Breffle, W. S., Smith, J. B., and Kalkstein, L. S. 1998. Analysis of differences in hot-weather-related mortality across 44 U.S. metropolitan areas. *Environmental Sciences Policy*, 1: 59–70.

Chinander, K. R., Kleindorfer, P. R., and Kunreuther, H. C. 1998. Compliance strategies and regulatory effectiveness of performance-based regulation of chemical accident risks. *Risk Analysis*, 18: 135–43.

Chock, D. P., and Sluchak, P. S. 1986. Estimating extreme values of air quality data using different fitted distributions. *Atmospheric Environment*, 20 (5): 989–93.

Clark, D. A. 1990. Verbal uncertainty expressions: A review of two decades of research. *Current Psychology: Research and Reviews*, 9: 203–35.

Clemen, R. T. 1989. Combining forecasts: A review and annotated bibliography. *International Journal of Forecasting*, 5: 559–83.

Clemen, R. T. 1991. *Making Hard Decisions: An Introduction to Decision Analysis*. PWS-Kent, Boston.

Clemen, R. T., and Reilly, T. 2001. *Making Hard Decisions with Decision Tools*. Duxbury, Thomson Learning, Pacific Grove, CA.

Coles, S. G., and Tawn, J. A. 1991. Modelling extreme multivariate events. *Journal of the Royal Statistical Society, Series B*, 53: 377–92.

Coles, S. G., and Tawn, J. A. 1994. Statistical methods for multivariate extremes: An application to structural design. *Applied Statistics*, 43 (1): 1–48.

Cooke, R. M. 1991. *Experts in Uncertainty: Opinion and Subjective Probability in Science*. Oxford University Press, Oxford.

Cooke, R., and Jager, E. 1998. A probabilistic model for the failure frequency of underground gas pipelines. *Risk Analysis*, 18: 511–27.

Cosmides, L., and Tooby, J. 1996. Are humans good intuitive statisticians after all? Rethinking some conclusions from the literature on judgment under uncertainty. *Cognition*, 58: 1–73.

Cox, D. R. 1972. Regression models and life tables. *Journal of the Royal Statistical Society, Series B*, 34: 187–220.

Crawford, S. L., DeGroot, M. H., Kadane, J. B., and Small, M. J. 1992. Modeling lake chemistry distributions: Bayesian methods for estimating a finite-mixture model. *Technometrics*, 34 (4): 441–53.

David, H. A. 1981. *Order Statistics*. Wiley, New York.

Davoudian, K., Wu, J.-S., and Apostolakis, G. 1994a. Incorporating organizational factors into risk assessment through the analysis of work processes. *Reliability Engineering and System Safety*, 45: 85–105.

Davoudian, K., Wu, J.-S., and Apostolakis, G. 1994b. The work process analysis model (WPAM). *Reliability Engineering and System Safety*, 45: 107–25.

DeGroot, M. H. 1970. *Optimal Statistical Decisions*. McGraw Hill, New York.

DeGroot, M. H., and Schervish, M. J. 2002. *Probability and Statistics*, 3rd ed. Addison-Wesley, Boston.

Delaunay, D. 1987. Extreme wind speed distribution for tropical cyclones. *Journal Wind Engineering and Industrial Aerodynamics*, 28: 61–8.

Di Toro, D. M. 1980. Statistics of advective dispersive system response to run-off. In Y. A. Yousef, ed., *Proceedings, Urban Stormwater and Combined Sewer Overflow Impact on Receiving Water Bodies*. EPA-600/9-80-056. U.S. Environmental Protection Agency, Municipal Environmental Research Lab, Cincinnati.

Dong, W., and Wong, F. S. 1986. From uncertainty to approximate reasoning: Part 1: Conceptual models and engineering interpretations. *Civil Engineering Systems*, 3: 143–50.

Duane, J. T. 1964. Learning curve approach to reliability monitoring. *IEEE Transactions on Air and Space*, 2: 563–6.

Eagleson, P. S. 1972. Dynamics of flood frequency. *Water Resources Research*, 8, 878–98

Everitt, B. S., and Hand, D. H. 1981. *Finite Mixture Distributions*. Chapman and Hall, London.

Ferson, S. 1993. Using fuzzy arithmetic in Monte Carlo simulation of fishery populations. In T. J. Quinn, ed., *Management of Exploited Fish: Proceedings of the International Symposium on Management Strategies for Exploited Fish Populations*. AK-SG-93-02. Alaska Sea Grant College Program.

Ferson, S. 1995. Quality assurance for Monte Carlo risk assessments. In *The Third International Symposium on Uncertainty Modeling and Analysis and Annual Conference of the North American Fuzzy Information Processing Society*. IEEE Computer Society Press, Los Alamitos, CA.

Ferson, S. 2002. *RAMAS Risk Calc 4.0 Software: Risk Assessment with Uncertain Numbers*. Lewis Publishers, Boca Raton, FL.

Ferson, S., and Ginzburg, L. R. 1996. Different methods are needed to propagate ignorance and variability. *Reliability Engineering and System Safety*, 54: 133–44.

Ferson, S., and Long, T. F. 1995. Conservative uncertainty propagation in environmental risk assessments. In J. S. Hughes, G. R. Biddinger, and E. Mones, eds.,

Environmental Toxicology and Risk Assessment 3, ASTM STP 1218. American Society for Testing and Materials, Philadelphia.

Fiering, M. B., and Kindler, J. 1984. Surprise in water-resource design. *International Journal of Water Resources Development*, 2: 1–10.

Finkel, A. M. 1994. The case for "plausible conservatism" in choosing and altering defaults. *Science and Judgment in Risk Assessment*. National Academy Press, Washington, DC.

Fisher, R. A., and Tippett, L. H. C. 1927. Limiting forms of the frequency distribution of the largest or smallest numbers of a sample. *Proceedings of the Cambridge Philosophical Society*, 24.

Fischhoff, B. 2000. Informed consent for eliciting environmental values. *Environmental Science & Technology*, 34: 1439–44.

Fischhoff, B., Slovic, P., Lichtenstein, S., Read, S., and Combs, B. 1978. How safe is safe enough? A psychometric study of attitudes towards technological risk. *Policy Sciences*, 9: 127–52.

Fleming, K. N., Mosleh, A., and Deremer, R. K. 1986. A systematic procedure for the incorporation of common-cause events into risk and reliability models. *Nuclear Engineering and Design*, 93: 245–75.

Fleming, K. N., Mosleh, A., and Kelley Jr., A. P. 1983. On the analysis of dependent failures in risk assessment and reliability evaluation. *Nuclear Safety*, 24: 637–57.

Foufoula-Georgiou, E., and Guttorp, P. 1987. Assessment of a class of Neyman-Scott models for temporal rainfall. *Journal of Geophysical Research*, 92 (D8): 9679–82.

Fowler, T. G., and Sorgard, E. 2000. Modeling ship transportation risk. *Risk Analysis*, 20: 225–44.

Frank, M. V. 2000. Probabilistic analysis of the inadvertent reentry of the Cassini Spacecraft's radioisotope thermoelectric generators. *Risk Analysis*, 20: 251–60.

Freudenburg, W. 1988. Perceived risk, real risk: Social science and the art of probabilistic risk assessment. *Science* 241, 44–9.

Frohwein, H. I. 1999. *Risk of Extreme Events in Multiobjective Decision Trees*. Ph.D. Thesis, Department of Systems Engineering, University of Virginia, Charlottesville, Virginia.

Frohwein, H. I., and Lambert, J. H. 2000. Risk of extreme events in multiobjective decision trees. Part 1. Severe events. *Risk Analysis*, 20: 113–23.

Frohwein, H. I., Haimes, Y. Y., and Lambert, J. H. 2000. Risk of extreme events in multiobjective decision trees. Part 2. Rare events. *Risk Analysis*, 20: 125–34.

Frohwein, F., Lambert, J., and Haimes, Y. 1999. Alternative measures of risk of extreme events in decision trees. *Reliability Engineering and System Safety*, 66 (1): 69–84.

Galambos, J. 1987. *The Asymptotic Theory of Extreme Order Statistics.*, 2nd ed. Robert E. Krieger Publishing, Malabar, FL.

Genest, C., and Zidek, J. V. 1986. Combining probability distributions: A critique and an annotated bibliography. *Statistical Science*, 1: 114–35.

Gerrard, M. B. 2000. Risk of hazardous waste sites versus asteroid and comet impacts: Accounting for the discrepancies in U.S. resource allocation. *Risk Analysis*, 20: 895–904.

Gerrard, M. B., and Barber, A. W. 1997. Asteroids and comets: U.S. and international law and the lowest-probability, highest consequence risk. *New York University Environmental Law Journal*, 6 (1): 4–49.

Gigerenzer, G. 2002. *Calculated Risks: How to Know When Numbers Deceive You*. Simon & Schuster, New York.

Gigerenzer, G., and Hoffrage, J. 1995. How to improve Bayesian reasoning without instruction: Frequency formats. *Psychological Review*, 102: 684–704.

Gleick, J. 1987. *Chaos: Making a New Science*. Viking, New York.

Gnedenko, B. V. 1943. Sur la distribution limité du terme maxium d'une serie aléatoire. *Annals of Mathematics*, 44: 423–53.

Goldman, S., 1953. *Information Theory*. Prentice-Hall, New York.

Gonzalez-Vallejo, C. C., and Wallsten, T. S. 1992. Effects of probability mode on preference reversal. *Journal of Experimental Psychology: Learning, Memory, and Cognition*, 18: 855–64.

Greeno, J. G. 1970. Evaluation of statistical hypotheses using information transmitted. *Philosophy of Science*, 37, 279–94.

Greenwood, J. A., Landwehr, J. M., Matalas, N. C., and Wallis, J. R. 1979. Probability weighted moments: Definition and their relation to parameters of several distributions expressible in inverse form. *Water Resources Research*, 15: 1049–54.

Griesmeyer, J. M., and Okrent, D. 1980. On the development of quantitative risk acceptance criteria. *An Approach to Quantitative Safety Goals for Nuclear Power Plants*. NUREG-0739. U.S. Nuclear Regulatory Commission, Washington, DC.

Griffith, R. M. 1949. Odds adjustments by American horse race bettors. *American Journal of Psychology*, 62: 290–4.

Gross, L. J., and Small, M. J. 1998. River and floodplain process simulation for subsurface characterization. *Water Resources Research*, 34: 2365–76.

Guey, C. N., and Heising, C. D. 1986. Development of a common cause failure analysis method: The inverse stress-strain interference (ISSI) technique. *Structural Safety*, 4: 63–77.

Gumbel, E. J. 1958. *Statistics of Extremes*. Columbia University Press, New York.

Haimes, Y. Y. 1988. Alternatives to the precommensuration of costs, benefits, risks, and times. In D. D. Rouman and Y. Y. Haimes, eds., *The Role of Social and Behavioral Sciences in Water Resources Planning and Management*. American Society of Civil Engineers, New York.

Haimes, Y. Y. 1991. Total risk management. *Risk Analysis*, 11: 169–71.

Haimes, Y. Y. 1998. *Risk Modeling, Assessment, and Management*. Wiley, New York.

Haimes, Y. Y. 1999. Editorial: The role of the Society for Risk Analysis in the emerging threats to critical infrastructure. *Risk Analysis*, 19: 153–7.

Haimes, Y. Y., and Hall, W. A. 1974. Multiobjectives in water resources systems analysis: The surrogate worth trade-off method. *Water Resources Research*, 10: 615–24.

Haimes, Y. Y., and Jiang, P. 2001. Leontief-based model of risk in complex interconnected infrastructures. *ASCE Journal of Infrastructure Systems*, 7 (1): 1–12.

Haimes, Y. Y., and Li, D. 1991. A hierarchical-multiobjective framework for risk management. *Automatica*, 27: 579–84.

Haimes, Y. Y., and Steuer R., eds. 2000. *Research and Practice in Multiple Criteria Decision Making*. Springer, Heidelberg, Germany.

Haimes, Y. Y., Lambert, J. H., and Li, D. 1992. Risk of extreme events in a multiobjective framework. *Water Resources Bulletin*, 28 (1): 201–9.

Haimes, Y. Y., Li, D., Karlsson, P., and Mitsiopoulos, J. 1990. Extreme events: Risk management. In M. G. Singh, ed., *Systems and Control Encyclopedia*, Supplementary Volume 1. Pergamon Press, Oxford.

Haimes, Y. Y., Matalas, N. C., Lambert, J. H., Jackson, B. A., and Fellows, J. F. R. 1998. Reducing the vulnerability of water supply systems to attack. *Journal of Infrastructure Systems*, 4 (4): 164–77.

Hakel, M. 1968. How often is often? *American Psychologist*, 23: 533–4.

Halpern, D. F., Blackman, S., and Salzman, B. 1989. Using statistical risk information to assess oral contraceptive safety. *Applied Cognitive Psychology*, 3: 251–60.

Harr, M. E. 1987. *Reliability-Based Design in Civil Engineering*. McGraw-Hill, New York.

Hart, S., Sharir, M., and Pneuli, A. 1983. Termination of probabilistic concurrent programs. *ACM Transactions on Programming Languages and Systems*, 5 (3): 356–80.

Hartonas-Garmhausen, V. 1998. *Probabilistic Symbolic Model Checking with Engineering Models and Applications*. Ph.D. Thesis, Department of Engineering and Public Policy, Carnegie Mellon University, Pittsburgh.

Hartonas-Garmhausen, V., Kurfess, T., Clarke, E. M., and Long, D. 1995. Automatic verification of industrial designs. In *Proceedings of the Workshop on Industrial Strength Formal Specification Techniques*. IEEE Computer Society Press, Los Alamitos, CA.

Heising, C. D., and Guey, C. N. 1984. A comparison of methods for calculating system unavailability due to common cause failures: The beta factor and multiple dependent failure fraction methods. *Reliability Engineering*, 8: 101–16.

Heising, C. D., and Luciani, D. M. 1987. Application of a computerized methodology for performing common cause failure analysis: The MOCUS-BACFIRE beta factor (MOBB) code. *Reliability Engineering*, 17: 193–210.

Hogarth, R. M. 1975. Cognitive processes and the assessment of subjective probability distributions. *Journal of the American Statistical Association*, 70: 271–91.

Hora, S. C., and Iman, R. L. 1987. Bayesian analysis of learning in risk analyses. *Technometrics*, 29 (2): 221–8.

Horowitz, J., and Barakat, S. 1979. Statistical analysis of the maximum concentration of an air pollutant: Effects of autocorrelation and non-stationarity. *Atmospheric Environment*, 13: 811–18.

Hosking, J. R. M., 1990. L-moments: Analysis and estimation of distributions using linear combinations of order statistics. *Journal of the Royal Statistical Society, Series B*, 52 (1): 105–24.

Hosking, J. R. M., and Wallis, J. R. 1987. Parameter and quantile estimation for the generalised Pareto distribution. *Technometrics*, 29 (3): 339–49.

Hosking, J. R. M., and Wallis, J. R. 1995. A comparison of unbiased and plotting-position estimators of L-moments. *Water Resources Research*, 31: 2019–25.

Hosking, J. R. M., Wallis, J. R., and Wood, E. F. 1985. Estimation of the generalized extreme value distribution by the method of probability-weighted moments. *Technometrics*, 27 (3): 251–61.

Hsueh, K.-S., and Mosleh, A. 1996. The development and application of the accident dynamic simulator for dynamic probabilistic risk assessment of nuclear power plants. *Reliability Engineering and System Safety*, 52: 297–314.

Interagency Advisory Committee on Water Data. 1982. *Guidelines for Determining Flood Flow Frequency*. Bulletin 17B of the Hydrology Subcommittee. Office of Water Data Coordination, U.S. Geological Survey, Reston, VA.

Jain, D., and Singh, V. P. 1987. Estimating parameters of the EV1 distribution for flood frequency analysis. *Water Resources Bulletin*, 23 (1): 59–71.

Jaynes, E. T. 1995. *Probability Theory: The Logic of Science*. Electronically published at http://bayes.wustl.edu/pub/Jaynes/book.probability.theory.

Jenkinson, A. F., 1955. The frequency distribution of the annual maximum (or minimum) value of meteorological elements. *Quarterly Journal of the Royal Meteorological Society*, 81: 158–71.

Johnson, C. W. 1993. A probabilistic logic for the development of safety-critical, interactive systems. *International Journal of Man-Machine Systems*, 39 (2): 333–51.

Johnson, N. L., Kotz, S., and Balakrishnan, N. 1995. *Continuous Univariate Distributions*, vol. 2. Wiley, New York.

Johnson, V. E., and Albert, J. H. 1999. *Ordinal Data Modeling*. Springer-Verlag, New York.

Julius, J., Jorgenson, E., Parry, G. W., and Mosleh, A. M. 1995. A procedure for the analysis of errors of commission in a Probabilistic Safety Assessment of a nuclear power plant at full power. *Reliability Engineering and System Safety*, 50: 189–201.

Kagan, Y., and Knopoff, L. 1977. Earthquake risk predictions as a stochastic process. *Physics of the Earth and Planetary Interiors*, 14: 97–107.

Kalkstein, L. S., and Greene, J. S. 1997. An evaluation of climate/mortality relationships in large U.S. cities and the possible impacts of climate change. *Environmental Health Perspectives*, 105 (1): 84–93.

Kaplan, R., Hammel, B., and Schimmel, L. 1986. Patient information processing and the decision to accept treatment. *Journal of Social Behavior and Personality*, 1: 113–20.

Kaplan, S., and Garrick, B. J. 1981. On the quantitative definition of risk. *Risk Analysis*, 1: 11–27.

Karlsson, P. O., and Haimes, Y. Y. 1988a. Risk based analysis of extreme events. *Water Resources Research*, 24: 9–20.

Karlsson, P. O., and Haimes, Y. Y. 1988b. Probability distributions and their partitioning. *Water Resources Research*, 24: 21–9.

Katz, R. W., and Brown, B. G. 1992. Extreme events in a changing climate: Variability is more important than climate. *Climatic Change*, 21: 289–302.

Kaufmann, A., and Gupta, M. M. 1991. *Introduction to Fuzzy Arithmetic*. Van Nostrand Reinhold, New York.

Keeney, R. L., and Raiffa, H. 1976. *Decisions with Multiple Objectives*. Wiley, New York.

Keigwin, L. D., and Jones, G. A. 1994. Western North Atlantic evidence for millennial-scale changes in ocean circulation and climate. *Journal of Geophysical Research*, 99 (12): 397–410.

Kinnison, R. R. 1985. *Applied Extreme Value Statistics*. MacMillan, New York.

Kottegoda, N. T., and Rosso, R. 1997. *Statistics, Probability, and Reliability for Civil and Environmental Engineers*. McGraw-Hill, New York.

Kotz, S., and Nadarajah, S. 2000. *Extreme Value Distributions: Theory and Applications*. Imperial College Press, London.

Kuchenhoff, H., and Thamerus, M. 1996. Extreme value analysis of Munich air pollution data. *Environmental and Ecological Statistics*, 3 (2): 127–41.

Kumamoto, H., and Henley, E. H. 1996. *Probabilistic Risk Assessment and Management for Engineers and Scientists*. IEEE Press, New York.

Lambert, J. H., and Li, D. 1994. Evaluating risk of extreme events for univariate loss functions. *Journal of Water Resources Planning and Management*, 120 (3): 382–99.

Lambert, J. H., Matalas, N., Ling, C. W., Haimes, Y. Y., and Li, D. 1994. Selection of probability distributions for risk of extreme events. *Risk Analysis*, 14: 731–42.

Lambert, J. H., Haimes, Y. Y., Li, D., Schooff, R., and Tulsiani, V. 2001. Identification, ranking, and management of risks in a major system acquisition. *Reliability Engineering and System Safety*, 72 (3): 315–25.

Landwehr, J. M, Matalas, N. C., and Wallis, J. R. 1979. Estimation of parameters and quantiles of Wakeby distributions. *Water Resources Research*, 15: 1361–79 (with a correction on p. 1672).

Langford, I. H., and Bentham, C. 1995. The potential effects of climate change on winter mortality in England and Wales. *International Journal of Biometeorology*, 38: 141–7.

Larsen, U. 1990a. The effects of monthly temperature fluctuations on mortality in the United States from 1921 to 1985. *International Journal of Biometeorology*, 34: 136–45.

Larsen, U. 1990b. Short-term fluctuations in death by cause, temperature, and income in the United States 1930 to 1985. *Social Biology*, 37 (3/4): 172–87.

Lave, L. B., and Balvanyos, T. 1998. Risk analysis and management of dam safety. *Risk Analysis*, 18: 455–62.

Leadbetter, M. R. 1991. On a basis for "peak over a threshold" modeling. *Statistics and Probability Letters*, 12: 357–62.

Leadbetter, M. R., Lindgren, G., and Rootzen, H. 1982. *Extremes and Related Properties of Random Sequences and Processes*. Springer-Verlag, New York.

Lehman, S. J., and Keigwin, L. D. 1992. Sudden changes in North Atlantic circulation during the last deglaciation. *Nature*, 356: 757–62.

Leonard, T., and Hsu, J. S. J. 1999. *Bayesian Methods: An Analysis for Statisticians and Interdisciplinary Researchers*. Cambridge University Press, Cambridge.

Leveson, N. G. 1995. *SafeWare: System safety and computers*. Addison-Wesley, Reading, MA.

Levine, R. D., and Tribus, M. 1978. *The Maximum Entropy Formalism*. MIT Press, Cambridge, MA.

Lichtenstein, S., and Newman, J. R. 1967. Empirical scaling of common verbal phrases associated with numerical probabilities. *Psychonomic Sciences*, 9: 563–4.

Lichtenstein, S., Slovic, P., Fischhoff, B., Layman, M., and Combs, B. 1978. Judged frequency of lethal events. *Journal of Experimental Psychology: Human Learning and Memory*, 4: 551–78.

Longstaff, T. A., Chittister, C., Pethia, R., and Haimes, Y. Y. 2000. Are we forgetting the risks of information technology? *Computer*, 33 (12): 43–51.

Loomis, J. B., and DuVair, P. H. 1993. Evaluating the effect of alternative risk communication devices on willingness to pay: Results from a dichotomous choice contingent valuation experiment. *Land Economics*, 69: 287–98.

Lopez Gomez, U. A. 1990. *Communicating Very Low Probability Events*. Ph.D. Thesis, Department of Engineering and Public Policy, Carnegie Mellon University, Pittsburgh.

Lowrance, W. 1976. *Of Acceptable Risk*. Kaufmann, Los Altos, CA.

Macwan, A., and Mosleh, A. 1994. A methodology for modeling operator errors of commission in probabilistic risk assessment. *Reliability Engineering and System Safety*, 45: 139–57.

Manna, Z., and Pnueli, A. 1992. *The Temporal Logic of Reactive and Concurrent Systems-Specification*. Springer-Verlag, Heidelberg.

Martz, H. F., and Waller, R. A. 1982. *Bayesian Reliability Analysis*. Wiley, New York.

Mathiesen, M., Goda, Y., Hawkes, P. J., Mansard, E., Martin, M. J., Pelthier, E., Thomson, E. F., and Van Vledder, G. 1994. Recommended practice for extreme wave analysis. *Journal of Hydraulic Research*, 32 (6): 803–14.

Merrick, J. R. W., van Dorp, J. R., Harrald, J. R., Mazzuchi, T. A., Grabowski, M., and Spahn, J. E. 2000. A systems approach to managing oil transportation risk in Prince William Sound. *Systems Engineering*, 3: 128–42.

Metcalfe, A. V. 1997. *Statistics in Civil Engineering*. Arnold, London (co-published by Wiley, New York).

Mitsiopoulos, J. A., Haimes, Y. Y., and Li, D. 1991. Approximating catastrophic risk through statistics of extremes. *Water Resources Research*, 27: 1223–30.

Monod, J. 1971. *Chance and Necessity: An Essay on the Natural Philosophy of Modern Biology*. Knopf, New York.

Morgan, M. G., and Henrion, M. 1990. *Uncertainty: A Guide to Dealing with Uncertainty in Quantitative Risk and Policy Analysis*. Cambridge University Press, Cambridge.

Morgan, M. G., and Keith, D. W. 1995. Subjective judgments by climate experts. *Environmental Science and Technology*, 29: 468A–77A.

Moses, F. 1998. Probabilistic-based structural specifications. *Risk Analysis*, 10. 445–54.

Mosleh, A., Bier, V. M. and Apostolakis, G. 1987. *Methods for the Elicitation and Use of Expert Opinion in Risk Assessment: Phase I – A Critical Evaluation and Directions for Future Research*, NUREG/CR-4962. U.S. Nuclear Regulatory Commission, Washington, DC.

Mosleh, A., Goldfeiz, E., and Shen, S. 1997. The ω-factor approach for modeling the influence of organizational factors in probabilistic safety assessment. In D. I. Gertman, D. L. Schurman, and H. S. Blackman, eds., *Proceedings of the 1997 IEEE Sixth Conference on Human Factors and Power Plants: Global Perspectives of Human Factors in Power Generation*. Institute of Electrical and Electronics Engineers, New York.

Mosleh, A., Hsueh, K. S., and Macwan, A. P. 1991. *A Simulation Based Approach to Modeling Errors of Commission*. UMNE 91-001. Materials and Nuclear Engineering Department, University of Maryland, College Park, MD.

Muir, L. R., and El-Shaarawi, A. H. 1986. On the calculation of extreme wave heights. *Ocean Engineering*, 13: 93–118.

Munera, H. A., Canal, M. B., and Munoz, M. 1997. Risk associated with transportation of spent nuclear fuel under demanding security constraints: The Columbian experience. *Risk Analysis*, 17: 381–9.

National Research Council. 1985. *Safety of Dams: Flood and Earthquake Criteria*. National Academy Press, Washington, DC.

National Research Council. 1996. *Understanding Risk: Informing Decisions in a Democratic Society*. National Academy Press, Washington, DC.

Natural Environment Research Council. 1975. *Flood Studies Report, Vol. 1. Hydrological Studies*, National Environment Research Council, London.

Nogee, P., and Lieberman, B. 1960. The auction value of certain risky situations. *Journal of Psychology*, 49: 167–79.

Nordquist, J. M. 1945. Theory of largest values, applied to earthquake magnitudes. *Transactions American Geophysical Union*, 26: 29–31.

North, M. 1980. Time-dependent stochastic models of floods. *Journal of the Hydraulics Division of the American Society of Civil Engineers*, 106: 649–55.

Olsen, J. R., Lambert, J. H., and Haimes, Y. Y. 1998. Risk of extreme events under nonstationary conditions. *Risk Analysis*, 18: 497–510.

Pannullo, J. E., Li, D., and Haimes, Y. Y. 1993. On the characteristics of extreme values for series systems. *Reliability Engineering and System Safety*, 40: 101–10.

Pate-Cornell, M. E. 1984. Fault trees vs. event trees in reliability analysis. *Risk Analysis*, 4: 177–89.

Pate-Cornell, M. E. 1990. Organization aspects of engineering system safety: The case of offshore platforms. *Science*, 250: 1210–17.

Pate-Cornell, M. E., and Bea, R. G. 1992. Management errors and system reliability: A probabilistic approach and application to offshore platforms. *Risk Analysis*, 12: 1–18.

Pate-Cornell, M. E., and Fischbeck, P. S. 1993. PRA as a management tool: Organizational factors and risk-based priorities for the maintenance of the tiles of the space shuttle orbiter. *Reliability Engineering and Systems Safety*, 40: 239–57.

Pate-Cornell, M. E., Lakats, L. M., Murphy, D. M., and Gaba, D. M. 1997. Anesthesia patient risk: A quantitative approach to organizational factors and risk management options. *Risk Analysis*, 17: 511–23.

Patz, J. A. 1996. Health adaptations to climate change: Need for farsighted integrated approaches. In J. B. Smith et al., eds., *Adapting to Climate Change*. Springer, New York.

Peebles, C. 2000. *Asteroids, A History*. Smithsonian Institution Press, Washington, DC.

Pepper, S. 1981. Problems in the quantification of frequency expressions. In D. Fiske, ed., *New Directions for Methodology of Social and Behavioral Sciences: Problems with Language Imprecision*. Jossey-Bass, San Francisco.

Pet-Armacost, J. J., Sepulveda, J., and Sakude, M. 1999. Monte Carlo sensitivity analysis of unknown parameters in hazardous materials transportation risk assessment. *Risk Analysis*, 19: 1173–84.

Petrakian, R., Haimes, Y. Y., Stakhiv, E. Z., and Moser, D. A. 1989. Risk analysis of dam failure and extreme floods. In Y. Y. Haimes and E. Z. Stahkiv, eds., *Risk Analysis and Management of Natural and Man-Made Hazards*. American Society of Civil Engineers, New York.

Phien, H. N., and Hira, M. A. 1983. Log Pearson type-3 distribution: Parameter estimation. *Journal of Hydrology*, 64 (3): 25–37.

Pnueli, A., and Zuck, L. 1993. Probabilistic verification. *Information and Computation*, 103 (1): 1–29.

Poston, T., and Stewart, I. 1978. *Catastrophe Theory and its Applications*. Pitman, London.

Powers, G. J., and Lapp, S. A. 1976. Computer-aided fault tree synthesis. *Chemical Engineering Progress*, 72 (April): 89–93.

Powers, G. J., and Tompkins Jr., F. C. 1974. Fault tree synthesis for chemical processes. *AIChE Journal*, 20 (2): 376–87.

Pratt, J., Raiffa, H., and Schlaifer, R. 1995. *Introduction to Statistical Decision Theory*. MIT Press, Cambridge, MA.

Press, S. J. 1989. *Bayesian Statistics: Principles, Models, and Applications*. Wiley, New York.

Preston, M. G., and Baratta, P. 1948. An experimental study of the auction-value of an uncertain outcome. *American Journal of Psychology*, 60: 183–93.

Purdy, G. 1993. Risk analysis of the transportation of dangerous goods by road and rail. *Journal of Hazardous Materials*, 33: 229–59.

Raiffa, H., and Schlaifer, R. 1961. *Applied Statistical Decision Theory*. Harvard University, Cambridge, MA.

Ramaswami, A., and Small, M. J. 1994. Modeling the spatial variability of natural trace element concentrations in groundwater. *Water Resources Research*, 30: 269–82.

Reiss, R.-D., and Thomas, M. 1997. *Statistical Analysis of Extreme Values*. Birkhauser Verlag, Boston.

Remo, J. L., ed. 1997. Near-earth objects: The United Nations International Conference. *Annals of the New York Academy of Sciences*, 822.

Resendiz-Carillo, D., and Lave, L. B. 1987. Optimizing spillway capacity with an estimated distribution of floods. *Water Resources Research*, 23 (1): 2043–9.

Resnick, S. I. 1987. *Extreme Values, Regular Variation and Point Processes*. Springer, New York.

Rice, S. O. 1944. Mathematical analysis of shot noise. *Bell Systems Technical Journal*, 23: 282–332; also 24: 46–156.

Roberts, E. M. 1979a. Review of the statistics of extreme values with application to air pollution data. Part I: Review. *Journal of the Air Pollution Control Association*, 29: 632–7.

Roberts, E. M. 1979b. Review of the statistics of extreme values with application to air pollution data, Part II: Applications. *Journal of the Air Pollution Control Association*, 29:, 733–40.

Robinson, J. S., and Sivapalan, M. 1997. Temporal scales and hydrological regimes: Implications for flood frequency scaling. *Water Resources Research*, 33: 2981–99.

Romei, S. F., Haimes, Y. Y., and Li, D. 1991. Exact determination and sensitivity analysis of a risk measure of extreme events. *Information and Decision Technologies*, 18: 265–82.

Rossi, F., Fiorentino, M., and Versace, P. 1984. Two-component extreme value distribution for flood frequency analysis. *Water Resources Research*, 20 (7): 847–56.

Royall, R. 2000. On the probability of observing misleading statistical evidence (with discussion). *Journal of the American Statistical Association*, 95 (451): 760–80.

Sandman, P. M., Weinstein, N. D., and Miller, P. 1994. High risk or low: How location on a 'risk ladder' affects perceived risk. *Risk Analysis*, 14: 35–45.

Sangal, B. P., and Biswas, A. K. 1970. The 3-parameter lognormal distribution and its applications in hydrology. *Water Resources Research*, 6 (2): 505–15.

Schlesinger, G. N. 1990. The stars are singing as they shine: The hand that made us is divine. In A. Gotfryd, H. Branover, and S. Lipskar, eds., *Fusion: Absolute Standards in a World of Relativity*. Feldheim Publishers, Jerusalem.

Shackle, G. L. S. 1972. *Imagination and the Nature of Choice*. Edinburgh University Press, Edinburgh.

Shafer, G. 1976. *A Mathematical Theory of Evidence*. Princeton University Press, Princeton, NJ.

Shannon, C. E. 1948. A mathematical theory of communication. *The Bell System Technical Journal*, 27: 623–56.

Sharit, J. 2000. A modeling framework for exposing risks in complex systems. *Risk Analysis*, 20: 469–82.

Shiloh, S., and Sagi, M. 1989. Effect of framing on the perception of genetic recurrence risks. *American Journal of Medical Genetics*, 33: 130–5.

Siegrist, M. 1997. Communicating low risk magnitudes: Incidence rates expressed as frequency versus rates expressed as probability. *Risk Analysis*, 17, 507–10.

Simpson, R. H. 1944. The specific meanings of certain terms indicating differing degrees of frequency. *Quarterly Journal of Speech*, 30: 328–30.

Simpson, R. H. 1963. Stability in meanings for quantitative terms: A comparison over 20 years. *Quarterly Journal of Speech*, 49: 146–51.

Singh, K. P. 1986. Flood estimate reliability enhancement by detection and modification of outliers. In B. C. Yen, ed., *Stochastic and Risk Analysis in Hydraulic Engineering*. Water Resources Publications, Littleton, CO.

Slovic, P. 1987. Perceptions of risk. *Science*, 236: 280–5.

Small, M. J., and Fischbeck, P. S. 1999. False precision in Bayesian updating with incomplete models. *Human and Ecological Risk Assessment*, 5 (2): 291–304.

Small, M. J., and Morgan, D. J. 1986. The relationship between a continuous time renewal model and a discrete Markov chain model of precipitation occurrence. *Water Resources Research*, 22: 1422–30.

Small, M. J., Sutton, M. C., and Milke, M. W. 1988. Parametric distributions of regional lake chemistry: Fitted and derived. *Environmental Science and Technology*, 22 (2): 196–204.

Smidts, C., Shen, S. H., and Mosleh, A. 1997. The IDA cognitive model for the analysis of nuclear power plant operator response under accident conditions. Part I: problem solving and decision making model. *Reliability Engineering and System Safety*, 55: 51–71.

Smith, J. A., and Karr, A. F. 1983. A point process model of summer rainfall occurrences. *Water Resources Research*, 19: 95–103.

Smith, J. A., and Karr, A. F. 1985. Statistical inference for point process models of rainfall. *Water Resources Research*, 21 (1): 73–9.

Smith, J. A., and Karr, A. F. 1986. Flood frequency analysis using the Cox regression model. *Water Resources Research*, 22: 890–6.

Smith, J. A., Bradley, A. A., and Baeck, M. L. 1994. The space-time structure of extreme storm rainfall in the Southern Plains. *Journal of Applied Meteorology*, 33 (12): 1402–17.

Smith, R. L., and Shively, T. S. 1995. Point process approach to modeling trends in tropospheric ozone based on exceedances of a high threshold. *Atmospheric Environment*, 29 (23): 3489–99.

Smoyer, K. E. 1998. Putting risk in its place: Methodological considerations for investigating extreme event health risk. *Social Science and Medicine*, 47: 1809–24.

Solari, G. 1996. Statistical analysis of extreme wind speed. In D. P. Lalas and C. F. Ratto, eds., *Modeling of Atmosphere Flow Fields*. World Scientific Publishing, Singapore.

Spash, C. L. 2000. Multiple value expression in contingent valuation: Economics and ethics. *Environmental Science & Technology*, 34 (8): 1433–8.

Specter, H., and Braun, C. 1994. Safety goals, safety policy, and economic opportunities. *American Nuclear Society Executive Conference on Policy Implications of Risk-Based Regulation*. Washington, DC.

Sprowls, R. C. 1953. Psychological-mathematical probability in relationships of lottery gambles. *American Journal of Psychology*, 66: 126–30.

Stedinger, J. R. 1980. Fitting lognormal distribution to hydrologic data. *Water Resources Research*, 16: 481–90.

Stedinger, J. R., Vogel, R. M., and Foufoula-Georgiou, E. 1993. Frequency analysis of extreme events. In D. Maidment, ed., *Handbook of Hydrology*, McGraw-Hill, New York.

Stiber, N. A., Pantazidou, M., and Small, M. J. 1999. Expert system methodology for evaluating reductive dechlorination at TCE sites. *Environmental Science and Technology*, 33 (17): 3012–20.

Stone, D. R., and Johnson, R. J. 1959. A study of words indicating frequency. *Journal of Educational Psychology*, 50: 224–7.

Stoppard, T. 1967. *Rosencrantz and Guildenstern Are Dead*. Grove Press, New York.

Surman, P. G., Bodero, J., and Simpson, R. W. 1987. The prediction of the numbers of violations of standards and the frequency of air pollution episodes using extreme value theory. *Atmospheric Environment*, 21 (8): 1843–8.

Taqqu, M. S. 1987. Random processes with long-range dependence and high variability. *Journal of Geophysical Research*, 92 (D8): 9683–86.

Tawn, J. A. 1992. Estimating probabilities of extreme sea levels. *Applied Statistics*, 41: 77–93.

Thom, H. J. S. 1954. Frequency of maximum wind speeds. *Proceedings of the American Society of Civil Engineers*, 80: 104–14.

Titterington, D. M., Smith, A. F. M., and Makov, U. E. 1985. *Statistical Analysis of Finite Mixture Distributions*. Wiley, Chichester.

Tung, Y. K., and Mays, L. W. 1981. Reducing hydrologic parameter uncertainty. *ASCE Journal of Water Resources Planning and Management Division*, 107 (WR1): 245–62.

U.S. Nuclear Regulatory Commission. 1975. *Reactor Safety Study: An Assessment of Accident Risks in U.S. Commercial Nuclear Power Plants*. WASH-1400. U.S. NRC, Washington, DC.

U.S. Nuclear Regulatory Commission. 1997. *The Use of PRA in Risk-Informed Applications*. Draft Report for Comment, NUREG-1602. U.S. NRC, Washington, DC.

U.S. Nuclear Regulatory Commission. 1998a. *Risk-Informed Inservice Testing*. Standard Review Plan (SRP) Chapter 3.9.7. U.S. NRC, Washington, DC.

U.S. Nuclear Regulatory Commission. 1998b. *Standard Review Plan for Trial Use For the Review of Risk-Informed Inservice Inspection of Piping*. SRP Chapter 3.9.8. U.S. NRC, Washington, DC.

U.S. Nuclear Regulatory Commission. 1998c. *Risk-Informed Decisionmaking: Technical Specifications*. SRP Chapter 16.1. U.S. NRC, Washington, DC.

U.S. Nuclear Regulatory Commission. 1998d. *Use of Probabilistic Risk Assessment in Plant-Specific, Risk-Informed Decisionmaking: General Guidance*. SRP Chapter 19.0. U.S. NRC, Washington, DC.

U.S. Nuclear Regulatory Commission. 1998e. *An Approach for Using Probabilistic Risk Assessment in Risk-Informed Decisions on Plant-Specific Changes to the Licensing Basis*. Regulatory Guide 1.174. U.S. NRC, Washington, DC.

U.S. Nuclear Regulatory Commission. 1998f. *An Approach for Plant-Specific, Risk-Informed Decisionmaking: Inservice Testing*. Regulatory Guide 1.175. U.S. NRC, Washington, DC.

U.S. Nuclear Regulatory Commission. 1998g. *An Approach for Plant-Specific, Risk-Informed Decisionmaking: Graded Quality Assurance*. Regulatory Guide 1.176. U.S. NRC, Washington, DC.

U.S. Nuclear Regulatory Commission. 1998h. *An Approach for Plant-Specific, Risk-Informed Decisionmaking: Technical Specifications*. Regulatory Guide 1.177. U.S. NRC, Washington, DC.

U.S. Nuclear Regulatory Commission. 1998i. *An Approach For Plant-Specific Risk-Informed Decisionmaking: Inservice Inspection of Piping*. Regulatory Guide 1.178. U.S. NRC, Washington, DC.

van Dorp, J. R., Merrick, J. R. W., Harrald, J. R., Mazzuchi, T. A., and Grabowski, M. 2001. A risk management procedure for the Washington state ferries. *Risk Analysis*, 21: 127–42.

Vanmarcke, E. 1983. *Random Fields: Analysis and Synthesis*. MIT Press, Cambridge, MA.

Vardi, M. 1985. Automatic verification of probabilistic concurrent finite-state programs. *Proceedings of the 26th IEEE Symposium on Foundations of Computer Science*. Institute of Electrical and Electronics Engineers, New York.

Vaughan, D. 1997. *The Challenger Launch Decision: Risky Technology, Culture, and Deviance at NASA*. University of Chicago Press, Chicago.

Vaurio, J. K. 1984. Learning from nuclear accident experience. *Risk Analysis*, 4: 103–15.

Vaurio, J. K. 1994. Estimation of common cause failure rates based on uncertain event data. *Risk Analysis*, 14: 383–7.

Vaurio, J. K. 1995. The probabilistic modeling of external common cause failure shocks in redundant systems. *Reliability Engineering and System Safety*, 50: 97–107.

Vaurio, J. K. 1998. An implicit method for incorporating common-cause failures in systems analysis. *IEEE Transactions on Reliability*, 47 (2): 173–80.

Vaurio, J. K. 2001. Modelling and quantification of dependent repeatable human errors in system analysis and risk assessment. *Reliability Engineering and System Safety*, 71 (2): 179–188.

Vere-Jones, D. 1970. Stochastic models for earthquake occurrence. *Journal of the Royal Statistical Society, Series B*, 32: 1–62.

Vere-Jones, D. 1978. Earthquake prediction – A statistician's view. *Journal of Physics of the Earth*, 26: 129–46.

Vere-Jones, D. 1992. Statistical methods for the description and display of earthquake catalogs. In A. T. Walden and P. Guttorp, eds., *Statistics in the Environmental and Earth Sciences*. Edward Arnold, London.

von Winterfeldt, D., and Edwards, W. 1986. *Decision Analysis and Behavioral Research*. Cambridge University Press, Cambridge.

Walley, P. 1991. *Statistical Reasoning with Imprecise Probabilities*. Chapman and Hall, London.

Wallis, J. R., and Wood, E. F. 1985. Relative accuracy of log Pearson III procedures. *Journal of Hydraulic Engineering*, 111 (7): 1043–56.

Wallsten, T. S., and Budescu, D. V. 1995. A review of human linguistic probability processing: General principles and empirical evidence. *The Knowledge Engineering Review*, 10: 43–62.

Wang, G., Lambert, J., and Haimes, Y. 1999a. Stochastic minimax decision rules for risk of extreme events. *IEEE Transactions on Systems, Man and Cybernetics A*, 29 (6): 533–41.

Wang, G., Lambert, J., and Haimes, Y. 1999b. Stochastic ordering of extreme value distributions. *IEEE Transactions on Systems, Man and Cybernetics A*, 29 (6): 696–701.

Webb, R. S., Rind, D. H., Lehman, S. J., Healy, R. J., and Sigman, D. 1997. Influence of ocean heat transport on the climate of the Last Glacial Maximum. *Nature*, 385: 695–9.

Weiss, G. 1977. Shot noise models for the generation of synthetic streamflow data. *Water Resources Research*, 13: 101–8.

Wilks, S. S. 1948. Order statistics. *Bulletin of the American Mathematical Society*, 54: 6–50.

Wilks, S. S. 1962. *Mathematical Statistics*. Wiley, New York.

Winkler, R. L. 1981. Combining probability distributions from dependent information sources. *Management Science*, 27: 479–88.

Winkler, R. L., and Clemen, R. T. 1992. The combination of forecasts. *Journal of the Royal Statistical Society, Series A*, 146: 150–7.

Youngblood, R. W. 1998. Applying risk models to formulation of safety cases. *Risk Analysis*, 18: 433–44.

Zadeh, L. A. 1965. Fuzzy sets. *Information and Control*, 8: 338–53.

Zadeh, L. A. 1978. Fuzzy sets as a basis for a theory of possibility. *Fuzzy Sets and Systems*, 1, 3–28.

Zhang, Y., and Der Kiureghian, A. 1994. First-excursion probability of uncertain structures. *Probabilistic Engineering Mechanics*, 9: 135–43.

Zimmermann, H.-J. 1991. *Fuzzy Set Theory and its Applications*, 2nd rev. ed. Kluwer, Dordrecht, The Netherlands.

For Further Reading

The following texts are recommended for further reading . . .
on technical and statistical methods for characterizing extreme and rare events:

Bierlant, J., Teugels, J. L., and Vynckier, P. 1996. *Practical Analysis of Extreme Values*. Leuven University Press, Leuven, Belgium.

Haimes, Y. Y. 1998. *Risk Modeling, Assessment, and Management*. Wiley, New York.

Kotz, S., and Nadarajah, S. 2000. *Extreme Value Distributions: Theory and Applications*. Imperial College Press, London.

Paté-Cornell, M. E. 1996. Uncertainties in risk analysis: Six levels of treatment. *Reliability Engineering and System Safety*, 54: 95–111.

Reiss, R.-D., and Thomas, M. 1997. *Statistical Analysis of Extreme Values*. Birkhauser Verlag, Boston.

on the role of extreme and rare events in decision making and policy:

Clemen, R. T., and Reilly, T. 2001. *Making Hard Decisions with Decision Tools*. Duxbury, Thomson Learning, Pacific Grove, CA.

Freudenburg, W. 1988. Perceived risk, real risk: Social science and the art of probabilistic risk assessment. *Science* 241, 44–9.

Heimann, C. F. L. 1997. *Acceptable Risks: Politics, Policy, and Risky Technologies*. University of Michigan Press, Ann Arbor, MI.

Kunreuther, H. 1997. Rethinking society's management of catastrophic risks. *Geneva Papers on Risk and Insurance*, 83: 151–76.

Page, T. 1978. A generic view of toxic chemicals and similar risks, *Ecology Law Quarterly*, 7: 207–44.

5

Environmental Risk and Justice

Mary R. English

1. INTRODUCTION

Traditionally, risk was the province of scientists and engineers; justice, the province of ethicists. Their conjunction received little systematic study. Over the past few decades, however, they have been linked in political action as well as academic deliberation, particularly on issues concerning environmental risks to human health and safety. By now, the compelling need to consider whether environmental risks contribute to injustice is clear.

   The purpose of this chapter is to provide a comprehensive overview of environmental risk and justice. Specifically, the paper has four objectives: (1) to elucidate the relationship between environmental risk and justice; (2) to examine developments in environmental justice, as that term is popularly conceived; (3) to consider a *broader* conception of environmental justice – one that includes but extends beyond anthropocentric concerns; and (4) to suggest work needed within risk assessment, social science, and ethics, in order to improve our collective ability to manage environmental

Writing this paper has enabled me to bring together a number of ideas that, while seemingly disparate, are part of a larger picture of environmental risk and justice. I have tried to sketch that larger picture, rather than concentrate exclusively on "environmental justice" as that term is popularly conceived. Many thanks to the Society for Risk Analysis for providing this opportunity and the financial support to carry it out. Thanks also to Roger Kasperson for his suggestions on the paper's outline; to Rae Zimmerman and the editors of these papers, Tim McDaniels and Mitchell Small, for their guidance; to three anonymous reviewers for their perceptive suggestions on an earlier draft; and to David Steele, a graduate student in Sociology at the University of Tennessee, for his assistance in locating some key references. In addition, while this paper may or may not be well received within the "environmental justice community," I would like to extend my thanks to the staff and members of the U.S. Environmental Protection Agency's National Environmental Justice Advisory Council, on which I served for several years, for their dedication, warmth, and caring. Being part of the council was an invaluable experience.

risks in a just manner. The paper draws primarily upon experiences in the United States but refers to international and global concerns as well.

To begin, six basic questions are considered. These questions focus on factors central to environmental risk and justice:

the principles of justice invoked;
the origin of the risk;
how the risk is managed;
the distribution of risks and benefits;
the severity of the risk; and
the risk decision process.

By understanding how these factors play out in the abstract, actual experiences with environmental risk and justice can be better understood.

1.1. Principles of Justice: What Conception of Justice Is Applied?

Justice can be thought of in several different ways. It can be thought of as equality: as like treatment for like cases – the guiding principle behind much of the U.S. legal system. It also can be thought of as fair process: as following procedural rules that, by being fair, will by definition result in fair outcomes (Rawls, 1971). The latter provides for the possibility of tempering equality with principles such as a compensatory principle: special allowances in instances of past harm or grave need. Alternatively, justice – especially justice as imposed by a system of governance – can be thought of in libertarian terms, as the minimal safeguards needed to ensure citizens' protection against force, theft, fraud, and so on (Nozick, 1974). Or justice can be thought of less formally and more personally: not as a set of universal principles, but as each person having the opportunity to pursue his or her self-defined and context-specific conception of the good (Walzer, 1983).

Despite their differences, all of these conceptions of justice focus on the individual. In contrast, other conceptions of justice place greater emphasis on the well-being of society as a whole: for example, act-utilitarianism – roughly speaking, the view that the rightness or wrongness of an action depends on the total goodness or badness of its consequences (Smart and Williams, 1973), or communitarianism – a call for increased social responsibility and a renewed sense of community (Etzioni, 1993; Sandel, 1996). While seemingly not prominent in Western societies, especially in the United States, these societal conceptions of justice do, in fact, come into play. For example, the cost-benefit analyses used to justify some public policy decisions are grounded in a utilitarian conception of justice.

1.2. Risk Origin: Is the Hazard Natural or Anthropogenic?

If a giant asteroid hits the earth, that's nobody's fault. The same cannot be said for global warming.

The Beginner's Guide to the U. N. Framework Convention on
Climate Change, 1994 *(http://www.unfccc.de/resource/beginner.html)*

Within the broad literature on risk, there are two fairly distinct strands: those that deal with technological risks (e.g., risks posed by nuclear power plants or hazardous waste sites) and those that deal with natural disasters (e.g., disasters caused by hurricanes or earthquakes). For example, *Risk Analysis: An International Journal* features articles on the former, whereas *Natural Hazards: Journal of the International Society for the Prevention and Mitigation of Natural Hazards* features articles on the latter.

Technological risks are anthropogenic. In contrast, natural disasters have been seen as acts of God or fate. As humans demonstrate their ability to alter large natural phenomena such as global climate patterns, even "natural" disasters are regarded as partly anthropogenic. Traditionally, however, their origins have been seen as out of human hands.

As discussed later, the management of both technological risks and natural disasters may raise concerns and contribute to feelings of injustice. Because technological risks are obviously and unquestionably anthropogenic, however, they are more likely to trigger a sense of injustice, all other things being equal. This is especially the case if they are assumed involuntarily (see Section 1.5).

1.3. Risk Management: Is the Risk Managed Well or Badly?

In the now-classic "red book" on characterizing the potential adverse health effects of human exposure to toxic substances, risk assessment was partitioned from risk management. Risk assessment culminates in quantitatively or qualitatively estimating the incidence and magnitude of an adverse effect on a given population, whereas risk management, drawing upon risk assessment, culminates in developing policy options and evaluating their health, economic, social, and political implications (National Research Council, 1983). In a more recent report, however, the line between risk assessment and risk management is drawn less brightly: It is argued that an understanding of both is needed for a full understanding of risk (National Research Council, 1996). It also has been argued that risk assessment, like risk management, is permeated with normative assumptions (Cranor, 1997). The term *risk management* thus is used loosely here, to incorporate a variety of activities undertaken by a variety of agents – corporations, regulators, emergency management agencies, and so on – to prevent, minimize, or mitigate the adverse effects of hazards.

Regardless of whether a hazard is natural or anthropogenic, principles of justice may be violated if risks posed by the hazard are managed badly. The U.S. Presidential/Congressional Commission on Risk Assessment and Risk Management convened in the mid-1990s identified six stages in its "risk management framework": formulating the problem at hand, analyzing risks associated with the problem, examining options, making decisions, implementing the decisions, and evaluating implementation effectiveness (Presidential/Congressional Commission on Risk Assessment, 1997). If those responsible for any given stage fail to perform their duties or do not perform them even-handedly, those adversely affected may feel a sense of injustice. For example, if a hurricane notification system fails to give early warnings of an impending storm, all those at risk may feel they have been treated unjustly. If the system gives early warnings but in English only, non-English speakers may believe that they have been treated unjustly.

The outcomes of risk management thus have implications for justice; so too does the *process* of risk management. At the centerpiece of the Commission's framework is "stakeholder collaboration," which, the Commission argues, should occur at each stage of risk management. If, with a given risk, stakeholder collaboration in problem formulation, risk analysis, option identification, decision making, and so on is advocated but then not carried out, or is carried out with some people but not with others, those who hold conceptions of justice grounded in fair process are likely to feel affronted. For example, if air quality problems with coal-fired plants are formulated by regulators in consultation with electric utilities but no one else, environmental groups, people living in the affected areas, and the utility rate-payers may believe that, simply by being excluded from the collaborative process, they have been treated unfairly.

Process issues are considered further in Section 1.6, after considering two fundamental outcome issues: the distribution of risks and benefits and the severity of the risks.

1.4. The Distribution of Risks and Benefits: Who Gets What?

Distribution of risks is at the heart of many claims of environmental injustice. As discussed in Sections 2, 3, and 4, the environmental justice movement, which seeks to lighten disproportionately heavy environmental burdens on minority and low-income communities, is particularly concerned with risk distribution. Unfair distribution of risks has been a rallying cry for communities of all stripes, however, especially when controversial facilities (e.g., facilities to treat or dispose of radioactive or chemically hazardous wastes) are proposed. The familiar "not in my backyard" (NIMBY) slogan arose largely out of risk distribution concerns, although, as the need for such facilities was questioned, this slogan often became "not in anyone's backyard" (NIABY).

To achieve distributive justice in siting potentially noxious facilities, various allocative schemes have been explored in the United States and other nations: for example, simultaneously siting numerous small facilities within the context of a regional siting strategy (Morrell, 1984) or nationally determining needs for new facilities and then allocating responsibility for siting these facilities among the states (Gerrard, 1994). Schemes such as these are based on a conception of justice as equal treatment. Inevitably, however, they fail in the details: The new facilities to be distributed across the landscape are not equally burdensome, and some people will still bear their brunt more than others. Reductio ad absurdum: People bury their own hazardous wastes in their own backyards – clearly, a distributionally infeasible and environmentally unsound solution.

One popular alternative has been to identify siting criteria and then choose the most well-suited site. This approach appeals to a very different principle of justice: "from each according to his ability." It has been employed in many attempts to site controversial facilities, but without much success. The "best site" approach is easily challenged: The criteria may be disputed, and "best" is very difficult to prove (English, 1992). Casting about for solutions, those attempting to site controversial facilities have tried other approaches, such as deliberative, consensual processes (see Section 1.6) and compensatory benefits (Kunreuther, 1995).

Benefits are part of the risk equation. Technological risks (and also some "natural disaster" risks) come with benefits – for example, the benefits of nuclear medicine, with radioactive waste as a by-product. The risks may not fall squarely on technology providers and consumers, however; they may be shifted spatially and/or temporally. For example, the risks of permanent, centralized storage of spent nuclear fuel from commercial nuclear power plants will not necessarily fall on electricity consumers; they will fall on workers handling the fuel, on the host area to the permanent waste repository, on areas along the transportation routes to the repository, and on people who will live near the repository many millennia into the future.

To balance the equation of risks and benefits, compensation in the form of up-front money, tax revenue, new schools and parks, and so on is sometimes offered to communities that agree to potentially hazardous facilities. This approach, which is like giving high pay for high-girder work, is intended to make the risk more palatable. Like its workplace equivalent, it may be used to entice people to assume risks voluntarily. This leads to questions of what "voluntary" and "informed consent" means, particularly when consent is sought from a community, not simply an individual (English, 1991). Obtaining consent is even more problematic when risk decisions are centralized (MacLean, 1986).

Using compensation to balance the scales of justice has led to a few siting successes, especially if the risks are perceived as not unacceptably high, the community is economically in need, the benefits are sought-after, and a participatory deliberative process is used (Rabe, 1994). Even then, however, benefits are not a surefire answer. They still may be seen as bribes by some people within the prospective host community, and a local consensus to accept the facility may not be reached (English, 1992).

Using compensation is even less plausible with other types of environmental risks – that is, if the risks are spatially diffuse (as with global climate change) or temporally extended (as with radioactive waste disposal). Then, lacking the softening effect of compensatory benefits, the pull between competing conceptions of justice becomes starkly obvious. On the one hand, it can be argued that individuals, known and unknown, living and not yet born, have rights not to be burdened unduly with risks not of their own making. On the other hand, it can be argued that the welfare of society as a whole is more important than the welfare of a relatively few individuals – in effect, a collective benefit-cost analysis. With risks that extend far into the future, discount rates applied to both costs and benefits are a key (and controversial) issue.

1.5. The Severity of the Risk: How Bad Is It Really?

Risk, stripped to its essence, is a function of the probability of an adverse event occurring and the consequence if it does. The most trivial risks are low-probability, low-consequence; the most grave are high-probability, high-consequence. While people can agree on these theoretical poles, the range in between – where most risks fall – often provokes disagreement. A main reason is that risk can't be stripped to its essence. Risk is a construct colored by many things: by individual and collective knowledge and beliefs about the nature of the risk; by forms of organization and communication; and by values (about the payoffs as well as the perils of risk taking) and vantage points (about how exposed one will be to a risk) (Douglas and Wildavsky, 1982; Krimsky and Golding, 1992; Luhmann, 1993).

It has been documented that – regardless of dry statistics about probability and consequences – seemingly similar risks may be perceived differently. Dreaded and latent risks, such as the risk of cancer, are more likely to be perceived as high risks than familiar and immediate risks, such as the risk of an automobile accident. In addition, by now it is clear that risks assumed involuntarily are much more likely to be perceived as unacceptable than voluntary risks (Slovic, 1987; for a discussion of "toxic dread" and the stress, uncertainty, and distrust that it engenders, see also Edelstein, 1988, and Erikson, 1990).

As the discussion of risk management implies (see Section 1.3), the line between voluntary and involuntary risks is not always bright. Even with seemingly voluntary risks, if people have been kept in the dark, deceived, or manipulated, they may well maintain that the risks they have assumed weren't truly voluntary and that they have been treated unjustly. Smokers' lawsuits against the tobacco industry are a case in point. Moreover, our collective knowledge of risks is continually evolving. As we learn more about hazards, exposure routes, and effects, we reevaluate both the probabilities and the consequences of adverse events. This reevaluation may lead to an upgrading or downgrading of a risk – or even to recognition of positive, hormetic responses to low doses of otherwise toxic substances (Calabrese et al., 1999). In addition, as discussed further in Section 4.1, we have only recently begun to move beyond a focus on single-risk, single-exposure events to comprehensive attention to the cumulative effects of protracted low-dose exposures as well as the synergistic effects of combinations of diverse environmental conditions.

This evolution of knowledge casts doubt on our past understanding of risks and, by extension, on our current understanding. When taken together with differences in values (about the tradeoffs inherent in risks), as well as differences in vantage points (about whether a risk will be personally experienced), it is often virtually impossible to reach agreement on "How bad is it, really?" Yet this question is at the heart of environmental risk and justice.

If an environmental risk is extremely trivial – for example, if it involves only a minor inconvenience – few people would care, even if it was imposed unjustly. If it is extremely grave – for example, if a dam upstream from a city is about to break – most people would agree that it should be dealt with immediately. It is in the broad, gray area in between, where disputes about the severity of a risk arise, that justice is most likely to be an issue.

1.6. The Risk Decision Process: Who Decides What?

According to recent research (English, 1999; Tonn, English, and Travis, 2000), there are at least six different modes of environmental decision making, each varying in who makes the decision, under what conditions, and with what consequences:

1. With "emergency action" decisions, emergency managers make rapid decisions concerning urgent situations with potentially high consequences.
2. With "routine procedures" decisions, administrative and technical staff, following preestablished protocols, make decisions concerning familiar situations whose individual consequences are not high.
3. With "elite corps" decisions, senior officials reach agreement on an issue whose consequences (including political consequences) are potentially high.

4. With "analysis-centered" decisions, technical and policy analysts make carefully deliberated recommendations concerning complex issues whose consequences, taken in the aggregate over time, are potentially high.
5. With "conflict management" decisions, controversial issues with potentially high consequences are addressed by bringing together people representing various sides of the conflict.
6. With "collaborative learning" decisions, complex issues fraught with uncertainty and potentially very high consequences are tackled through open, deliberative, iterative processes.

These six modes help to elucidate that risk decisions can be made in a variety of ways, with different types of input by those responsible for managing the risk as well as those affected by it. These modes are, however, theoretical constructs. An actual decision process may involve two or more modes concurrently or over time. Moreover, an evolution is occurring in which modes are favored. While the first four modes (emergency action, routine procedures, elite corps, and analysis-centered) remain dominant for many environmental risk decisions, the latter two modes (conflict management and collaborative learning) increasingly are favored, especially for complicated, controversial issues with potentially severe risks. Those who might otherwise have used an analysis-centered or elite corps mode are now being urged to pursue an analytic-deliberative process, engaging interested and affected parties in the steps leading up to a decision, including the steps to characterize risks (National Research Council, 1996).

The shift toward more open, participatory ways to make decisions on environmental risks has taken place gradually in many nations. In the United States, for example, with the 1946 Administrative Procedure Act, which set forth federal requirements concerning public notice, public hearings, and public right to comment, the camel was able to get his nose in the tent. Nevertheless, this law mainly provided formal means for after-the-fact criticism of public agency decisions. It has taken several decades to evolve toward the right to participate in *reaching* decisions.

Now, with more widespread acceptance of the idea that those affected by an environmental risk decision should be both allowed and encouraged to participate in the decision process, from problem scoping onward, advice is plentiful on how this should be accomplished. In *Fairness and Competence in Citizen Participation* (Renn, Webler, and Wiedemann, 1995), for example, various approaches are discussed and evaluated, with the underlying theme that meaningful citizen participation is requisite for decision-making competence and legitimacy. Similarly, there are calls for transparency in public policy making in order to ensure that facts, values, and arguments are clearly presented to members of the public participating in decision processes (Andersson et al., 1999).

Underlying this shift toward more participatory decision processes is a reconceptualization of justice and environmental risk decisions. Whereas a consequentialist view of justice previously held sway, there is a new emphasis on procedural justice. Procedural justice is widely seen as both a good in itself and as a way to ensure more just outcomes. As was illustrated in a comparison of Austrian and U.S. siting processes (Linerooth-Bayer and Fitzgerald, 1996), notions about both fair process and fair outcomes vary. Nevertheless, as will become clear in the following four sections, an interweaving of fair process and fair outcomes is a hallmark of environmental justice today, on both a national and a global scale.

2. THE ENVIRONMENTAL JUSTICE MOVEMENT: BACKGROUND

The following discussion centers on the environmental justice movement in the United States. The movement is not unique to the United States, however. Roughly parallel developments have occurred in other countries, with growing awareness of disproportionate environmental burdens on people who often are disadvantaged in other ways as well.

2.1. Popular Conception of Environmental Justice

In the United States, *environmental justice* has become the popular term for the need to improve the environmental conditions of low-income communities and, especially, communities of color. The problem of environmental justice is a classic example of how environmental problems are socially constructed. (For a discussion of the social constructionist perspective, see Hannigan, 1995.) Virtually unrecognized a few decades ago, environmental justice has been raised to widespread national consciousness through the awareness, anger, and passion of people speaking for environmentally beleaguered communities that have, historically, been subjected to social, economic, and, especially, racial discrimination and oppression. In both the identity of the oppressed and the rhetorical, political, and legal tactics for alleviating oppression, environmental justice (as the term has come to be used) and civil rights have been closely intertwined.

Environmental injustices have not been borne solely by African-Americans. In 1978, New York State health authorities and then-President Carter announced that a chemical emergency existed in the largely blue-collar, Caucasian neighborhood of Love Canal in Niagara Falls, New York. This event, which came to a boil because of pressure from residents, helped to trigger a 1980 federal act to clean up waste sites (the Comprehensive Environmental Response, Compensation, and Liability Act, or Superfund) (Levine, 1982). Lois Gibbs, the president of the Love Canal Homeowners Association, went on to found a national networking organization, the Citizen's Clearinghouse for Hazardous Wastes, which included but was not limited to grassroots groups in communities of color. Community organizations in predominantly white, predominantly poor areas such as Appalachia also had formed during the 1970s to contest environmental degradations and other hardships imposed by corporations that owned and exploited large tracts of land, minerals, and forests (Gaventa, 1980).

But when the term *environmental justice* gained currency, approximately a decade ago, it was closely linked with civil rights issues. As discussed further in Section 3.2, Title VI of the 1964 Civil Rights Act refers to discrimination on the ground of race, color, or national origin. It does not mention economic discrimination. The problem of environmental justice, with its strong link to civil rights, thus has been seen by some as one imposed mainly on African-Americans; for this reason, it sometimes has been termed "environmental racism" (Bullard, 1993). Scholarly and political debates continue over whether environmental justice is, in fact, mainly a problem experienced by blacks: in other words, whether African-Americans, poor or not, live with heavier environmental burdens than other racial and ethnic minorities or people who are white but poor.

Nevertheless, both the term *environmental justice* and the movement have been inclusive. The President's 1994 executive order addressing environmental justice uses the term expansively, to refer to low-income communities as well as communities of color. (See Section 2.3 for a discussion of this order.) Latinos, Asian-Americans, and the obviously poverty-stricken and downtrodden, including but not limited to migrant workers – all have been drawn into the U.S. environmental justice movement, even as African-Americans have remained central to it. Native Americans also are included under the environmental justice umbrella, although, because of treaties between tribal nations and the federal government, they have somewhat different legal recourses.

The nature of the environmental justice problem gradually has been expanded as well. In the 1980s, as will be discussed further later, the location of waste facilities was the prime focus. Since then, attention has turned to the many other environmental insults that, individually or collectively, can burden a community: factories with toxic emissions, improperly managed mines and other resource extraction enterprises, highways and their vehicular traffic, and so forth. The lack of rapid solutions and slow, inadequate regulatory enforcement at the federal and state level (e.g., protracted and insufficient cleanups of some Superfund sites contaminated with hazardous wastes) is seen as part of the problem. The failure to adequately monitor and provide help for adverse health effects caused by environmental toxics (e.g., neurological problems in children ingesting or inhaling lead) is also part of the problem. Parenthetically, it might be said that shifting from a tight focus on waste facility siting to a broader spectrum of concerns may have been strategically wise in that it may have helped to minimize divisiveness within the environmental justice movement. Arguably, waste disposal sites such as sanitary landfills are needed, and someone has to live near them; the same can't be said for many other environmental justice concerns.

As an understanding has evolved of who and what are at issue with environmental justice, a widely shared (if still fuzzy) understanding of the term's meaning has evolved. A decade ago, the problem of disproportionate environmental burdens was referred to by some, especially government officials, as the need for "environmental equity" (U.S. Environmental Protection Agency, 1992). Pushed by the movement, however, the term *equity* was soon banished, and *justice* is now used in most circles. Equity could be interpreted simply as equal treatment under the laws – for example, as having no more than an equal likelihood of a new noxious facility nearby. What was sought, however, was not only equity but also a recognition of an accumulation of past wrongs, resulting in tipped scales of justice. To balance the scales, it would not be enough to be treated equally in the future; preexisting disproportionate environmental burdens would need to be mitigated and would need to be taken into account in future decisions. (For further discussion of the distinction between equity and justice, see Bryant, 1995.) *Justice* evokes a sense of the broad social and economic factors that have contributed to disproportionate environmental burdens, and – while imprecise – the term has a unifying resonance that invokes other widely held principles. One key principle is that of participation in decision making.

Fundamental to the conception of environmental justice as it has developed in the United States is the right of people affected by environmental risk decisions to participate in those decisions – especially the decisions of public agencies, but also those of corporations. On the view of environmental justice advocates, those making environmental risk decisions should treat affected communities as their equal partners; they should recognize community knowledge; they should encourage active participation using cross-cultural formats and exchanges; and they should maintain honesty and integrity when articulating goals, expectations, and limitations (National Environmental Justice Advisory Council, 1996a).

Underlying this conception is the principle that procedural fairness is a necessary but not sufficient condition for environmental justice. It is necessary, not only because of a belief in the right of people to have a voice in decisions that affect them, but also because of a lack of trust in those who, in the past, have made decisions resulting in disproportionate environmental burdens. But participation in the decision process increasingly is seen as not sufficient because of a growing skepticism about whether public participation can affect outcomes without pursuing other tactics as well (e.g., lawsuits and political lobbying). There also is a growing sense that when research funds are disbursed or staff are hired to address environmental justice issues, members of the "environmental justice community" should be tapped. They should not be simply the subjects of work within the halls of power or learning; they should be engaged in (and paid for) the work.

This last point suggests a difficult issue for those within or consulting with the environmental justice movement. Embedded in the movement is a potent but elusive concept: that of the environmental justice community. A community is a social entity characterized by interaction, interdependence, and common ties (Hillery, 1955; see also English and Zimlich, 1997). As with some other communities (e.g., the scientific community), the environmental justice community is a broad community of interest. However, it is also composed of smaller, place-based communities. This sets up a potential tension between those who belong to the environmental justice community by virtue of their skin color or ethnic background together with their commitment to the cause of environmental justice and those who directly encounter disproportionate environmental burdens in their daily lives. As is mentioned later, this tension has grown as the movement has developed.

2.2. Development of the Movement

In January 1990, the Conference on Race and the Incidence of Environmental Hazards was held by the University of Michigan School of Natural Resources. At the conference, a group consisting of social scientists and civil rights leaders (informally called the Michigan Coalition) organized and subsequently wrote to the Administrator of the U.S. Environmental Protection Agency (EPA), requesting action on environmental risk in minority and low-income communities. EPA responded by forming an Environmental Equity Workgroup, which, in May 1992, issued *Environmental Equity: Reducing Risk For All Communties* [sic], and, in late 1992, EPA established an Office of Environmental Equity (later changed to the Office of Environmental Justice). In the meantime, the First National People of Color Environmental Leadership Summit was held in Washington, DC, in 1991. Attended by more than 650 people, many of them community activists and members of grassroots organizations, the conference resulted in a set of seventeen principles to make explicit the movement's purpose and goals.

While the early 1990s could be said to mark the beginning of the officially recognized environmental justice movement, it had, in fact, been gathering force for a number of years. Otherwise, agencies such as EPA might not have been motivated to respond. A harbinger was heard as early as 1971, when the Council on Environmental Quality's *Annual Report to the President* noted that racial discrimination combined with poverty hampered people's ability to improve their environments. In 1979, a class-action lawsuit, *Bean v. Southwestern Waste Management*, was filed in Texas, opposing the planned location of a solid waste landfill in a residential, predominantly

African-American neighborhood. Although unsuccessful, the lawsuit and a related study (Bullard, 1983) drew attention to the link between civil rights and environmental decisions. In 1982, a signal event occurred: massive resistance by local activists, civil rights leaders, and members of Congress to the siting of a hazardous waste landfill in predominantly African-American Warren County, South Carolina.

The Warren County confrontation triggered a new awareness of potential environmental injustices. It also triggered various studies concerning siting inequities. The U.S. General Accounting Office, in a 1983 study, found that three-fourths of commercial waste sites located in the southeastern U.S. were to be found in predominantly poor African-American communities (U.S. General Accounting Office, 1983). Subsequently, similar cases were made in other notable studies, such as *Toxic Wastes and Race in the United States: A National Report on the Racial and Socioeconomic Characteristics of Communities with Hazardous Waste Sites* (1987), which was conducted by the United Church of Christ Commission for Racial Justice; and *Dumping in Dixie: Race, Class, and Environmental Quality* (1990) by Robert Bullard, a sociologist and long-standing leader within the environmental justice movement.

Initially, the movement was a loosely woven fabric of community groups, regional or national advocacy groups (often with community groups as members), academics, and political/legal activists. While the elements of the fabric remain largely the same today, the pattern has changed. Particularly with greater official recognition and increased status resulting from the National Environmental Justice Advisory Council established in July 1993 (see Section 2.3), the leadership of the movement became more sharply defined, even as the movement became more inclusive. Office doors of senior management at EPA and other agencies (e.g., the National Institute of Environmental Health Sciences, the Agency for Toxic Substances and Disease Registry) opened to key movement leaders who had risen to the forefront. In a few cases, movement leaders had been or became employees of the establishment being challenged.

With this evolution in the movement, the opportunity to influence agency and corporate decisions, especially environmental risk decisions, has increased. Movement members, working through formal advisory groups such as the National Environmental Justice Advisory Council and the Children's Health Protection Advisory Board at EPA, as well as through informal consultation, have altered the agendas of those making decisions. They have even captured the attention of the National Academies, hardly a hotbed of radical thinking: The Institute of Medicine recently released *Toward Environmental Justice: Research, Education, and Health Policy Needs* (1999).

There has been a concern within the movement, however, that the talk about environmental justice by public agency and corporate officials is mostly lip service, and that environmental justice is still largely unrealized. For example, a group called the Interim National Black Environmental and Economic Justice Coordinating Committee formed in late 1999 to prevent backsliding by EPA on enforcement of Title VI of the 1964 Civil Rights Act (see Section 3.2) and to rally activists against perceived efforts by industry and conservative groups to discredit the environmental justice movement (*Environmental Policy Alert*, January 25, 2000).

There also has been a concern (although often only murmured) that the movement has become stratified and disconnected from its roots. On the one hand, there are its leaders – those who stand out in the environmental justice community and have access to the halls of power – on the other hand, there are the affected people – those whose daily lives are spent in disproportionately burdened communities. Although settings such as meetings of the National Environmental Justice Advisory Council always include opportunity for comment from affected community members and the importance of this moving testimony is recognized, a gulf remains between those at the council table and those speaking to the council. One reason may be that, in order to change national policy, the movement leaders must speak in the language of policy and politics.

2.3. Recent Indications of the Movement's Influence

On Earth Day 1993, shortly after taking office, President Clinton made a commitment to address environmental inequity and discrimination. Following up on this commitment, he signed an executive order (Executive Order 12898) entitled "Federal Actions to Address Environmental Justice in Minority Populations and Low-Income Populations" on February 11, 1994. The order requires federal agencies to develop agencywide strategies for identifying and addressing "disproportionately high and adverse human health or environmental effects of [their] programs, policies, and activities on minority populations and low-income populations" (Sec. 1–103); to conduct their programs affecting human health and the environment in a manner that does not have discriminatory effects; to include minority and low-income populations whenever practicable and appropriate when conducting environmental human health research; and to collect and analyze, whenever practicable, information assessing and comparing environmental and human health risks borne by populations identified by race, national origin, or income. It also requires federal agencies to provide minority and low-income people with the opportunity for participating in the development of the agencies' environmental justice strategies and research programs.

The order, as a document internal to the executive branch, does not have the force of law; nevertheless, it was intended to guide how agencies carry out their administrative duties. It also has affected – on paper, at least – how the 1969 National Environmental Policy Act (NEPA) is implemented. In a memo accompanying the order, federal agencies were directed to analyze the effects on minority and low-income communities when a NEPA analysis is required. Nearly four years later, in December 1997, the Council on Environmental Quality, which is responsible for NEPA compliance, released a document providing guidance for incorporating environmental justice considerations into NEPA documents (U.S. Council on Environmental Quality, 1997).

In the 1990s, a number of state governments considered comparable measures to ensure that state agencies adhere to environmental justice principles. Many of these initiatives have taken the form of legislative bills; many of them have not passed (National Conference of State Legislatures, 1995). Nevertheless, in the late 1990s, with lawsuits brought under Title VI of the 1964 Civil Rights Act and with the threat of losing federal funds from EPA (see Section 3.2 for a discussion of Title VI), states have become somewhat less lethargic about environmental justice. Still, the question remains whether so far all of the talk has resulted in significant change in risk management – specifically, in the distribution and severity of risks.

3. CURRENT CONTROVERSIES CONCERNING RISK AND ENVIRONMENTAL JUSTICE

In the United States over the past few years, two controversial environmental justice issues have stood out. The first concerns research on whether low-income communities and communities of color typically do, in fact, have disproportionate environmental burdens. The second concerns how Title VI of the 1964 Civil Rights Act (hereafter referred to as Title VI) should be used to prevent further environmental burdens on racial and ethnic minorities. While the two issues are seemingly distinct one is an academic dispute and the other is a legal/regulatory dispute the former has implications for the latter, and the two issues are linked by their methodological complexity.

3.1. Environmental Risk Distribution

For more than a decade, a controversy has been percolating: Do communities of people who are low-income, from a racial or ethnic minority (in the United States) or both, actually have more numerous and severe environmental burdens than other communities? As yet, this controversy has not been put to rest.

On the one hand, numerous studies have shown disparities in the distribution of environmental hazards (White, 1998). Disparities based on race were found in 87 percent of the studies; disparities based on income, in 74 percent of the studies. According to Benjamin Goldman, these disparities were found in both urban and rural areas, in all regions of the country, and in a variety of circumstances – workplace conditions, the location of hazardous waste facilities, and so forth (Goldman, 1993).

Such studies have been criticized for their methods, however. Targets of criticism include equating membership in a community with exposure to an environmental harm; implying rather than making explicit the risk presented by the facilities or other environmental hazards in question; failing to determine whether the hazard was antecedent to the community; using lenient (low-ratio) percentages for the compositions of communities qualifying as "low-income" or "community of color"; and defining the communities too broadly in geographic terms (Boerner and Lambert, 1994; Foreman, 1998). On the last point, it has been found that using census tracts rather than larger areas such as zip codes or counties can yield very different results. A study of twenty-five metropolitan areas with commercial hazardous waste facilities indicated that, using census tracts as the unit of analysis for demographic characteristics, the facilities were slightly more likely to be located in predominantly white working-class neighborhoods than in minority neighborhoods (Anderton, Anderson, and Oakes, 1994). Similarly, a 1995 study of municipal landfills by the U.S. General Accounting Office found that, more often than not, the percentage of minority and low-income people within one mile of a landfill was lower than their percentages in the rest of the county (U.S. General Accounting Office, 1995).

At issue in this dispute are the following questions:

1. How should environmental burdens be counted?
2. How should a potential environmental justice (EJ) community be identified and analyzed?
3. How do EJ communities compare with other communities, in terms of the heaviness of their environmental burdens?

The following discussion gives a flavor of the complex issues at hand.

HOW SHOULD ENVIRONMENTAL BURDENS BE COUNTED? The term *environmental burdens* is in flux. For example, as discussed earlier, awareness has grown of the discriminatory effects of planning for highways and other modes of transportation (Bullard and Johnson, 1997). With this growing awareness, the definition of environmental burdens has been expanded. In a startling move, the director of EPA's Office of Environmental Justice even suggested that the definition of environmental justice should be expanded to include the right to esthetically and culturally pleasing surroundings (Barry Hill, memorandum, quoted in *The Reinvention Report*, March 10, 1999).

To date, however, most research has focused on one or a few classes of facilities that stand out as potential environmental burdens – for example, hazardous waste treatment and disposal facilities, other facilities with toxic releases, and Superfund sites. Typically, point-source environmental risks – especially those on which databases are maintained and readily available – are used as the focal point for research. Simply being in the same database does not mean that all facilities are equivalent, however. Some may be significant sources of pollution; others may have no off-site releases. In addition, if there are off-site releases, a simple radius around the facility may not accurately represent the affected area; instead, the plume of the releases may be downstream or downwind. Moreover, if an EJ community has become established in the vicinity of an environmentally burdensome facility, not vice versa, it can be argued that the facility was not "imposed" on the community; rather that broader social and economic forces drove the community to locate in an undesirable area (Been, 1994). Taking all of these complexities into account, especially when comparing a number of facilities and their surrounding communities, is extraordinarily difficult.

HOW SHOULD A POTENTIAL EJ COMMUNITY BE IDENTIFIED AND ANALYZED? In identifying and then analyzing a potential EJ community, two linked issues arise: the scope of its geographic area and (typically) the number of low-income and racial or ethnic minority individuals as a percentage of its total population. Regarding the former issue, it has been established that the geographic unit of analysis can have profound effects on research findings and their interpretations (Glickman et al., 1995). Various researchers have urged that the unit of analysis be carefully chosen to capture as accurately as possible the affected community (Greenberg, 1993; Zimmerman, 1993, 1994; National Research Council, 1999a). Often, the unit of analysis is chosen because of census data availability (e.g., counties, "places," "minor civil divisions," zip code areas, census tracts, census blocks). While this data driven necessity often can't be avoided, especially in large-sample studies, it has been recommended that data be analyzed at a number of geographic levels, and also that, when possible, field examinations should be conducted of the community potentially at risk (Zimmerman, 1994).

Once the geographic unit of analysis has been established, data on its demographic composition can be assembled. If "low-income and racial or ethnic minority" is used to define a potential environmental justice community, comparisons of the area's demographics may be made with national statistics (e.g., percent below official, federally established poverty levels; percent within particular racial or ethnic categories). Alternatively, comparisons may be made with other, smaller, and more proximate units within which the area is situated – for example, comparisons of the area's demographic statistics with state or county statistics. It has been noted, however, that "race and ethnic classifications are often used to simplify the reporting of data, but may, in fact, be inappropriate surrogates for more complex social attributes" (Zimmerman, 1994, p. 665). Similarly, research using the linked National Health Interview Survey–National Death Index has found that "the least socioeconomically advantaged groups – Mexican, Native, and African Americans – are able to reduce their mortality gaps with Caucasian and Asian Americans when social characteristics [such as marital status, family size, income, and household crowding] are controlled" (Rogers et al., 1996, p. 1434).

As with the geographic unit of analysis of a potential EJ community, simplifying assumptions in assessing the community's demographic composition may be unavoidable, especially when large-sample studies are undertaken. Nevertheless, these findings point to the need to select carefully which attributes will be studied. Moreover, when evaluating the environmental burdens of a potential EJ community, both the importance and the difficulty of meaningful environmental risk exposure studies cannot be overstated. (For further discussion of this point, see Section 4.1.)

HOW DO THE ENVIRONMENTAL BURDENS OF POTENTIAL EJ COMMUNITIES COMPARE WITH THOSE OF OTHER COMMUNITIES? Because of the methodological difficulties noted here, it is easy to challenge research that systematically and quantitatively tries to compare the environmental burdens of potential EJ communities with those of other communities. For this reason, the qualitative research that has been conducted on environmental burdens of low-income communities and communities of color is at least as important. Often conducted through interviews and case studies, this research documents the incidence of environmental burdens within these communities. (See, for example, Hurley, 1995.) Similarly, situation-specific evidence presented at meetings such as those of the National Environmental Justice Advisory Council makes a strong case for the need for environmental justice. Qualitative research and personal case histories cannot, however, provide the means to draw comparisons conclusively; instead, they anecdotally help to substantiate that environmental risk problems are, in general, worse in low-income communities and communities of color.

3.2. Title VI and Environmental Justice

Title VI of the 1964 Civil Rights Act, as amended, states that:

No person in the United States shall, on the ground of race, color, or national origin, be excluded from participation in, be denied the benefits of, or be subjected to discrimination under any program or activity receiving Federal financial assistance.

During the 1990s, EPA received an increasing number of complaints alleging discriminatory effects due to pollution control permits issued by states, and lawsuits were brought to bring this point home. For example, residents of the predominantly African-American city of Chester, Pennsylvania, through an organization called Chester Residents Concerned for Quality Living (CRCQL), brought a lawsuit against the state's Department of Environmental Protection, alleging discrimination under Title VI for granting a permit for yet another waste facility in the city. The suit eventually reached the U.S. Supreme Court, where it was declared moot because the specific reason for the suit (the permit) had become a dead issue. Nevertheless, the case established that groups such as CRCQL had standing to sue under Title VI.

States are responsible for administering programs enforcing federal laws such as the Clean Air Act and the Clean Water Act; their permitting agencies receive EPA financial assistance for these and other programs. In February 1998, EPA's Office of Civil Rights issued interim guidance describing how the office intended to process Title VI complaints regarding environmental permits issued by recipients of EPA financial assistance. (Low-income populations were not included in the Title VI language; hence, they were not included in this guidance.)

The interim guidance made clear that, following a prior Supreme Court ruling interpreting Title VI, proof of discriminatory intent was not needed; proof of discriminatory *effects* was enough to provide grounds for a Title VI complaint, unless it could be shown that there was no less-discriminatory alternative. The guidance established a framework for processing complaints. These included procedures for acceptance of the complaint; assessment of disparate impacts (at which point, if the Office of Civil Rights concluded that there was no disparate impact, the case would be dismissed); rebuttal, mitigation, or justification in response to an initial finding of disparate impacts; notice of determinations of noncompliance; informal resolution where practicable at any point in the process; voluntary compliance; and penalties (including withdrawal of funds and the possibility of litigation) in the event of noncompliance.

In response to this interim guidance, many states and industries cried "Foul!" and aligned to combat the interim guidance. The states, to whom the guidance was directed, were especially irate. Arrayed on the other side was the environmental justice community, which urged EPA's Office of Civil Rights to hang tough, although they too had technical criticisms of the guidance.

Many objections were raised concerning ambiguities in the guidance. Other objections, especially from the states, concerned whether EPA had gone too far in its attempt to guide state and local decisions. A key objection concerned the methodology for determining whether the permit(s) at issue would create (or add to) a disparate impact on a racial or ethnic population group.

The methodology laid out by the Office of Civil Rights in its February 1998 guidance entailed five steps: (1) identifying the population affected by the permit that triggered the complaint, using proximity as a reasonable indicator of where impacts are concentrated; (2) determining the racial and/or ethnic composition of the affected population; (3) determining whether other permitted facilities under the permitting agency's jurisdiction should be included in the universe of facilities to be analyzed, and determining the populations affected by those facilities; (4) conducting a disparate impact analysis by comparing the racial or ethnic characteristics of the affected population with those of the nonaffected population; and (5) using arithmetic or statistical analyses to determine whether the disparate impact is significant.

Questions and objections to the proposed methodology were heard from a number of quarters, including a fourteen-member panel of EPA's National Advisory Council for Environmental Policy and Technology. Core issues included how the "affected community" should be defined; whether social, economic, and cultural impacts should be considered along with health impacts; and what degree of disparity was needed to justify a determination of a Title VI violation.

In EPA's 1999 congressional appropriation, EPA was prohibited from using funds to implement the interim guidance for Title VI complaints filed after October 21, 1998. EPA then began a lengthy process of revising the guidance, using an advisory committee and consultation with representatives of various interested and affected parties. The extent to which Title VI has a major impact on environmental risk decision making remains to be seen.

3.3. Risk Analysis, Politics, and Justice

These two key controversies confronting the environmental justice movement in the United States, both of which concern the systematic analysis of risk distribution, present dilemmas for risk analysts and movement leaders alike. The double-barreled question – *how* a risk is distributed, and *who* is affected by it – is methodologically complex; these complexities become all the more intricate when cumulative risks from multiple sources must be taken into account. While risk analysts may find these complexities intellectually intriguing, unraveling them – or even agreeing upon plausible analytic methods that can be widely applied – is dauntingly difficult. And while movement leaders would like convincing factual evidence to support their claims for systemic and widespread disproportionate environmental burdens, this evidence is, similarly, dauntingly difficult to document.

Research and analysis on environmental risk distribution thus has become a necessity but also a quagmire. Gut intuition, anecdotes, rhetoric, and politics initially can be at least as effective as dispassionate statistical analysis, but, without the latter, these other strategies may not result in a durable movement. As has been pointed out in a recent discussion of the environmental justice movement (Sizemore, 2000), the movement, while still powerful, may be on the wane. Like other social movements, it appears to have passed through three phases of growth: first, emergence, development, and initial claims-making; second, framing, frame extension, and frame alignment; and third, external validation and domain expansion. But, particularly without convincing research methods and results to bulwark this validation, the movement may be entering a fourth phase, one of contention and recession. Even if this is the case, however – and it is not yet clear whether it is – the environmental justice movement has altered our collective thinking about environmental risks.

4. ROLE OF THE U.S. ENVIRONMENTAL JUSTICE MOVEMENT IN ELUCIDATING ENVIRONMENTAL RISK ISSUES

Putting aside these controversies concerning risk distribution and also the question of whether the environmental justice movement has prompted significantly different risk decision outcomes, the movement still has made a difference in the United States It has established the problem of environmental justice as an issue that must be considered when environmental risk decisions are made. Twenty years ago, environmental justice was a nascent concept; today, it has become a common part of our thinking about risk. Moreover, the environmental justice movement has helped raise to the foreground different ways to think about risks. Twenty years ago, risk analysts, regulators, and exposed individuals tended to think of environmental risks in nonaggregated terms, particularly across various kinds of risks. Today, as is discussed later, environmental risks and their potential effects are seen more holistically and with greater attention to differences among exposed individuals. While the environmental justice movement has not been the only champion of these changes, it has helped to foster them.

4.1. Health Risks Arising from Cumulative Exposures

For a number of years, the possibility of adverse health effects from chronic, low-dose exposure to toxic substances has been recognized. The science is difficult, however. For example, questions must be tackled about whether the toxic substance bioaccumulates and whether a linear, no-threshold model for adverse effects should be used (National Research Council, 1994). Moreover, the risks of multiple sources of exposure to the same toxic substance are often difficult to assess because of the complexity of multiple means of uptake in different individuals. Perhaps for this reason, while the cumulative effects of toxic exposure have been studied, they traditionally have received somewhat less attention than acute, episodic exposure to a toxic substance from a single source.

Further complicating the environmental health effects picture is the possibility of synergistic interaction among different toxic substances – in other words, "a response to a mixture of toxic substances that is greater than that suggested by the component toxicities" (National Research Council, 1994, p. 227, quoting U.S. Environmental Protection Agency, 1988). Individuals are exposed to many different substances, both simultaneously and over time. Some of these may be due to lifestyle choices (e.g., cigarette smoking); others may be due to environmental conditions (e.g., asbestos). As new products using new chemical formulas proliferate, the difficulty of anticipating adverse health effects of combined exposures to different substances becomes extraordinarily difficult. Yet it is becoming apparent that in some instances synergistic interactions do exist (National Research Council, 1994), although they may be unrecognized or underestimated by laypersons (Hampson et al., 1998).

While no one is exempt from multiple environmental exposures to toxic substances, it can be argued that low-income populations and racial or ethnic minorities are, on average, more likely to be exposed to greater quantities, over longer time periods, and in more settings – outside, at home, and at work. For this reason, the need for analysis of multiple and cumulative exposures was recognized in Executive Order 12898; for this reason also, attempts have been made to estimate multiple toxic exposures within low-income communities and communities of color. For example, in 1997, EPA's Region III office developed a two-part chemical indexing system using a geographic information system platform: one part to identify chemical releases and their relative toxicities in a particular geographic area; the second part to demographically characterize potentially exposed populations (U.S. Environmental Protection Agency, 1997).

4.2. Differences Among Population Groups

The environmental justice movement also has highlighted that the "average adult white male" should not be used as the standard by which to judge exposure likelihood. By now, it has become commonplace to recognize that different population groups have different dietary habits because of necessity and/or culture. For example, some people grow fruits and vegetables in the soil outside their homes; some, especially those with lifestyles rooted in strong cultural traditions, rely on fishing or hunting for much of their sustenance (Harris and Harper, 1997). These formerly healthy sources of food may now be contaminated with toxic substances such as lead, mercury, polychlorinated byphenols, and so forth. While the level of contamination may pose insignificant risks in the "typical" diet, the risks are much graver when the contaminated foods are consumed in large quantities.

In addition to "nonstandard" exposure scenarios, differences in dose-response sensitivities are now more well-recognized than they were two decades ago. Pregnant women and their fetuses, young children, the elderly, people with preexisting conditions such as asthma: all of these sub-populations are receiving greater attention in the scholarly literature (see, for example, Crump et al., 1998) and the manuals of regulators. For example, EPA published an extensive manual, *Sociodemographic Data Used for Identifying Potentially Highly Exposed Populations* (1999) as a companion to its *Exposure Factors Handbook* (1989). In addition, to routinize attention to the needs of particular groups, new offices such as EPA's Office of Children's Health Protection have been created.

Underlying this heightened attention is the recognition that, atypical or not, these population groups often have few alternatives to living the way they do. While some unhealthy practices may be discouraged, deterring people from getting food in traditional ways may be more difficult. Moreover, disadvantaged people often have a limited ability to improve their home and work environments or to move elsewhere. The environmental justice movement has helped open our collective eyes to differences, not only in proclivities but also in circumstances.

4.3. Risks to Economic, Social, or Psychological Well-Being

According to NEPA regulatory guidance issued in 1978 by the Council on Environmental Quality, whenever a federal action requires an environmental impact assessment, the action's economic and social impacts must be assessed as well (40 CFR 1508.14). This guidance not withstanding, until recently little attention had been paid in the United States to the socioeconomic impacts of environmental decisions, either within or outside the context of NEPA-mandated assessments. Over the past decade or so, however, a shift has occurred. When environmental risk decisions are made, the focus is no longer on health risks alone; the importance of risks to other aspects of human well-being has been more fully acknowledged.

Environmental decisions, particularly those involving the location of waste treatment/disposal sites and other controversial facilities, may have both positive and negative socioeconomic impacts on the surrounding community (National Research Council, 1999a). On the positive side, the new facility may create jobs for local workers and add to local tax revenue. But on the negative side, it may create a stigma effect, causing the community to be ignored as a possible location for desirable businesses. As Gregory, Flynn, and Slovic have noted, "Stigma goes beyond conceptions of hazard. It refers to something that is to be shunned or avoided not just because it is dangerous but because it overturns or destroys a positive condition" (1995, p. 220). The facility also may increase the cost of local services, if new roads or special arrangements for pollution monitoring and emergency management by the local government are needed, and it may cause nearby property values to decline. In addition, it may contribute to individual stress, depression, and feelings of powerlessness. Freudenburg and Jones (1991) have noted that technological risk is associated with a broad range of negative and potentially long-lived psychological impacts, sometimes but not always correlated with negative attitudes toward the risk source. The facility also may factionalize the community, by creating a rift between local proponents and opponents, and may alter the community fundamentally, by triggering a cascade of changes over time in its ratio of industrial to nonindustrial uses as well as its demographic size and composition.

The environmental justice movement has helped to publicize these non-health risks. For example, the environmental justice movement has stressed that the community's interests should be taken into account when reaching decisions about "brownfields" (derelict, potentially contaminated industrial and commercial sites). In other words, such projects should be thought of in comprehensive, community-oriented terms, as revitalization; not simply in limited, market-oriented terms, as reuse (National Environmental Justice Advisory Council, 1996b). Brownfields have been of particular interest and concern within the environmental justice movement because they often are found in or near to urban low-income communities and communities of color. Issues with brownfields illustrate that, insofar as members of these communities lack mobility, they must live with not only the health effects but also the other consequences of decisions about how such sites are remediated and reused.

Brownfields issues also illuminate an important paradox within the environmental justice movement: out of concern about permitting problems, corporations may decide not to site new industries in potential environmental justice communities. Yet it is widely recognized that these communities need economic as well as environmental well-being. How to attain both remains a challenge for environmental justice leaders and other leaders of impoverished communities. As the following section briefly illustrates, this struggle to improve environmental quality while also improving social and economic well-being is occurring, not only in U.S. communities, but also in other nations and internationally.

5. ENVIRONMENTAL JUSTICE: INTERNATIONAL AND GLOBAL PERSPECTIVES

While this paper is written mainly from the perspective of the United States, the United States does not exist in a vacuum. Environmental justice matters in other nations as well. This paper will not delve into the multitude of environmental problems and injustices experienced within other nations that arise from, for example, the actions of multinational corporations or policies and practices internal to these other countries. For discussions of environmental policies in countries such as Chile, China, India, Indonesia, Mexico, and Nigeria, with these nations' attempts to develop economically and meet the basic needs of growing populations while confronting environmental degradation, see, for example, Dwivedi and Vajpeyi (1995). For poignant and detailed accounts of the intersection between human rights and environmental abuses in developing countries especially, but also in industrialized nations, see Johnston (1994, 1997).

In addition, environmental justice on a global scale is an enormously difficult problem. In Sections 5.1 and 5.2, two examples are briefly discussed: the 1992 United Nations Conference on Environment and Development and the 1997 Kyoto Protocol on global climate change. Both examples concern, on the one hand, the rights of individual nations and, on the other hand, the responsibility of each nation to redress unsustainable practices and avoid imposing harm on others, including other nations as well as future generations. Both examples mirror at the international level some of the tensions between advantaged and disadvantaged populations commonly found at the subnational level.

5.1. The 1992 Earth Summit

The UN Conference on Environment and Development (UNCED, often referred to as the Earth Summit or the Rio conference) was held in Rio de Janeiro, Brazil, in June 1992. The culmination of a 1989 decision of the UN General Assembly, the Earth Summit built on prior efforts such as the UN Conference on the Human Environment held in Stockholm, Sweden, in June 1972 and the 1987 report of the World Commission on Environment and Development (the Brundtland report).

Prior to the Earth Summit, "Agenda 21" was released. a document of several hundred pages detailing strategies and program measures to halt environmental degradation and promote environmentally sustainable development in all counties. Agenda 21 resulted from extensive negotiations among governments, using as a basis proposals prepared by the UNCED Secretariat, which had, in turn, drawn upon UN agencies, experts, intergovernmental and nongovernmental organizations, regional conferences, national reports, and four sessions of the Preparatory Committee of the Conference (United Nations Conference on Environment and Development, 1992).

At the Earth Summit, more than 100 heads of state met. Agenda 21 was adopted by the assembled leaders, together with the Rio Declaration of Environment and Development (a set of principles that set the stage for the action program of Agenda 21); Principles for the Management, Conservation and Sustainable Development of Forests; the Framework Convention on Climate Change; and the Convention on Biological Diversity.

(In UN parlance, the term *convention* generally is used for formal multilateral treaties with a broad number of parties.) In December 1992, the Commission on Sustainable Development was created within the UN Economic and Social Council to ensure follow-up of the Earth Summit; to monitor implementation at the local, national, and international level of agreements reached at the Summit; and to lay plans for a five-year progress review. This five-year review ("Earth Summit + 5") was held as a special session of the UN General Assembly in June 1997; at the session, "Programme for the Further Implementation of Agenda 21," a document that had been prepared by the Commission on Sustainable Development, was adopted.

The core of the Rio Declaration of Environment and Development, as well as Agenda 21, is a set of twenty-seven principles concerning sustainable development and equity among nations. These principles are sweeping in scope. For example, the first five principles are:

1. Human beings are at the centre of concerns for sustainable development. They are entitled to a healthy and productive life in harmony with nature.
2. States have, in accordance with the Charter of the United Nations and the principles of international law, the sovereign right to exploit their own resources pursuant to their own environmental and developmental policies, and the responsibility to ensure that activities within their jurisdiction or control do not cause damage to the environment of other States or of areas beyond limits of national jurisdiction.
3. The right to development must be fulfilled so as to equitably meet developmental and environmental needs of present and future generations.
4. In order to achieve sustainable development, environmental protection shall constitute an integral part of the development process and cannot be considered in isolation from it.
5. All States and all people shall cooperate in the essential task of eradicating poverty as an indispensable requirement for sustainable development, in order to decrease the disparities in standards of living and better meet the needs of the majority of the people of the world.

The Rio Declaration pulls its punches on two key issues, however: the consumption practices and exploitation of global resources in developed countries, and the birth control practices (or lack thereof) and burgeoning populations in developing countries. Both issues are mentioned in the list of principles and elsewhere in Agenda 21:

In view of the different contributions to global environmental degradation, States have common but differentiated responsibilities. The developed countries acknowledge the responsibility that they bear in the international pursuit of sustainable development in view of the pressures their societies place on the global environment and of the technologies and financial resources they command (Principle 7).

To achieve sustainable development and a higher quality of life for all people, States should reduce and eliminate unsustainable patterns of production and consumption and promote appropriate demographic policies (Principle 8).

But in the delicate balance between rights and responsibilities, Agenda 21 tilts toward rights on both issues. It does not go further than exhorting nations and their populations to mend their ways. Similarly, the International Conference on Population and Development held in Cairo, Egypt, in September 1994 affirmed that the preferred route to stabilizing population growth is through more equal opportunity for women and improved quality of life, not through government-imposed limits on family size.

On both issues – resource consumption and population control – the apparent hope is that ultimately, a sustainable and just world can be obtained through enlightenment, voluntary lifestyle changes, and attractive alternatives (e.g., "green" products and readily available family planning services), rather than through coercion. In contrast, as is discussed later, at least one global environmental risk issue – global climate change – has now become urgent enough in the minds of many to be treated with a firmer hand.

5.2. The 1997 Kyoto Protocol

In December 1997, an important international conference on climate change took place in Kyoto, Japan. This conference was held within the broader context of the UN Framework Convention on Climate Change (UNFCCC), which had been adopted at the UN headquarters in New York City in May 1992 and was then opened for signatures at the June 1992 Earth Summit in Rio de Janeiro. By June 1993, the UNFCCC had received 166 signatures. It entered into force in March 1994, with the provision that governments that had not signed on could do so at any time. By December 1999, the UNFCCC had received 181 instruments of ratification. (According to UN practices, a signature indicates the signatory government's good-faith intent to continue the treaty-making process; ratification [or acceptance or approval] expresses its consent to be bound by the treaty.)

The UNFCCC establishes a framework of general principles and a process through which governments meet regularly to address climate change issues. In itself, it does not specify specific actions to be taken. Instead, it recognizes that anthropogenic climate change is a problem, albeit one with considerable scientific uncertainty; it encourages scientific research; it sets an ultimate objective of preventing dangerous levels of human-induced interference with the global climate system; and it takes preliminary steps such as encouraging countries ratifying the Convention to share technology and cooperate in other ways to reduce emissions of "greenhouse gases"

(gases such as carbon dioxide, methane, and nitrogen oxide, which are much more prevalent because of human activities and are thought to be major contributors to global warming).

The Kyoto conference held in December 1997 was the third session of the parties to the UNFCCC. At this session, "teeth" were added to the UNFCCC through the Kyoto Protocol, an agreement that specifies greenhouse gas emission limits and reduction commitments of each of the nations that are parties to the agreement. (As used by the United Nations, a protocol contains specific substantive obligations; it implements the broad objectives of a prior framework or umbrella convention.) The protocol was adopted at the Kyoto conference on December 11, 1997. Parties to the UNFCCC were invited to sign the protocol beginning March 1998; as of January 2000, eighty-four parties had signed. For the Protocol to be entered into force, it must be ratified by at least 55 parties to the UNFCCC accounting for a total of at least 55 percent of the carbon dioxide emissions listed in Annex I to the Protocol.

Of the 34 countries listed in Annex I (i.e., the countries, most of them industrialized, that submitted data prior to December 1997), the United States, with 36 percent of the total, is by far the largest producer of carbon dioxide emissions. Under the Kyoto Protocol, the United States and other industrialized nations would cut their greenhouse gas emissions to below 1990 baseline levels by the period 2008–12. Before the United States and other nations ratify the Protocol and commit to these reductions, several difficult issues must be resolved. These include questions about (1) how emission cuts will be monitored and what penalties will be imposed if reductions aren't met; (2) what role land-based activities (e.g., forestry, agriculture) should play in determining credit for greenhouse gas emission reductions; (3) to what extent industrialized nations will be allowed to take credit for sponsoring emission-prevention or emission-reduction projects in developing countries; and (4) to what extent emission credit trading between nations will be allowed, or whether tangible emission reductions will be required within a nation for its Protocol obligations to be met (*Environment Reporter*, January 21, 2000).

The latter two questions, especially, raise sensitive issues about the respective obligations of industrialized and developing countries. The UNFCCC, in its principles, puts the major responsibility for battling climate change on the wealthy industrialized countries, since these countries have contributed the greatest burden of greenhouse gas emissions. Many of these emissions are closely tied to energy use, because the fossil fuel combustion used in many forms of energy production is a major contributor to greenhouse gases. Moreover, the UNFCCC recognizes both that poorer nations have a right to economic development, including development with greenhouse gases as a by-product, and that these nations may be the most vulnerable to the effects of climate change, in terms of their ecosystems (some of which are highly susceptible to the impacts of climate change) as well as their economies (some of which depend heavily on the sale of oil or coal).

Nevertheless, for the industrialized nations to meet their emission reduction targets, the easiest, cheapest, quickest way may be either to sponsor emission prevention and reduction projects in developing countries or to purchase some of their emission allowances. The infrastructures of the industrialized countries (e.g., for lighting, heating, cooling, transportation, and industrial manufacturing processes) are mammoth. To "turn these ships around" is likely to require considerable time and enormous expense. In contrast, the infrastructures of developing countries are, as the term *developing* implies, still rudimentary in many cases. The opportunity to realize a bigger and faster "bang for the buck" in greenhouse gas reduction or prevention is much greater in these countries, especially those that are undergoing rapid development. This opportunity may be realized if funds are provided by industrialized countries either directly, through project sponsorship, or indirectly, through purchasing emission allowances.

Under such arrangements, however, the industrialized countries may then be seen as buying their way out of a global problem largely of their own making, and transferring to developing countries the ultimate responsibility to solve the problem. As David Rose aptly noted two decades ago, "Preaching of conservation by the industrialized countries sounds to many LICs [less industrialized countries] like curtailment of LIC energy. *Sie pregen Wasser und trinken Wein* [they're preaching water and drinking wine], so to speak" (Rose, 1980, p. 257). Further questions about the justness of such arrangements arise when, rather than a transfer of new funds, arrangements such as "debt for nature" are used. Many developing countries have large debts; writing off part of this debt in exchange for guarantees of protection of wilderness areas, including forests that serve as carbon sinks, is, in effect, a purchase of emission allowances. But using yesterday's loans to pay for today's obligations may be regarded as unjust.

Underlying these questions about the justness of various arrangements for confronting global environmental risks is a long and unsavory history. Some nations have exploited other nations' natural and human capital, profiting off the weaker status and dire need of these other nations, as well as (in some cases) the unscrupulous greed of their economic and political leaders. As with environmental justice within a nation, environmental justice among nations may not be realized with arrangements that simply provide for equal treatment. Instead, contexts of deeper, often longstanding economic and social injustices must be factored in when assessing whether a proposed arrangement to address an environmental risk is indeed just – especially if that risk is likely to fall most heavily on those who are disadvantaged or who historically have been exploited.

6. A BROADER CONCEPTION OF ENVIRONMENTAL JUSTICE

So far, this chapter has used the term *environmental justice* in its popular sense, as justice to *humans* when human actions create or contribute to environmental risks. But the reader who has come this far is asked to stretch further, by thinking of environmental justice in a more expansive sense. In fact, the term *environmental justice* can be and has been used less restrictively (see, for example, Wenz, 1988), with the targets of concern extending not only to humans but also to other sentient species (e.g., monkeys, dogs); not only to sentient species but also to other living organisms (e.g., trees, flowers); not only to living organisms but also to other aspects of the environment (e.g., rivers, mountains).

Often referred to as environmental ethics, the scholarship and activism on nonhuman environmental concerns (see, for example, the journal *Environmental Ethics*) has largely been distinct from the scholarship and activism on environmental justice in its popular, human-oriented sense. They have been two separate camps. They are not battling, but they are not communicating. Yet this compartmentalization of different senses of environmental justice does us all a disservice. It fails to recognize the interdependence of human and ecological well-being. For this reason, this paper concludes with what might seem to be a tangent – a digression into a different and broader conception of environmental justice than the one discussed in previous sections. This digression will be brief, without adequate acknowledgment of the extensive scholarly tradition of ecocentric environmental thinking and action. The point here is to illuminate that there is, indeed, a connection between these two camps; one that bears further exploration.

These more-encompassing views of environmental justice are known by various names, and each takes a somewhat different slant. The following discussion does not try to analyze the roots of or nuances among these different interpretations. Instead, a simplifying dichotomy – on the one hand, the popular conception of environmental justice, which focuses mainly on humans, and on the other hand, an ecocentric conception of environmental justice, which does *not* focus exclusively on humans – will be used to facilitate a brief discussion of three fundamental issues:

1. What is distinctively different about an ecocentric conception of environmental justice?
2. Do human values and views inevitably come into play?
3. In terms of their risk management implications, how do the two conceptions of environmental justice diverge and converge?

6.1. Distinctive Characteristic of an Ecocentric Conception

The key distinguishing characteristic of an ecocentric conception of environmental justice is that humans are not the "be-all and end-all." In other words, even if *no* human, present or future, is harmed in *any* way by an environmental risk, the natural environment still may be harmed, and that harm may be a wrong. As discussed in Section 6.2, it is conceptually difficult to extricate humans from an understanding of what it means to harm the environment. Nevertheless, the theoretical possibility must be admitted for the ecocentric conception of environmental justice to be comprehensible.

6.2. The Natural Environment: Inherent Value and Inherent Rights?

The natural environment can have value in three different ways (Beatley, 1994, adapted from Hare, 1987). First, it can have *instrumental* value to humans – as, for example, natural resources that can improve people's lives. Second, it can have *intrinsic* value to humans: as something valued by people for itself, not for the benefits it brings to them. And third, it can have *inherent* value: as something that has value regardless of whether it is used or cherished by humans. The line between the first two types of value is not always clearcut: If a mountain is appreciated for nonuse reasons such as its beauty, is this value instrumental or intrinsic? For present purposes, however, this ambiguity doesn't matter; what matters is that the first two kinds of value arise because of, and only because of, a human conception of the desirable. They are both anthropocentric. In contrast, the third type of value is ecocentric; the value inheres in the object, not because of humans. While the semantic distinction between *intrinsic* and *inherent* may be fabricated (they are synonymous in *Webster's Third New International Dictionary*) and has provoked extensive debate (see, for example, Callicott, 1985; Fieser, 1993), the conceptual distinction captures an important difference between the anthropocentric and the ecocentric viewpoint. One says that value exists only when it is conferred by humans; the other rejects this notion.

Nevertheless, while the distinction between intrinsic and inherent value is theoretically important, it may be impossible in practice to draw this distinction. How can a person comprehend the value of nature qua nature without bringing a human perspective to bear? Even Native Americans and other indigenous peoples, who have brought to populist environmental justice movements a heightened sense of the value of the natural environment as a good in itself, base their valuations on their own cultural perspectives. Because of these practical difficulties, the latter two categories often are collapsed, with the terms *intrinsic* and *inherent* value used interchangeably. But the distinction does point to a subtle difference worth preserving. Nevertheless, rather than thinking of the instrumental, intrinsic, and inherent value of the natural environment as mutually exclusive categories, it may be more appropriate to think of a continuum of values, from those derived solely from the potential for human consumption and use to those derived solely from the essential nature of the thing itself.

A similar problem occurs in considering the rights of natural objects. Thirty years ago, Christopher Stone in his landmark *Should Trees Have Standing? Toward Legal Rights for Natural Objects* (1972) proposed that natural objects such as trees, mountains, and rivers should, like humans (or organizations of humans), have legal rights. Fifteen years later, Stone refined his theory of rights for natural objects in *Earth and Other Ethics: The Case for Moral Pluralism* (1987). The former work was concerned largely with developing an argument that, because natural objects have rights but cannot speak for themselves, environmental or public interest organizations should be given legal standing to speak on their behalf. The latter work was concerned with elaborating what it means for a natural object to have a "right."

Rights, in human terms, can take the form of negative rights (e.g., the right not to be physically harmed by others) and liberties (e.g., the right to pursue happiness). They also may, arguably, take the form of claim rights (e.g., the right to sustenance), although whether claim rights can in fact be called "rights" has been debated, since – it is sometimes argued – a right implies a corresponding obligation. In *Earth and Other Ethics*, Stone develops a concept, not of a set of universally applicable rights, but of moral pluralism wherein not only humans but also other living species and natural objects deserve moral consideration, although not always the same *kind* of moral consideration. (That is where moral pluralism, as opposed to moral monism, enters in.) As Stone says, "the position I want to espouse is one whose development Monism has discouraged, and for which there is consequently little support even in the animal rights literature: that (animal) life does count, but that not all life counts equally, or comes under the same rules and considerations" (1987, pp. 136–7).

The moral pluralism view is elastic in ways that a strict, rights-based view is not. It extends moral consideration beyond humans, but not in an "all-or-nothing" fashion. It is, however, ineluctably anthropocentric: moral consideration requires that someone do the considering.

6.3. Points of Divergence and Convergence Between the Two Conceptions

At this point, it is useful to reiterate the first principle of the Rio Declaration, quoted in Section 5.1:

Human beings are at the centre of concerns for sustainable development. They are entitled to a healthy and productive life in harmony with nature.

As a counterpoint, consider the oft-quoted statement of Aldo Leopold – one of the fathers of modern ecology – in *A Sand County Almanac*:

A thing is right when it tends to preserve the integrity, stability, and beauty of the biotic community. It is wrong when it tends otherwise (1966 [1949], p. 262).

Taken at face value, the first statement is blatantly anthropocentric. Humans and only humans (it is implied) matter in environmental decisions. Granted, in the Rio Declaration, as in many other strongly anthropocentric documents, the needs and wants of future generations are recognized as part of the moral calculus (although how they should be incorporated is often disputed). But these are future generations of humans, not other species. Justice – the entitlement to "a healthy and productive life in harmony with nature" – applies only to humans, on this view, and environmental decisions are made only with humans in mind.

In contrast, the second statement is strongly ecocentric. People are not all that matter, on this view; in fact, their interests should not always be put first when environmental decisions are made. In the first statement, humans are the summum bonum; in the second, the natural world (of which humans are only a part) is. Seen in this way, the anthropocentric view and the ecocentric view diverge radically.

If the term *anthropocentrism* is taken literally, however, it does not have to mean speciesism – that is, an exclusive focus on the welfare of humans. It *may* mean that, but it may mean a broader view more akin to that articulated by Christopher Stone in *Earth and Other Ethics*. Humans, especially those present today, may be at the center of our concerns, but they do not necessarily describe the *universe* of our concerns. Instead, as in the "concentric circle" perspective of environmental justice proposed by Peter Wenz, moral relationships are defined pluralistically: "The closer our relationship is to someone or something, the greater the number of obligations in that relationship, and/or the stronger our obligations in that relationship" (Wenz, 1988, p. 316). (The appropriate extent of moral pluralism has been the subject of debate – see, for example, Wenz, 1993 – but will not be considered here.)

In this less restrictive form of anthropocentrism, close relationships are emphasized, but so too is a broader conception of environmental justice. This conception is akin in spirit to Ernest Partridge's argument for self-transcendence: for people's need to "identify themselves as part of larger, ongoing, and enduring processes, projects, institutions, and ideals" (Partridge, 1981, p. 217). This broader conception is still anthropocentric in its prioritization of concerns, all other things being equal. In addition, as is recognized by Stone, Wenz, and other advocates of moral pluralism, tradeoffs must be made contextually, through dialogue among humans. In that sense too, the conception is anthropocentric. But it recognizes, as a narrow form of anthropocentrism does not, a broader scope of concern and a broader view of the tradeoffs to be made.

As Brian Norton notes in *Why Preserve Natural Variety?* (1987), an inclusive form of anthropocentrism is one based not only on human demand values but also on human transformative values. He concludes that "little is gained by emphasizing the split between nonanthropocentrists and the inclusive version of anthropocentrism.... The two positions would seem to have very similar policy implications" (Norton, 1987, p. 237).

6.4. Implications of a Broader Conception of Environmental Justice

When justice is sought in deliberative decision processes on environmental risk, a tilt away from a narrow form of anthropocentrism could sometimes make a telling difference. The concerns of those who have borne disproportionate environmental burdens (as well as disproportionate social and economic burdens) would still figure importantly in this more expansive form of anthropocentrism. These concerns would not be moral trumps, however; they would simply be high cards. Ideally, they would be congruent with an ecological perspective of environmental justice. When they were not, complex, difficult decisions would be required.

In the "Principles of Environmental Justice" adopted at the First National People of Color Environmental Leadership Conference in October 1991, the first principle was:

Environmental Justice affirms the sacredness of Mother Earth, ecological unity and the interdependence of all species, and the right to be free from ecological destruction. (http://www.brookings.org/gs/envjustice/ejprinciples.htm)

Nevertheless, much of the focus of the U.S. environmental justice movement has been on perceived risks to health and safety from disproportionate environmental burdens in low-income communities and communities of color. The movement is reluctant to prioritize its concerns. To do so would jeopardize its large but loose coalition and its authenticity as a populist voice for disadvantaged people. However, that voice speaks most passionately when it is raised against big, powerful corporate or government organizations generating toxic substances. Whether these toxic substances are, in fact, always the most pressing environmental problem facing communities within the environmental justice movement, other communities of people, or biotic communities is another question – one that merits probing and careful analysis.

Associate Supreme Court Justice Stephen Breyer, in *Breaking the Vicious Circle: Toward Effective Risk Regulation* (1993), refers to "the problem of the last 10 percent" – in other words, the problem of achieving rigidly strict regulatory standards, which can entail big costs for small returns. In a similar vein, Christopher Foreman, in his challenge to the conventional wisdom of the U.S. environmental justice movement (*The Promise and Peril of Environmental Justice*, 1998), urges that:

Those who would advance social justice in the environmental realm, or through environmental policy tools, must begin thinking of better ways to discriminate – between the very important and the less important, between the deserving and the undeserving, between what is empirically defensible and what is not. The pursuit of justice requires the making of distinctions, not just the pursuit of rights and the flexing of advocacy muscle (p. 136).

Given the focus of Foreman's book, it would seem that he is speaking primarily of risks to human health and safety, but the same eloquent statement could be applied to a broader conception of environmental justice as well. To achieve this broader conception, collective deliberations are needed that rely, less on ardent advocacy, and more on "discursive ethics" – a dialogic process whose promise lies in determining genuinely social values (National Research Council, 1999b).

In addition, further thought is needed about how the popular conception of environmental justice and a broader, more inclusive conception can be reconciled. In this regard, work has begun. For example, Mark Sagoff (1988), among others, emphasizes that we must think less in terms of consumers and the marketplace and more in terms of citizens and public values; and Holmes Rolston, III (1994) adroitly discusses both the tension and the connection between various anthropocentric values, including those involving human rights to development and the natural values of this planet that we call home. Yet much remains to be done to forge links between the two camps mentioned here – on the one hand, a narrowly anthropocentric conception of environmental justice, as popularly conceived; on the other hand, a largely ecocentric conception of environmental ethics.

7. CONCLUSION

A few interlocking themes have run throughout this discussion. One theme concerns identifying the appropriate spatial and temporal breadth (human and ecological) for a given environmental risk issue. A second theme concerns finding the right balance between, on the one hand, unduly cautious and expensive risk prevention or mitigation and, on the other hand, blithely proceeding with risks that have potentially severe consequences. A third theme concerns achieving justice of process as well as outcome. A fourth theme concerns understanding and, insofar as possible, treating the root social and economic causes that contribute to environmental injustices, as that term is popularly understood. A fifth theme concerns reconciling the popular conception of environmental justice with a less narrowly anthropocentric conception of environmental justice. A sixth theme concerns the importance of discourse in achieving this broader conception.

In each of these six thematic areas, much more work is needed by risk analysts, social scientists, and ethicists – when appropriate, working in concert with individuals and communities that are unduly exposed to environmental risks. For example, there is a need for more sophisticated methods to assess risks to human life and health, backed by more extensive epidemiological studies; for better ecological risk assessment data and methods; for empirical studies that can provide improved understanding of disproportionate environmental risks, including risks to material and mental well-being as well as to human life and health; for further philosophical research on ways to integrate anthropocentric and ecocentric perspectives on environmental risks; and for more policy research on ways to improve environmental risk decision processes. There also is an ongoing need for concerted interdisciplinary and international research on global environmental risks that, while fraught with uncertainty, have potentially grave consequences.

Global environmental risks serve as a paradigmatic example that in the end, environmental risk and justice cannot be extricated. No single discipline holds the answers. At best, researchers can serve as skilled professionals while remaining sensitive to their limitations as well as to the bigger picture. In so doing, they will be better equipped to contribute to an evolving, collaborative effort to attain justice in the management of environmental risks.

References

Andersson, K., Balfors, B., Schmidtbauer, J., and Sundqvist, G. 1999. *Transparency and Public Participation in Complex Decision Processes*. Research Report. Royal Institute of Technology, Department of Civil and Environmental Engineering, Stockholm.

Anderton, D. L., Anderson, A. B., and Oakes, J. M. 1994. Environmental equity: The demographics of dumping. *Demography*, 31: 229–48.

Beatley, T. 1994. *Ethical Land Use*. Johns Hopkins University Press, Baltimore.

Been, V. 1994. Locally undesirable land uses in minority neighborhoods: Disproportionate siting or market dynamics? *Yale Law Journal*, 103: 1383–422.

Boerner, C., and Lambert, T. 1994. *Environmental Justice?* Policy Study Number 21. Washington University, Center for the Study of American Business, St. Louis.

Breyer, S. 1993. *Breaking the Vicious Circle: Toward Effective Risk Regulation*. Harvard University Press, Cambridge, MA.

Bryant, B. 1995. Introduction. In B. Bryant, ed., *Environmental Justice: Issues, Policies and Solutions*, pp. 1–7. Island Press, Washington, DC.

Bullard, R. D. 1983. Solid waste sites and the black Houston community. *Social Inquiry*, 53: 273–88.

Bullard, R. D. 1990. *Dumping in Dixie: Race, Class, and Environmental Quality*. Westview Press, Boulder, CO.

Bullard, R. D., ed. 1993. *Confronting Environmental Racism: Voices from the Grassroots*. South End Press, Boston.

Bullard, R. D., and Johnson, G. S., eds. 1997. *Just Transportation*. New Society Publishers, Stony Creek, CT.

Calabrese, E. J., Baldwin, L. A., and Holland, C. D. 1999. Hormesis: A highly gen-eralizable and reproducible phenomenon with important implications for risk assessment. *Risk Analysis*, 19 (2): 261–81.

Callicott, J. B. 1985. Intrinsic value, quantum theory, and environmental ethics. *Environmental Ethics*, 7 (3): 257–75.

Cranor, C. F. 1997. The normative nature of risk assessment. *Risk: Health, Safety & Environment*, 8 (2): 123–36.

Crump, K. S., Kjellstrm, T., Shipp, A. M., Silvers, A., and Steward, A. 1998. In-fluence of prenatal mercury exposure upon scholastic and psychological test performance: Benchmark analysis of a New Zealand cohort. *Risk Analysis*, 18 (6): 701–13.

Douglas, M., and Wildavsky, A. 1982. *Risk and Culture*. University of California Press, Berkeley.

Dwivedi, O. P., and Vajpeyi, D. K., eds. 1995. *Environmental Policies in the Third World: A Comparative Analysis*. Greenwood Press, Westport, CT.

Edelstein, M. R. 1988. *Contaminated Communities: The Social and Psychological Impacts of Residential Toxic Exposure*. Westview Press, Boulder, CO.

English, M. R. 1991. Risk and consent. In C. Zervos, ed., *Risk Analysis*, pp. 547–54. Plenum Press, New York.

English, M. R. 1992. *Siting Low-Level Radioactive Waste Disposal Facilities*. Quorum Books, New York.

English, M. R. 1999. Environmental decision making by organizations: Choosing the right tools. In K. Sexton, A. A. Marcus, K. W. Easter, and T. D. Burkhardt, eds., *Better Environmental Decisions*, pp. 57–75. Island Press, Washington, DC.

English, M. R., and Zimlich, M. A. 1997. *The "Community" in Community-Based Environmental Protection*. Synopsis Report. University of Tennessee, Waste Man-agement Research and Education Institute, Knoxville, TN.

Erikson, K. 1990. Toxic reckoning: Business faces a new kind of fear. *Harvard Business Review*, 68 (1) (January–February 1990): 118–26.

Etzioni, A. 1993. *The Spirit of Community*. Simon & Schuster, New York.

Fieser, J. 1993. Callicott and the metaphysical basis of ecocentric morality. *Environ-mental Ethics*, 15 (2): 171 80.

Foreman Jr., C. H. 1998. *The Promise and Peril of Environmental Justice*. Brookings Institution Press, Washington, DC.

Freudenburg, W. R., and Jones, R. E. 1991. Criminal behavior and rapid community growth: Examining the evidence. *Rural Sociology*, 56 (4): 619–45.

Gaventa, J. 1980. *Power and Powerlessness: Quiescence and Rebellion in an Appalachian Valley*. University of Illinois Press, Urbana.

Gerrard, M. B. 1994. *Whose Backyard, Whose Risk*. MIT Press, Cambridge, MA.

Glickman, T. S., Golding, D., and Hersh, R. 1995. GIS-based environmental equity analysis: A case study of TRI facilities in the Pittsburgh area. In G. E. G. Beroggi and W. A. Wallace, eds., *Computer Supported Risk Management*, pp. 95–114. Kluwer Academic Publishers, Dordrecht, The Netherlands.

Goldman, B. A. 1993. *Not Just Prosperity: Achieving Sustainability with Environmental Justice*. National Wildlife Federation, Vienna, VA.

Greenberg, M. 1993. Proving environmental inequity in siting locally unwanted land uses. *Risk – Issues in Health and Safety*, 4 (3): 235–52.

Gregory, R., Flynn, J., and Slovic, P. 1995. Technological stigma. *American Scientist*, 83 (3): 220–3.

Hampson, S. E., Andrews, J. A., Lee, M. E., Foster, L. S., Glasgow, R. E. and Lichtenstein, E. 1998. Lay understanding of synergistic risk: The case of radon and cigarette smoking. *Risk Analysis*, 18 (3): 343–50.

Hannigan, J. A. 1995. *Environmental sociology: A social constructionist perspective.* Routledge, New York.

Hare, R. M. 1987. Moral reasoning about the environment. *Journal of Applied Philosophy*, 4 (1): 3–14.

Harris, S. G., and Harper, B. L. 1997. A Native American exposure scenario. *Risk Analysis*, 17 (6): 789–95.

Hillery, Jr., G. A. 1955. Definitions of community: areas of agreement. *Rural Sociology*, 20 (June): 111–23.

Hurley, A. 1995. *Environmental Inequalities: Class, Race, and Industrial Pollution in Gary, Indiana, 1945–1980*. University of North Carolina Press, Chapel Hill.

Institute of Medicine, Committee on Environmental Justice. 1999. *Toward Environmental Justice: Research, Education, and Health Policy Needs*. National Academy Press, Washington, DC.

Johnston, B. R., ed. 1994. *Who Pays the Price? The Sociocultural Context of Environmental Crisis*. Island Press, Washington, DC.

Johnston, B. R., ed. 1997. *Life and Death Matters: Human Rights and the Environment at the End of the Millennium*. AltaMira Press, Walnut Creek, CA.

Krimsky, S., and Golding, D., eds. 1992. *Social Theories of Risk*. Praeger, Westport, CT.

Kunreuther, H. 1995. Voluntary siting of noxious facilities: The role of compensation. In O. Renn, T. Webler, and P. Wiedemann, eds., *Fairness and Competence in Citizen Participation*, pp. 283–95. Kluwer Academic Publishers, Dordrecht, The Netherlands.

Leopold, A. 1966 [1949]. *A Sand County Almanac*. Oxford University Press, New York.

Levine, A. G. 1982. *Love Canal: Science, Politics, and People*. D. C. Heath and Company, Lexington, MA.

Linnerooth-Bayer, J., and Fitzgerald, K. B. 1996. Conflicting views on fair siting processes: Evidence from Austria and the U.S. *Risk: Health, Safety & Environment*, 7 (2): 119–34.

Luhmann, N. 1993. *Risk: A Sociological Theory*. Aldine de Gruyter, New York.

MacLean, D. 1986. Risk and consent: philosophical issues for centralized decisions. In D. MacLean, ed., *Values at Risk*, pp. 17–30. Rowman & Allanheld, Totowa, NJ.

Morrell, D. 1987. Siting and the politics of equity. *Hazardous Waste*, 1: 555–71. Reprinted in R. W. Lake, ed. (1987), *Resolving Locational Conflict*, pp. 117–36. Center for Urban Policy Research, New Brunswick, NJ.

National Environmental Justice Advisory Council, Public Participation and Accountability Subcommittee. 1996a. *The Model Plan for Public Participation*. U.S. EPA, Office of Environmental Justice, Washington, DC.

National Environmental Justice Advisory Council, Waste and Facility Siting Subcommittee. 1996b. *Environmental Justice, Urban Revitalization, and Brownfields: The Search for Authentic Signs of Hope*. U.S. Environmental Protection Agency, Office of Environmental Justice, Washington, DC.

National Research Council, Committee on the Institutional Means for Assessment of Risks to Public Health. 1983. *Risk Assessment in the Federal Government: Managing the Process*. National Academy Press, Washington, DC.

National Research Council, Committee on Risk Assessment of Hazardous Air Pollutants. 1994. *Science and Judgement in Risk Assessment*. National Academy Press, Washington, DC.

National Research Council, Committee on Risk Characterization, Paul C. Stern and Harvey V. Fineberg, eds. 1996. *Understanding Risk*. National Academy Press, Washington, DC.

National Research Council (NRC) Committee on Health Effects of Waste Incineration. 1999a. *Waste Incineration and Public Health*. Washington, DC: National Academy Press.

National Research Council (NRC), Committee on Noneconomic and Economic Value of Biodiversity. 1999b. *Perspectives on Biodiversity: Valuing Its Role in an Everchanging World*. National Academy Press, Washington, DC.

National Conference of State Legislatures, The Environmental Justice Group. 1995. *Environmental Justice: A Matter of Perspective*. National Conference of State Legislatures, Denver.

Norton, B. G. 1987. *Why Preserve Natural Variety?* Princeton University Press, Princeton, NJ.

Nozick, R. 1974. *Anarchy, State, and Utopia*. Basic Books, New York.

Partridge, E. 1981. Why care about the future? In E. Partridge, ed., *Responsibilities to Future Generations*. Prometheus Books, Buffalo, NY.

Presidential/Congressional Commission on Risk Assessment and Risk Management. 1997. *Framework for Environmental Health Risk Management*. Final Report, Volume 1.

Rabe, B. G. 1994. *Beyond NIMBY: Hazardous Waste Siting in Canada and the United States*. Brookings Institution, Washington, DC.

Rawls, J. 1971. *A Theory of Justice*. Harvard University Press, Cambridge, MA.

Renn, O., Webler, T., and Wiedemann, P., eds. 1995. *Fairness and Competence in Citizen Participation*. Kluwer Academic Publishers, Dordrecht, The Netherlands.

Rogers, R. G., Hummer, R. A., Namm, C. B., and Peters, K. 1996. Demographic, socioeconomic, and behavioral factors affecting ethnic mortality by cause. *Social Forces* 74 (4): 1419–38.

Rolston III, Holmes. 1994. *Conserving Natural Value*. Columbia University Press, New York.

Rose, D. J. 1980. Summary. In R. A. Bohm, L. A. Clinard, and M. R. English, eds., *World Energy Production and Productivity*, pp. 253–63. Ballinger Publishing Company, Cambridge, MA.

Sagoff, M. 1988. *The Economy of the Earth: Philosophy, Law, and the Environment*. Cambridge University Press, New York.

Sandel, M. J. 1996. *Democracy's Discontent*. Harvard University Press, Cambridge, MA.

Sizemore, D. 2000. Environmental racism's life course: Domain expansion, contention, and recession. Working paper presented at the Midwest Sociological Society's Annual Meeting (April 2000), Chicago, IL.

Slovic, P. 1987. Perception of risk. *Science*, 236: 280–5.

Smart, J. J. C., and Williams, B. 1973. *Utilitarianism for and against*. Cambridge University Press, New York.

Stone, C. D. 1987. *Earth and Other Ethics: The Case for Moral Pluralism*. Harper & Row, New York.

Stone, C. D. 1974 [1972]. *Should Trees Have Standing? Toward Legal Rights for Natural Objects*. William Kaufmann, Los Altos, CA.

Tonn, B., English, M., and Travis, C. 2000. A framework for understanding and improving environmental decision making. *Journal of Environmental Planning and Management* 43 (2): 165–85.

United Church of Christ, Commission for Racial Justice. 1987. *Toxic Wastes and Race in the United States: A National Report on the Racial and Socioeconomic Characteristics of Communities with Hazardous Waste Sites*. United Church of Christ, New York.

United Nations Conference on Environment and Development. 1992. *The Global Partnership for Environment and Development: A Guide to Agenda 21*. United Nations 92-1-100481-0. United Nations, New York.

U.S. Council on Environmental Quality. 1971. *Annual Report to the President*. U.S. Council on Environmental Quality, Washington, DC.

U.S. Council on Environmental Quality. 1997. *Environmental Justice: Guidance Under the National Environmental Policy Act*. U.S. Council on Environmental Quality, Washington, DC.

U.S. Environmental Protection Agency, Office of Research and Development. 1988. *Technical Support Document on Risk Assessment of Chemical Mixtures*. EPA-600/8–90/064. U.S. Environmental Protection Agency, Washington, DC.

U.S. Environmental Protection Agency, Environmental Equity Workgroup. 1992. *Environmental Equity: Reducing Risk For All Communties* [sic], Volumes 1 and 2. EPA230-R-92-008 and -008A. U.S. Environmental Protection Agency, Washington, DC.

U.S. Environmental Protection Agency, Region III. 1997. *Chemical Indexing System Part I: Chronic Index (revised)*, EPA/903/R-97/020, and *Chemical Indexing System Part II: Vulnerability Index*, EPA/903/R-97/021. U.S. Environmental Protection Agency, Washington, DC. (Draft documents.)

U.S. Environmental Protection Agency, Office of Research and Development, National Center for Environmental Assessment. 1997. *Exposure Factors Handbook*. EPA/600/P-95/002Fa,b,c. U.S. Environmental Protection Agency, Washington, DC.

U.S. Environmental Protection Agency, Office of Research and Development, National Center for Environmental Assessment. 1999. *Sociodemographic Data Used for Identifying Potentially Highly Exposed Populations*. EPA/600/R-99/060. U.S. Environmental Protection Agency, Washington, DC.

U.S. General Accounting Office. 1995. *Hazardous and Nonhazardous Waste: Demographics of People Living Near Waste Facilities*. GAO/RCED-95-84. U.S. General Accounting Office, Washington, DC.

U.S. General Accounting Office. 1983. *Siting of Hazardous Waste Landfills and Their Correlation with Racial and Economic Status of Surrounding Communities*. GAO/RCED 83-168. U.S. General Accounting Office, Washington, DC.

Walzer, M. 1983. *Spheres of Justice*. Basic Books, New York.

Wenz, P. 1993. Minimal, moderate, and extreme moral pluralism. *Environmental Ethics*, 15 (1): 61–74.

Wenz, P. S. 1988. *Environmental Justice*. State University of New York Press, Albany, NY.

White, H. L. 1998. Race, class, and environmental hazards. In D. E. Camacho, ed., *Environmental Injustices, Political Struggles*, pp. 61–81. Duke University Press, Durham, NC.

Zimmerman, R. 1993. Social equity and environmental risk. *Risk Analysis* 13, (6): 649–66.

Zimmerman, R. 1994. Issues of classification in environmental equity: How we manage is how we measure. *Fordham Urban Law Journal*, XXI (3): 633–69.

PART TWO

METHODS FOR RISK ASSESSMENT

6

Uncertain Risk

The Role and Limits of Quantitative Assessment

Alison C. Cullen and Mitchell J. Small

1. INTRODUCTION

The assessment and management of uncertain risks is an essential component of science, engineering, medicine, business, law, and other modern disciplines. I. Good argues that risk science based on the evaluation of probabilistic events has been useful in saving and improving lives throughout history, "since the assessment of uncertainty incorporates the idea of learning from experience which most creatures do" (Good, 1959). Learning from experience is only part of the challenge, since many decisions involve new or evolving risk scenarios in which humans have little or no experience. Risk management capabilities must include the ability to respond to risks that are known and well understood, and the ability to detect and anticipate those that are newly emerging (e.g., U.S. EPA SAB, 1995; NRC, 1999; U.S. EPA ORD, 2000). Uncertainty analysis provides a framework for characterizing the position of a problem along this continuum and the implications for appropriate response and management.

Uncertainty about risk arises as a result of a fundamental lack of understanding about how the world works, randomness in the outcome of natural and human processes, and an inability to measure, characterize, or even at times define key quantities of interest that affect, or are affected by, risk. Risks can vary across time, space, and from individual to individual in a population. Characterization of these variations is an essential part of a risk analysis; however, the basic knowledge or data to fully and effectively characterize risk, and its variability, are often lacking. Uncertainty analysis helps to reveal the nature of this incomplete knowledge and its potential impact on decisions.

In this chapter, the essential features of risk and uncertainty are reviewed to identify the role of quantitative assessment in helping to characterize

The authors wish to thank Gail Charnley, Lauren Zeise, and Alex Farrell for helpful comments.

and manage risks. The landscape of risk assessment is reviewed, noting the role that scientific uncertainty plays during the evolution of a risk problem through phases of discovery, increasing awareness, deliberation, management decision, and implementation. Probabilistic methods for uncertainty analysis are considered, and the historic growth of their use is documented. The role of probabilistic assessment in informing or guiding risk management decisions is discussed, including alternative risk-based or precautionary approaches for dealing with uncertainty. The role and limits of quantitative assessment of uncertainty are explored by examining three case studies: (1) siting and permitting municipal waste combustors; (2) addressing global climate change; and (3) protecting against species extinction.

2. BACKGROUND AND DEFINITIONS

Uncertainty analysis may be qualitative or quantitative. Qualitative descriptions of uncertainty seek to characterize the general state of knowledge about the risk and the overall weight of evidence concerning the nature and source of the hazard. This type of evaluation is routinely carried out early in a risk assessment, during the "hazard assessment" stage (NRC, 1983). Qualitative assessments may seek to identify alternative scenarios for the risk outcome, the possible implications of these scenarios, their severity and likelihood. Once specific measures of a risk outcome are identified – such as the likelihood of occurrence of a health endpoint for an individual, population incidence, lives lost, hectares of habitat destroyed, or dollars required for remediation – and methods and models are developed to predict these quantities, then quantitative uncertainty analysis can provide insight into the probability of occurrence of different values of these measures. Even then, qualitative evaluation retains an important role in characterizing the fundamental assumptions of the analysis and the state of the science supporting them.

Quantitative uncertainty analysis seeks to identify the relationship between risk model assumptions and inputs and the resulting uncertainty in predictions of risk. In the context of human health risk assessment, uncertainty analysis focuses on models and inputs that estimate exposure, dose, and the resulting health effects. Quantitative analysis may be used to examine the impact of person-to-person *variability* in factors such as ambient concentrations, time-activity patterns, ingestion and inhalation rates, susceptibility to disease, body weight, or age. It may also consider scientific *uncertainty*, for example, in the fate and transport of chemicals released to the environment, or in the dose-response relationship.

The need for careful delineation of variability and uncertainty, and proper recognition of the differing roles that these play in a quantitative risk assessment, are now widely recognized (e.g., Bogen, 1990, 1995; Hoffman and Hammonds, 1994; NRC, 1994; Burmaster and Wilson, 1996; Frey and Rhodes, 1996; Cullen and Frey, 1999; Hertwich, McKone, and Pease, 1999). Variability reflects the different levels of risk experienced by individuals, both before and after a mitigating action is taken. Variability reflects true differences between individuals, timeframes, and geographic locations. Uncertainty dictates that the true levels of risk faced by any individual or group are unknown. Uncertainty arises due to a lack of knowledge, the use of models that require inherent simplification of real-world processes and relationships, limits in the feasibility of measurement or other data collection, and the need to project future impacts and consequences that cannot be observed at the present time.

While variability and uncertainty are often addressed with the same tools and techniques, they have different impacts on the decision process. "Uncertainty forces decision makers to judge how probable it is that risks will be overestimated or underestimated for every member of the exposed population, whereas variability forces them to cope with the certainty that different individuals will be subjected to risks both above and below any reference point one chooses" (NRC, 1994). Uncertainty analysis may assist a decision maker in communicating or justifying a decision to the public, the media, a judge, or a jury, indicating "that the decision made was neither arbitrary nor ill-informed" (Finkel, 1990). Uncertainty analysis may provide information about the degree of, or lack of, conservatism in single point estimates of risk. It may assist with the prioritization of information gaps for additional study by revealing the relative impact of uncertainty and variability on the decision under consideration.

Sensitivity analysis is an important adjunct of uncertainty analysis, determining the impact of particular model inputs and assumptions on the estimated risk. Sensitivity analysis is often conducted as a precursor to uncertainty analysis, helping to identify those model assumptions or inputs that are important – if the models are not sensitive to a particular input or set of inputs, there is no need to examine these inputs as part of a more sophisticated uncertainty analysis. Sensitivity analysis is revisited in the subsequent phases of an uncertainty analysis to identify those inputs and assumptions that are significant contributors to the overall variance of the output and/or critical to pending decisions (for an example of the latter, see Merz, Small, and Fischbeck, 1992), thereby identifying the uncertainties *that matter*. In this manner, priorities can be established for further research and data collection efforts.

The introduction of explicit techniques for the treatment of uncertainty in risk assessment in U.S. regulatory policy has occurred over the past thirty years. An early example was the U.S. Atomic Energy Commission-sponsored "Reactor Safety Study," also known as the Rasmussen study (NRC, 1975), which together with reviews and follow-up studies led the Nuclear Regulatory Commission to develop probabilistic safety objectives to supplement qualitative safety goals already in place. In the Reactor Safety Study, event trees that systematically laid out scenarios with the potential to lead to reactor accidents were developed. Probabilities were assigned to these scenarios using a combination of quantitative information found in industrial records and expert judgment. The Study's reliance on subjective information elicited from experts led to criticism. In retrospect, the reliance on judgment was found to be generally sound and largely unavoidable; however, in its implementation, it resulted in "greatly understated uncertainty bands" (Lewis et al., 1979).

The last several decades have brought major advances in probabilistic risk assessment in civil engineering for building codes, and in aeronautical/astronautical and electrical engineering for ensuring the reliability of systems and circuits. These techniques have been used to identify weak points and opportunities for systems improvement. Still, many engineers (and their lawyers) remain uncomfortable explicitly computing the probability that their design will fail, thereby acknowledging that such a possibility is an inherent part of the design. It is expected that some time will be required to overcome this aversion, and the associated history of ignoring or avoiding explicit statements of uncertainty in the design, siting, permitting and operation of both industrial and public sector facilities (Morgan and Henrion, 1990; Lave, Resendiz-Carrillo, and McMichael, 1990).

Policies and practices regarding the treatment of uncertainty in environmental health risk assessment have been evolving for many years in government agencies, academic circles, and industrial fora. The initial guidance of the Red Book (NRC, 1983) addressed uncertainty in the risk characterization phase of a health risk assessment, noting the need to respond to the following:

- What are the statistical uncertainties in estimating the extent of health effects? How are these uncertainties to be computed and presented?
- What are the biologic uncertainties in estimating the extent of health effects? What is their origin? How will they be estimated? What effect do they have on quantitative estimates? How will the uncertainties be described to agency decision-makers? (p. 33)

The report went on to note that:

Little guidance is available on how to express uncertainties in the underlying data and on which dose-response assessments and exposure assessments should be combined to give a final estimate of possible risk. (p. 36)

The authors of the Red Book were clearly aware of the importance of uncertainty and the need to consider and communicate its implications. In practice, however, government agencies typically relied solely upon point estimates of risk in their early studies and applications of risk assessment.

More recently, publications containing guidance and encouragement for uncertainty analysis, mostly in the form of variability characterization, have come from various branches of government and the U.S. National Research Council (NRC). These include the NRC reports *Science and Judgment in Risk Assessment* (NRC, 1994) and *Understanding Risk: Informing Decisions in a Democratic Society* (NRC, 1996), several bills proposed by members of Congress, and a set of U.S. EPA memoranda on agency policy on risk assessment, risk characterization, and Monte Carlo analysis (Habicht, 1992; Browner, 1995; U.S. EPA Science Policy Council, 1997b). Probabilistic risk assessment has begun to be used by the U.S. Department of Agriculture and the U.S. Food and Drug Administration for evaluating food safety (see, for example, http://www.cfsan.fda.gov/~dms/fs-toc.html and http://www.foodsafety.gov/~fsg/fssyst2.html). Similarly, uses of probabilistic methods for risk assessment have been instituted in other countries (e.g., UK ILGRA, 1996, 1999; NRC, 2000). As these reports and guidance suggest, probabilistic approaches have gradually come to be accepted in recent years, especially for exposure assessment. The most recent U.S. EPA *Exposure Factors Handbook* (U.S. EPA, 1997a) includes distributions that describe variability in a number of exposure factors across target populations and information on the uncertainty of these estimates.

The adoption of probabilistic methods has occurred much more slowly in the area of dose-response assessment. This is so in part because the uncertainties in the overall dose-response assessment are so fundamental that our ability to grapple with them is limited. Even if a chemical has been evaluated using animal studies, the need for extrapolation across doses and species introduces uncertainties not especially amenable to analysis using many of the standard tools of uncertainty analysis. Further uncertainty is introduced in the application of individual chemical toxicity factors since compounds do not act alone in the environment, but rather together, resulting in exposures to a variety of complex mixtures. Even with good data on actual human exposures – and such datasets are rare – the low signal-to-noise ratios that typify epidemiological studies introduce considerable uncertainty when these data are introduced in practice.

While simpler, consensus or conservative approaches, such as the use of point (e.g., "upper 95% confidence") estimates for dose-response potency factors remain most common in current practice, such approaches leave major and fundamental uncertainties unaddressed. An example resulting from the format of traditional animal studies is the questionable relevance to human exposure of the timing of animal dosing. Despite the presence of great uncertainty many now argue persuasively for explicit uncertainty analysis, based on the finding that dose-response uncertainty is often the largest and most important source of error in an integrated health risk assessment (Bogen, 1990; McKone and Bogen, 1992; Cullen, 1995).[1] Indeed, the application of detailed, rigorous methods for uncertainty analysis in an exposure assessment, followed by use of point estimates for the dose-response calculation, may rightfully be considered akin to the classic search for lost keys under the lamppost – the key to resolving uncertainty in a risk assessment may not lie where the tools and estimates already exist but rather where our current knowledge is dimmest.

3. THE SOCIAL-SCIENTIFIC CONTEXT OF RISK AND UNCERTAINTY ANALYSIS

The social and cultural context of a risk, and the associated scientific studies undertaken to characterize it and inform risk management decisions, can greatly influence the way that uncertainty is studied, presented, and interpreted (Wynne, 1980; Clarke, 1988; MacKenzie, 1990; Jasanoff, 1993; NRC, 1996; Thompson and Dean, 1996). Heuristic tendencies toward overconfidence and a systematic underestimation of uncertainty in model predictions by those who develop and apply risk models are well documented (Kahneman and Tversky, 1973; Fischhoff, Slovic, and Lichtenstein, 1982; Henrion and Fischhoff, 1986; Freudenburg, 1988; Morgan and Henrion, 1990; Small and Fischbeck, 1999). Some assessments may tend to highlight uncertainty, while others downplay it (Jasanoff, 1990; MacKenzie, 1990; NRC, 1996), especially when considering extreme outcomes (Patt, 1999). Historic adaptation by decision makers and the public to the systematic underreporting of uncertainty, combined with questions about the motivations and biases of some risk studies, can lead to interpretations of reported uncertainties by targeted audiences that differ considerably from those intended by the scientists reporting them (Wynne, 1987; Slovic, 1993; Peters, Covello, and McCallum, 1997; Petts, 1998; Hunt, Frewer, and Shepard, 1999; Word et al., 1999).

Despite skepticism (albeit healthy) toward highly confident estimates of risk, we often prefer simple, unambiguous answers from science. In some contexts, a "one-armed scientist" is preferred,[2] especially when creating or maintaining confidence in the decision (even if illusory) is as important as the decision itself (Jasanoff, 1987). Johnson and Slovic (1995) explore the quandary faced by organizations considering how to communicate their own degree of uncertainty about risk issues that they must manage. Their empirical studies with focus groups indicate that highly confident statements about the problem build public confidence in the competency of the agency but can lead citizens to question their honesty. Full and open expressions of uncertainty can make the agency appear more honest, but (alas) less competent. The challenge today is to conduct and communicate honest, competent risk and uncertainty analyses, with appropriately structured stakeholder and expert input and deliberation in problem formulation, study, and evaluation so that trust can be built for both the competency *and* the objectivity of the analysis (Jasanoff, 1993; NRC, 1996).

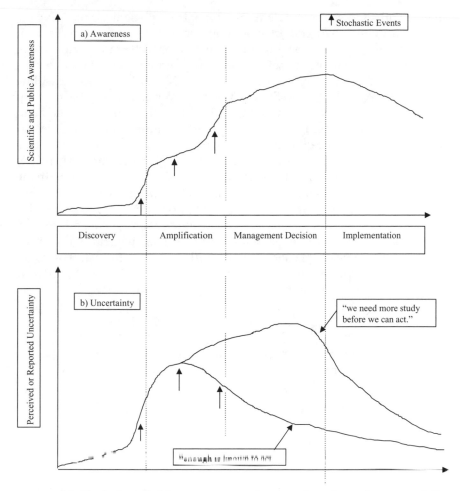

FIGURE 6.1. Evolution of risk awareness and uncertainty through stages of discovery, amplification, management decision, and implementation for a typical risk problem.

The evolution of awareness of a risk problem and the perceived uncertainty associated with it often follows a consistent recognizable pattern. Scientific and public awareness, and demands for risk management, undergo predictable growth and amplification through the processes of scientific discovery, information transmission, media coverage, and response to information (Kasperson et al., 1988; Renn et al., 1992; Kasperson and Kasperson, 1996). Risk awareness and perceived uncertainty often evolve through four major phases, illustrated in Figure 6.1:

1. *Discovery*. Here awareness of the risk emerges; then, depending on the magnitude of the problem, press coverage diffuses information

to all sectors. The problem may have been present for many years, but previously undetected with available measurement tools, or it may emerge as a result of a new technology, chemical formulation, greatly expanded use of an existing technology or product, or increased public awareness. For example, MTBE, used as a gasoline additive in the United States to improve air quality, has been recognized for a number of years to be highly mobile and persistent, and capable of resulting in the widespread contamination of surface and groundwaters (Squillace et al., 1996).

2. *Amplification.* Here the risk is the subject of growing concern in response to new study results, press releases, incidents, or accidents. In the case of MTBE, evidence of, and anger over, widespread water pollution impacts continued to accumulate (Johnson et al., 2000; see also http://fitzgerald.senate.gov/legislation/mtbe/mtbemain.htm).

3. *Management Decision.* The need for risk management is apparent, and the problem is characterized with enough information to identify an appropriate response. In the case of MTBE, EPA has now proposed elimination or very strict limitation on its use as a fuel additive. See the Federal Register Notice at http://www.epa.gov/fedrgstr/EPA-TOX/2000/March/Day-24/t7323.htm.

4. *Implementation.* Here the risk management strategy is implemented, and ongoing monitoring and study are conducted to ensure that the problem is properly and permanently resolved or addressed.

What is the role of uncertainty in reflecting, guiding, or promoting transitions of risk problems through these various phases? While each problem undergoes its own unique path through discovery, amplification, and management, the profile shown in Figure 6.1 is typical. When the awareness of a risk first emerges, the reported uncertainty is often low, and usually underestimated. This is to be expected; when dealing with surprise, our tools for analysis are quite limited (Shlyakhter, 1994; Hammitt and Shlyakhter, 1999; Casman, Morgan, and Downlatabadi, 1999). It is difficult to know much at all about an emerging risk, even your current level of uncertainty. During the awareness amplification phase, uncertainty – at least that reported and acknowledged – is amplified as well. This is especially common as new events or new data expose previously unrecognized aspects of the problem.

Risk management decisions require *sufficient* information to justify and prescribe an effective response strategy. However, different stakeholders often differ considerably in their assessment of what is sufficient. Those favoring rapid implementation of (possibly costly) control strategies typically argue that enough is already known, while those favoring delay argue that the uncertainties remain too great and that further study is needed.

Those familiar with the decade-long debate over the health and sustainability of salmon populations in the northwest of the United States during the 1990s will recognize these arguments. Only recently was a clear decision and transition to the risk management phase made, with the listing of multiple species under the Endangered Species Act (see Section 6.3) without full resolution of many of the attendant uncertainties. As suggested in Figure 6.1, once the pressure and high stakes of a decision have passed, perceptions of uncertainty among stakeholders often converge (though usually not completely) to some moderate level, as the risk management program is implemented. Whether this pattern will hold in the case of salmon recovery efforts remains to be seen.

4. ESTIMATING UNCERTAIN RISK

Given the context in which analyses of risk are embedded, it is not surprising that no single approach is uniformly applicable in dealing with variability and uncertainty. Still, it *is* important to acknowledge all sources of variability and uncertainty about risk and to justify the selection or omission of each of these sources in subsequent analyses. As is outlined later, choosing among the array of approaches from qualitative to quantitative assessments, or from simple bounding to sophisticated numerical analyses, requires analysts to think about the appropriateness of increasingly complex assessments given the decision under consideration and the information at hand.

While one approach to dealing with uncertainty has been to simply ignore it, a qualitative discussion of major data gaps and assumptions and their projected impact on the analysis should be within reach in even the most data-poor situation. For cases in which fundamental uncertainty owing to a lack of scientific knowledge dominates, a qualitative discussion may be most appropriate. At the next level, explicit evaluations of alternative assumptions or scenarios can be conducted to demonstrate model and decision sensitivity to various assumptions. This approach could involve the selection of representative scenarios to capture key features of possible future conditions – this approach has been especially useful in assessments of future climate change (IPCC, 1995, 2001a,b,c); it could also involve more-formal computational procedures to explore system behavior under plausible alternative conceptual models and assumptions (Bankes, 1993, 1994; Lempert, Schlesinger, and Bankes, 1996).

Quantitative procedures for uncertainty analysis include the evaluation of the uncertainties in model inputs and assumptions and calculation of the resulting uncertainty in model predictions. Simple quantitative estimates can be generated using analytic approximations (such as first- or second-order uncertainty analysis, based on Taylor series approximations) that consider sensitivity to, and uncertainty in, model parameters and inputs (see Cox and Baybutt, 1981, and Morgan and Henrion, 1990, for discussion of analytical approximation methods for uncertainty analysis). More sophisticated numerical methods such as Monte Carlo analysis can be used to simulate uncertainty and variability in model inputs and outputs. In addition, a number of new methods and conceptual approaches to probabilistic analysis have been developed, including fuzzy arithmetic, interval analysis, and statistics- or AI (artificial intelligence)-based methods.

In most quantitative risk assessment, it makes sense to start simply and to increase the level of complexity of the analysis where justified on the basis of available information, intermediate results, the importance of the pending decision, and time and resource constraints in developing the analysis to support the decision. An iterative progression is recommended, whether one is adhering to a precautionary, screening approach in which a set of qualitative tests will be used to gauge the acceptability of a risk, or one is applying a decision framework into which quantitative information about the risk will be directly input. Specific approaches to uncertainty are discussed as they arise in the following sections. In so doing, this paper provides a general overview of the suite of available methods for risk and uncertainty analysis; further detail and elaboration can be found in a number of papers, texts, and reports (Cox and Baybutt, 1981; Finkel, 1990; Morgan and Henrion, 1990; Pate-Cornell, 1996; NCRP, 1996; Thompson and Graham, 1996; Cullen and Frey, 1999; Bedford and Cooke, 2001; Isukapalli and Georgopoulos, 2001).

4.1. Objective and Subjective Information in the Risk Assessment Framework

The body of information identified to support a risk assessment may include empirical data, expert knowledge, judgment, opinions, beliefs, intuition, model results, analogies, or a combination of these. In short, it is likely to contain both objective and subjective elements. It is almost impossible to develop a general hierarchy of information for use in risk assessments because what constitutes value is context-dependent. Are three data points of extremely high quality, on exactly the right population, species, chemical, averaging time, and location, more or less valuable than 3000 data points taken out of context?

The field of probability theory includes distinct definitions of objective and subjective probability, though many of the same tools are used to represent and manipulate both. Objective probability, sometimes referred to as classical probability, is based on the theoretical outcomes of ideal, mathematically defined events or random processes. As shown in the early work of Laplace, event probabilities can be derived for such cases from a few basic theorems and axioms for the rules of probability. An understanding of the underlying processes that generate or influence a risk informs an analyst about the expected probabilities of various outcomes. Objective probabilities may also be established empirically by observing the relative frequency with which various outcomes occur in long-run trials. This is the frequentist's approach.

Subjective probability by contrast involves personal judgment, intuition, and other subjective factors. While references to subjective probability are relatively recent additions to the published record (Ramsey, 1931; De Finetti, 1937), the use and understanding of this type of information extends back much further. Indeed the genesis of subjective probability and the rule for treating it are traced to Thomas Bayes, a near contemporary of Laplace in the late 1700s; consequently, the subjectivist approach to probability is often referred to as the *Bayesian* method. Medical diagnosis, or projections of success and failure of a startup business, are examples of probabilistic assessments requiring some degree of subjective judgment. Depending on the nature of these assessments and their use of theory and observed data, the line between objective and subjective information can begin to blur.

The search for a definitive distinction between objective and subjective information has a long history. In a discussion of the characteristics of subjective and objective sources of information and knowledge, Cooke (1991) reminds us of the divided line framework proposed by Plato. Plato argues that knowledge is partitioned into four ordered categories. From lowest to highest he calls them *eikasia* (conjecture), *pistis* (belief), *dianoia* (correct reasoning from hypotheses as in mathematics), and *episteme* (knowledge). He draws a line dividing the lower two – the uncertain, fuzzy, and subjective – from the higher two – the rigorous and conclusive, which require scholarly competence. Although admittedly an analyst may find comfort in empirical data or apparently concrete measurements, after honest deliberation many, including Cooke, conclude that a schism of this type is neither useful nor real.

Bayesian methods provide a framework for refining risk estimates by combining multiple sources of information. A full Bayesian approach allows subjective assessments of probability to be informed, or "updated" with observed data, thus allowing a combination of the available subjective knowledge and objective information. For many, this is worrisome in that independent analysts can no longer be assured of arriving at the same conclusion. Still, it is rare that one faces a well-defined risk, with a large, perfectly applicable, high-quality data set. Evans et al. (1994a,b) suggest that the real choice is not whether or not to use judgment in risk analyses but rather where to draw the boundary of analysis. Some analysts draw a "tight" analytic boundary – answering only those questions informed by direct reference to classical or frequentist concepts of probability, but often leaving questions of relevance or interpretation unanswered. Others draw a looser boundary, including questions of relevance and interpretation in their formal analysis. In doing so, they rely on both objective and subjectivist definitions of probability. Such a "loose" analysis is willing to forgo some degree of objectivity for improved relevance.

4.2. Screening Analysis: Worst Case, Plausible Upper Bounds, Best Estimates

The first step in many quantitative risk assessments involves screening, an initial assessment of whether a hazard is present or whether a present hazard is of sufficient magnitude to be of concern. It is critical that a potential hazard first be defined clearly and the goals and purposes of the analysis understood. For scenarios of interest, a model or set of models is developed. They will link a possible exposure or natural occurrence or other chance event to some outcome of interest (often an adverse effect). Next it is advisable to gather and examine closely the available information pertaining to the problem. A rough comparison of the quality of the available data and the quality requirements for the results of the analysis up front may help avoid disappointment later. Screening assessments are sometimes quite crude depending on the quality and quantity of available information and the state of the science informing the analysis. A lack of either scientific understanding or of relevant information may lead to reliance on default assumptions, and it may be difficult to advance beyond a crude look without additional data collection.

Usually one or several point estimates are generated in a screening assessment. The relevant decision context may suggest a focus on worst-case upper bounds, representative central estimates, more plausible upper bounds, or perhaps several of these to gain a sense of uncertainty or variability based on a range of possible estimates. The results of a screening assessment may prompt one of several courses of action. If a worst case screening analysis indicates low hazard or risk relative to the level of concern in the eyes of the decision maker, steps to mitigate the risk might appear unjustified, and no further study may be needed. Conversely, immediate steps to mitigate risk might be prompted by high screening estimates of risk, especially if the costs of mitigation are not too high. High costs of action might constitute cause for additional, more detailed analysis in order to become more convinced that these expenditures are warranted. Often screening begins with a worst-case assessment and then moves to increasingly more realistic point estimates as warranted, if the risk is not discounted on the basis of insufficient concern (Pate-Cornell, 1996).

4.3. Beyond the Single Number: Probabilistic Analysis

Probabilistic risk assessment (PRA) often refers to the generation of distributions of exposure or risk representing uncertainty or variability or both.

TABLE 6.1. *Reasons to Pursue More Complex/Sophisticated Assessment*

To avoid costs of alternatives, substitutes or remedial activity (when they are
  higher than analysis costs)
To avert severe consequences of poor or biased estimates
To support decisions about whether to reduce actively exposure or do additional
  research
To allow more rigorous comparisons of alternatives
To provide a systematic consideration of all sources of variability and uncertainty
To provide a basis for comparison among impacted groups (e.g., for evaluation
  of equity concerns)
To prioritize opportunities for additional scientific research and to identify
  significant contributors to overall variance in risk, allow value of information
  analysis, and identify fruitful avenues for future research

TABLE 6.2. *Reasons Not to Pursue More Complex/Sophisticated Assessment*

To acknowledge a profound lack of understanding about the problem and its
  causative relationships
To avoid the cost of analysis (when it is higher than the costs of alternatives,
  substitutes, or remediation)
To acknowledge the results of screening indicating that the likelihood of
  unacceptable risk or hazard is low
To act immediately to reduce risk when safety is an immediate concern
To acknowledge that probabilities are prohibitively uncertain and/or
  indeterminate
To make explicit that there is little variability or uncertainty in the risk
To acknowledge a lack of desire, interest, guidance, or experience on the part of
  decision makers

This approach attempts to account for information from a range of sources
and for the quality and relevance of that information. The usefulness of pursuing a probabilistic analysis is affected by a number of factors (Tables 6.1
and 6.2), though the ability to perform such analysis may be limited by
available information.

Assume that a screening analysis has been pursued before launching
a more complex analysis. The problem has been defined, the purposes
of the analysis are understood, and information has been gathered and
scrutinized. A model or set of models is in place to describe the problem.
Here is where screening analysis and PRA diverge. In PRA, probabilities
are assigned to represent uncertainty (and/or variability) in the model(s)
and their inputs. These probabilities are propagated through the model(s),
and an output distribution describing the probability of various outcomes
is generated (see Figure 6.2).

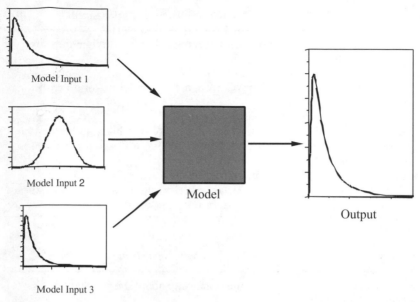

Model Input 1

Model Input 2

Model

Model Input 3

Output

FIGURE 6.2. Propagating uncertainty.

To illustrate, assume that the model output depicted in Figure 6.2 is the probability distribution of various degrees of risk faced by an individual. Some examination of this output can inform a decision about whether to invest more time and analysis in the project. Is the risk at a level of concern? Is the variance of the risk distribution prohibitively large considering the decision context? If proceeding with analysis, one could next assess the most significant contributors to overall variance in the risk distribution. Armed with information about the major contributors to variance, it may be helpful to weigh the cost or feasibility of obtaining better information against the consequences of making a decision on the basis of the current information and analytic results. At this point, the results may be deemed adequate to inform the decision, or the analysis may be continued iteratively with sequential refinement of model choices and inputs. Value-of-information techniques, discussed in more detail later, help to identify the improved knowledge or further data expected to be most useful for informing these subsequent model refinements.

4.3.1. Developing Distributions

A major challenge in applying probabilistic risk assessment techniques is representing variable and uncertain risk models and model inputs, often with probability distributions (Morgan and Henrion, 1990; Taylor, 1993; Evans et al., 1994a,b; Haimes and Lambert, 1994; NCRP, 1996; Cullen and Frey, 1999; Thompson, 1999; Ferson, 1996; Ferson, Ginzburg, and Akcakaya, in press). For cases where the model variables derive from known underlying mechanisms, a theoretical basis may be present for selecting a particular distribution type or form. The Central Limit Theorem dictates that variables resulting from the sum of many individual components tend to be normal, while those derived from multiplicative processes approach a lognormal distribution. Purely random processes (e.g., rare events occurring independently in time or space) can often be represented by a Poisson process, with an exponential distribution for the time (or distance) between events and a Poisson distribution for the number of events occurring over a given time (or spatial) interval. Even though such theoretical models can provide a good starting point for selecting candidate distributions, real-world conditions often lead to deviations from theoretical models, and an empirical approach to selecting and fitting the distribution must be adopted. When relevant, high-quality data are available for these cases, a variety of model identification and parameter estimation techniques can be utilized (Morgan and Henrion, 1990; Taylor, 1993; Burmaster and Thompson, 1998; Cullen and Frey, 1999). In the more common case of sparse data, elicitation techniques and related approaches are used to convert judgments into probabilities. In between, a combination of approaches may be pursued. In particular, Bayesian methods allow an elicited probability distribution, or one fit to a particular data set, to be updated using information from another data set (general texts on Bayesian methods include Lee, 1989; Press, 1989; Gelman et al., 1995; example applications to risk assessment can be found in Iman and Hora, 1989; Eddy, Hasselblad, and Shachter, 1990; Taylor, Evans, and McKone, 1993; Berry and Stangl, 1996; Wolfson, Kadane, and Small, 1996).

A complete discussion of the myriad approaches to information collection is well beyond the scope this paper; however, the critical role of judgment in supporting risk assessments warrants provision of a set of references to the field of expert elicitation. For most people, expert and lay alike, it is very difficult to convert understandings, beliefs, and intuition into probability distributions. Common heuristic biases that often overlay human judgment, and have been observed in many applications, include (NRC, 1996):

- A tendency to overestimate probabilities for events commonly encountered or frequently mentioned in work or personal environments (termed the "availability" bias);
- A tendency to be overly influenced by speculative or illustrative information presented to initially frame a problem ("anchoring and adjustment");

- A tendency to be overconfident, especially for complex problems with sparse data that may or may not be pertinent to the problem at hand ("overconfidence" and "representativeness"); and
- A tendency to ignore data or discount evidence that contradicts strongly held convictions ("disqualification").

Expert elicitation protocols have been developed in an attempt to counteract these heuristics and biases, though some degree of bias or misrepresentation should still be expected in elicited probabilities (Winkler, 1967; Spetzler and Holstein, 1975; Lichtenstein and Fischhoff, 1977; Slovic, Fischoff, and Lichtenstein, 1979; Morgan, Henrion, and Morris, 1980; Kahneman, Slovic, and Tversky, 1982; Merkhofer, 1987; Wolpert, 1989; Morgan and Henrion, 1990; Cooke, 1991; Wolfson, 1995; Chaloner, 1996; Kadane and Wolfson, 1998).

4.3.2. Propagating Variability/Uncertainty Through Models

Multiple uncertainties in model inputs and structure must be combined and evaluated to determine the overall uncertainty of model predictions. The tools used most frequently in the risk assessment field for this purpose are based on simulation. With these methods models are evaluated repeatedly, with each run of the model representing a possible outcome. The inputs to the model are simulated by sampling one value from each input distribution and then calculating the corresponding value of the model output. Through many iterations, a distribution of the output quantity is generated (Figure 6.2).

Possible methods for sampling the input distributions include the following:

- *Simple random or "Monte Carlo" sampling*, in which each sample of the model inputs is independent of all others. This method, which may be thought of as "independent sampling with replacement," is relatively easy to characterize in terms of the statistical properties of the resulting sample, but it often requires a large sample size to obtain accurate, representative estimates of the model input and output distributions.
- *Stratified random sampling*, in which the input distributions are divided into strata that are subsampled. Stratified sampling methods allow more accurate characterization of variability and/or uncertainty distributions with smaller sample sizes (this may be especially important when it is difficult to evaluate the model many times owing to computational requirements), but it is more difficult to describe the statistical properties of the sample. Available methods include Latin Hypercube sampling (McKay, Conover, and Beckman, 1979; Iman and Convover, 1982) and Hammersley, Halton, and Sobol sampling (Morokoff and Caflisch, 1994; Tezuka, 1995; Kalagnanam and Diwekar, 1997; Gentle, 1998).

- *Targeted or importance sampling*, in which a portion of the model input (and resulting output) space is sampled more often because this portion is associated with high risk conditions or events or is more relevant to possible risk management decisions (Morgan and Henrion, 1990).

Before embarking on a simulation exercise, it is also necessary to consider the presence of correlations among the model inputs, as these may significantly influence the accuracy or applicability of the output generated (Smith, Ryan, and Evans, 1992; Haas, 1999).

4.3.3. Methods for Distinguishing Variability and Uncertainty

PRA may be pursued as a one-dimensional exercise in which uncertainty and variability are combined without regard for their different consequences for a decision, although interpretation is limited as a consequence. Alternatively, either variability or uncertainty may be eliminated from the problem by looking at a series of narrowly defined scenarios in which only a subset of factors is allowed to vary across a range of possibilities. For a full discussion of these possibilities, see Cullen and Frey (1999).

For a full evaluation of the implications of variability and uncertainty, PRA can be carried out as a "two-dimensional" (or *n*-dimensional) analysis in which the sources of variability and uncertainty are handled separately (and often sequentially), with a resulting output distribution that is a multidimensional surface (Frey, 1992; Hoffman and Hammonds, 1994; Simon, 1999). A two-dimensional approach to Monte Carlo simulation allows for the disaggregation and evaluation of the consequences of variability and uncertainty. It can also provide a framework for exploring the interaction of uncertainty and variability. A risk model may be written as (Bogen and Spear, 1987):

Risk = f(**V**, **U**)

The estimate of **Risk** to a population of exposed individuals is a function of both variability in parameters **V**, which have different values for each member of the population, and uncertainty in parameters **U**, for which there is some lack of information about their true values (Frey, 1992). All three components of the model, **Risk**, **V**, and **U** are shown in bold to emphasize that they are vectors, representing risk across multiple individuals in a target population, with multiple sources of variability and uncertainty.

A risk model with both variable and uncertain components is illustrated in Figure 6.3. Frequency distributions are specified for all variable quantities and probability distributions are specified for all uncertain quantities (Cullen and Frey, 1999). Some model inputs are both uncertain and variable and thus are specified with multiple interacting distributions. A technique such as Latin Hypercube Sampling is then employed to generate two sets of samples: the frequency distribution representing each variable quantity

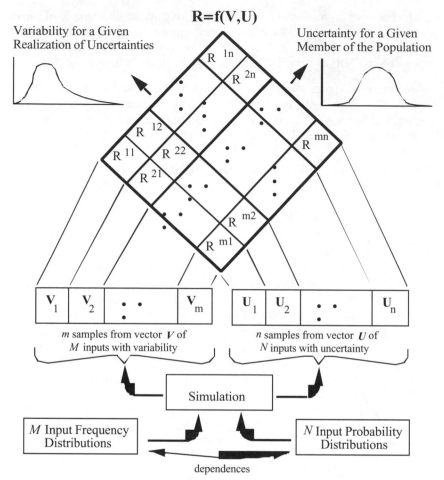

FIGURE 6.3. Two–dimensional simulation of uncertainty and variability (from Frey, 1992 with permission).

is simulated with a sample size of n, and the probability distribution representing each uncertain quantity is simulated with a sample size of m. The model is repetitively evaluated for each combination of samples from the variable and uncertain parameters, thereby generating the output surfaces (Frey, 1992; Frey and Rhodes, 1996).

An alternative approach to simulation evaluates all uncertain variables in an outer loop, followed by sampling from variable parameters in an inner loop. Each uncertain parameter is assigned a particular value before the variable parameters are simulated (for the given realization of uncertainty). This approach is consistent with variability/uncertainty characterizations based on mixture distributions, in which parametric distributions are used to represent variable quantities, but the *values of the parameters* for these distributions are uncertain. The sequential, hierarchical model is then specified as

Risk $= f(\mathbf{V})$

Variability: $\mathbf{V} \sim f_1(\boldsymbol{\theta})$

Uncertainty: $\boldsymbol{\theta} \sim f_2(\mathbf{U})$

where f_1 is the parametric distribution for the variable quantities $\mathbf{V}$, with parameters $\boldsymbol{\theta}$, while f_2 is the uncertainty distribution for these parameters. The distribution f_2 is simulated in the outer loop to determine $\boldsymbol{\theta}$; then $\mathbf{V}$ is simulated in the inner loop from f_1. An example application of this outer-inner loop simulation method for uncertainty and variability analysis is presented in Gurian et al. (2001a,b).

The determination of the distribution f_2 for $\boldsymbol{\theta}$ from an observed dataset can be accomplished using classical methods, based on sampling distributions (Brattin et al., 1996), the likelihood function (Burmaster and Thompson, 1998), or bootstrap methods (Frey and Burmaster, 1999). It can also be derived using Bayesian methods, starting with a prior distribution for f_2, updated as the data for the quantities of interest are considered (e.g., Lee, 1989; Press, 1989; Ashby and Hutton, 1996; Berry and Stangl, 1996; Lockwood et al., in press). This approach can also allow for probabilistic weighting and sampling of alternative model *structures* (i.e., different candidate parametric forms for f_1) (e.g., Wood and Rodriguez-Iturbe, 1975; Iman and Hora, 1989; Draper, 1995; Carrington, 1996, 1997).

4.4. Interpreting and Communicating Results

The communication and interpretation of PRA results is eased by graphical display and an understanding of probability. Because the analyst and decision maker are usually not the same individual, it is important to accompany results with the full set of assumptions and caveats encompassed in the analysis.

How can PRA results be interpreted to help identify the uncertainties that matter most and to point the analyst to further study or data collection activities that can be most beneficial in reducing these critical uncertainties? Most often only a relatively small subset of inputs is responsible for a majority of the variance in a model output. Morgan and Henrion (1990), Cullen and Frey (1999), and others describe the use of summary statistics, visual methods, regression approaches, and other sensitivity analysis tools to help find the most important input uncertainties. Broader approaches for risk communication and methods for testing the effectiveness of alternative presentations are discussed in Finkel (1990); Bostrom, Fischhoff, and Morgan (1992); Morgan et al. (1992); and Fischhoff et al. (1998).

After identifying model inputs and assumptions that contribute significantly to variance in the output, it is necessary to consider how to use this knowledge. Value of information (VOI) techniques are recommended for assessing the importance of the variability and uncertainty contributed by individual inputs to the expected value (or, conversely, the "loss") associated with a decision under uncertainty (Raiffa and Schlaifer, 1961; Raiffa, 1968; March and Simon, 1958; DeGroot, 1969; Hilton, 1981; Henrion, 1982; Evans, 1985; Finkel and Evans, 1987; Evans, Hawkins, and Graham, 1988; Taylor et al., 1993; Dakins et al., 1996; Thompson and Evans, 1997; Costello, Adams, and Polasky, 1998; Solow et al., 1998). VOI techniques seek to identify situations in which the cost of reducing uncertainty is outweighed by the expected benefit of the reduction. In short, VOI is helpful in identifying model inputs that are significant because (1) they contribute significantly to variance in the output and (2) they change the relative desirability of the available alternatives in the decision under consideration.

It is instructive to consider the results of probabilistic assessment along with the point estimates that would have been generated under alternative deterministic calculations (for a comparison of methods for generating these, see Burmaster and Thompson, 1997). While deterministic risk assessments are sometimes criticized for their cascading conservatism since they often combine upper-percentile estimates of exposure and risk factors, such comparison may yield surprising results. There are cases in which conservatism compounds dramatically for estimates of risk constructed from the upper percentiles of input parameters, as well as those in which the effect is less notable (Cullen, 1994). The degree to which the compounding of conservatism occurs in any specific situation is influenced by the structure of the risk or exposure model and the particulars of the relative contributions to overall variance from the individual inputs. For example, the assignment of upper percentiles to influential inputs will have more impact than the assignment of upper percentiles to other inputs. Finally, it should be stressed that the degree of health conservatism imparted by use of upper-percentile estimates in these cases is in fact unknown.

4.5. Sample Results

How are results from analyses of uncertain risk used to gain insight? In Figure 6.4, we show the results of an analysis of the overall variance in dose (sum of inhalation and ingestion doses) of PCBs to adult females living within 3 miles of a Superfund site (Cullen and Frey, 1999). Two-dimensional PRA is pursued in order to explore the relative significance of uncertainty and variability, as well as implications for decision makers. Figure 6.4 presents the percentiles of variability, with their associated uncertainties depicted as cumulative distribution functions.

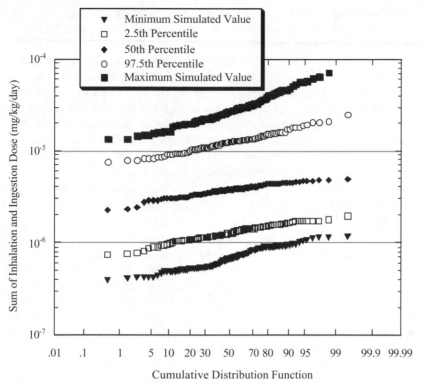

FIGURE 6.4. Uncertainty in percentiles of variability across an exposed population (from Cullen and Frey, 1999). *Each symbol represents a given percentile of variability across the exposed population (e.g., the 50th percentile, shown with a solid diamond, indicates the prediction for the median-dose individual); while the cumulative distribution function along the x-axis represents the uncertainty for a given percentile of the population. The median estimate of variability across the population can be determined by reading up from the value of 50 on the X-axis, the 90th percentile estimates by reading up from 90, etc. For the median, a plausible range from about 2×10^{-6} mg/kg/day as a lower-bound estimate, to about 6×10^{-6} mg/kg/day as an upper-bound estimate is indicated.*

Figure 6.4 indicates that variability, represented by the degree of spread across the 2.5th, 50th, and 97.5th percentiles, is greater in magnitude than uncertainty about any of those percentiles, represented by the range covered by any single line of plotted points. Note that on both an absolute and a relative basis, uncertainty increases at the upper percentiles, a result of our decreasing ability to predict exposure as we move out into the upper tails of variability (i.e., for highly exposed individuals). Uncertainty is, therefore, greater at the upper tail of the variability distribution where the greatest risk management concern is usually focused.

For the case shown in Figure 6.4, variability and uncertainty are explored for a randomly chosen individual in the population. Variability is found to contribute more significantly than uncertainty to overall variance. This conclusion rests on the assumption that the models for human intake via ingestion and inhalation are correct and appropriately applied, and that the measurements of PCB concentration from the site are representative of actual exposures. If the decision endpoint is defined in terms of the entire population (e.g., a decision based on population risk or incidence[3]), then the variability across individuals is integrated out, leaving just uncertainty. In other work, probabilistic sensitivity analysis revealed that for the exposure scenario considered (and assuming a decision focus on individual risk), variability plays a greater role than uncertainty (Cullen and Frey, 1999). For this case, the uncertainty that was present was found to result primarily from random sampling error in the selection of members of the population to include in the study, with much smaller contributions from dose measurement errors.

These results provide a starting point for thinking about subsequent analyses and the value of collecting additional information. For example, resources directed toward reducing random sampling error by increasing sample sizes would alter the results more than would improving measurement techniques through reduced measurement error. On the other hand, variability, the irreducible component, is the most significant contributor to the overall variance in dose. Specific next steps would depend on the decision(s) being supported by the analysis, for example, whether to target exposure-reduction management strategies to the population as a whole or to highly-exposed segments of the target population. The greater magnitude of variability (as compared to uncertainty) in this analysis suggests that efforts to identify the more highly exposed segments of the target population could, at this stage, be more productive than further efforts to reduce the overall uncertainty in the assessment.

4.6. New Approaches to Uncertainty Analysis

A number of new approaches for characterizing the uncertainty in risk estimates have emerged in recent years. Notable among these include methods based on fuzzy sets and classification techniques (Brieman et al., 1984; Eisenberg and McKone, 1998) using a combination of artificial intelligence (AI)/machine learning and statistical methods.

Fuzzy sets arise in two distinct contexts (Klir and Folger, 1988). The first occurs when we wish to classify a condition that is inherently vague and ambiguous such as a "good" water quality, a "healthy" ecosystem, or a "low" (or perhaps even an "acceptable") risk. When mapping from observed or modeled variables (water quality concentrations, species diversity or abundance measures, individual or population risks), there are fuzzy boundaries between good (or healthy or low) and other classifications that might be assigned, such as "fair," "poor," or "high." Membership in any one of these fuzzy sets can be expressed as a measure (between zero and one, summing to one across all possible sets) that is a function of the observed or modeled "hard" variables. The spread and degree of overlap in the membership functions reflect uncertainty in our scientific understanding, as well as differences between individuals (expert and lay alike) in what constitutes a good, fair, or poor condition. Used in this manner, fuzzy sets can help to bridge the gap between the quantitative output of a risk assessment and the more qualitative, judgmental features of a problem that are often used to summarize and value its outcome.

The second major context in which fuzzy sets arise occurs when the sets themselves are "crisp" (i.e., like the hard outputs or measures of an exposure or risk model), but uncertainty in membership is expressed by a fuzzy measure that reflects "plausibility" or "possibility." In this context, fuzzy sets are more controversial, competing with the more traditional methods of probability, with a number of similar concepts, but a different set of rules and calculus (see for example, http://www.amstat.org/sections/spes/jsm1997/5Finvited5.html). Fuzzy set models of either type have been shown to provide practical insights and effective operational tools in a number of technical domains, including medical diagnosis (Adlassnig, 1986), transportation (Nakatsuyam, Nagahashi, and Nishizuka, 1984; Larkin, 1985), meteorology (Cao and Chen, 1983), ecosystem assessment (Giering and Kandel, 1983), and stakeholder preferences for alternative site-cleanup outcomes (Apostolakis and Pickett, 1998). More in-depth discussion of specific applications of fuzzy methods in risk analysis is provided in a companion paper in this volume (Bier et al., 2003), and its role in supplementing the more-established procedures of probabilistic risk assessment continues to evolve.

New methods in AI, data-mining, and machine learning also provide a potentially rich set of tools for exploring and characterizing uncertainty in risk relationships and models. These tools can be used for initial exploration of complex datasets or for summary evaluation of risk model results. Rule-based expert systems and neural networks are examples of the methods that can be used for evaluations of this type (e.g., Julien, Fenves, and Small, 1992; Song and Hopke, 1996; Brasquet, Bourges, and Le Cloirec, 1999). A particularly useful tool that allows expert, probabilistic assessments to be updated with observed evidence is the Bayesian Belief Network (BBN). The BBN is a graphical causative network, or influence diagram, that defines the conditional probability relationships among a set of events (Pearl, 1988; Charniak, 1991). Bayes rule is used to propagate the effect of new information forward and backward through the network, updating the probability of each node.

Stiber, Pantazidou, and Small (1999) demonstrate use of BBNs for evaluating evidence in determining whether or not natural attenuation (versus active remediation) is a viable alternative for a site cleanup. They elicited models from twenty-two experts relating different chemical and environmental measurements that can be taken at a site to infer whether or not natural attenuation (in particular, reductive dechlorination of trichloroethene) can occur or is occurring. The analysis revealed different beliefs and assumptions among experts about key biochemical processes and the value of different measurements. This helps to provide insights on the current state-of-the-science and needs for further research. This type of evaluation is useful in exploring broader structural uncertainties in risk models, when causative relationships are still unclear. Druzdzel and colleagues provide a number of citations on BBNs (see http://www.pitt.edu/~druzdzel/publ.html), including available software. Further information on BBN software is available from Norsys (1998) and at

- http://www.hugin.dk/ and
- http://http.cs.berkeley.edu/~murphyk/Bayes/bnsoft.html.

5. FRAMEWORKS FOR DECISIONS ABOUT UNCERTAIN RISK

Risk assessment provides a structure for systematically exploring risks, but not necessarily the trade-offs associated with decisions about risks. The latter is the domain of risk management and risk-based decision making. The management of trade-offs between competing risks may often rely on risk assessment as a source of information but, more often than not, within a decision framework that incorporates other factors as well.

In decision making generally, it is necessary to gauge the acceptability of, or relative preference for, multiple alternatives. Environmental health regulatory decisions are sometimes considered as falling into multiple categories, for example, those determined on a strict health basis, those determined by technological feasibility, and those requiring the balancing of health and other considerations (Lave, 1978). The exclusive health basis for prioritizing risks, admitting no other considerations, has become rare. Examples included the historic Delaney clause ban (now overturned) on the use of any food additive known to be carcinogenic no matter how low the associated risk (NAS, 1987; Kammen and Hassenzahl, 1999) and statutory mandates for setting ambient particulate matter (PM) standards under the Clean Air Act (CAA). The standards for hazardous air pollutants (HAPs) under the 1990 CAA Amendments serve as examples of the application of a technology basis.

A balancing basis in which health risks are considered along with cost of protection, technological and political feasibility, risks of other alternatives, and social and cultural values is increasingly common. Many statutes, including the 1969 U.S. National Environmental Policy Act (NEPA), explicitly call for environmental impact statements that explore the full range of projected effects of proposed measures, including health, environmental, economic, and social impacts (Spensley, 1995). Precautionary approaches, discussed in more detail later in this section, in practice may also provide a basis for protecting public health within a broader context of selective balancing.

5.1. Balancing

Risks may be described quantitatively, but they are invariably interpreted according to our values, and acted on (or not) in the context of legal requirements, practical constraints, and other factors. Consider for a moment a set of decision makers, public or private, individuals, groups, or their representatives. They may pursue a process that includes iterative stages of problem framing, deliberation, data collection, analysis, preliminary decision, final decision, implementation, and ongoing monitoring of the effects and implications of decisions (NRC, 1996). When these decisions involve risk, they ultimately integrate science and values to decide, given practical constraints, about appropriate tolerance of risk and/or what constitutes acceptable levels of safety and protection (Pate, 1983).

In most cases, individual and social tolerance of a risk depend upon both its quantitative and qualitative characteristics (see Table 6.3). These characteristics have been explored in the literature on the perception of

TABLE 6.3. *Qualitative and Quantitative Characteristics of Risk*

Magnitude of risk
  Probability
  Severity
  Uncertainty/knowledge
Distribution of impacts
  Across species
  Across time
  Across one or more human populations
Time frame
  Catastrophic
  Chronic
Alternatives for avoidance
  Feasibility – technical, political, etc.
  Cost
  Legality
Newness/familiarity/dread/voluntariness/reversibility
Process

Adapted from Lave, 1978; Slovic et al., 1979; Pate, 1983; NRC, 1996; CRAM, 1997; Raffensperger and Tickner, 1999.

risk, decision analysis, and, more recently, the adoption of precautionary approaches (Slovic et al., 1979; Lave, 1978; Raffensperger and Tickner,
1999). These are deliberately introduced and acknowledged at this point
for two reasons. First, approaches to assessing risk are out of context without an awareness of the many factors that influence the acceptability of risk
in a management decision. Furthermore, the acceptability of a risk-based
decision is affected by the decision *process* as well, such as whether fair
and equitable stakeholder participation and informed consent have been
achieved (Pate, 1983; NRC, 1996; CRAM, 1997). More recently, the precautionary principle has been suggested as an approach that can address some
of the limitations of risk assessment. Since risk management must consider
characteristics such as those outlined in Table 6.3, well beyond those em-
bodied in a traditional risk assessment, a precautionary approach to risk
management may de facto allow such considerations to gain standing as
factors in the decision.

5.2. Precaution

Recent press comparing and contrasting risk assessment frameworks with
the precautionary principle warrants a brief treatment of the topic here.
Many of those promoting use of the precautionary principle have done
so viewing it as an alternative to risk assessment (Andrews, 1997). The
precautionary principle suggests that when an activity raises threats of
harm to human health or the environment, precautionary measures should
be taken even if some of the cause-and-effect relationships are not fully
established scientifically. This principle has been summarized as "better
safe than sorry" and "a stitch in time saves nine." First employed in the
late 1970s as *vorsorge* (German for foresight), the precautionary principle
has a wide following in Europe and increasingly has garnered attention in
the United States. In its most pure application, the precautionary principle
does not call for a balancing of risk with cost or other factors, but rather
full avoidance of an unknown risk. In practice, given the reality of risk-
risk and risk-cost trade-offs in any choice among alternatives, it is often
applied along with consideration of costs and other factors. For example,
in the United Kingdom and Sweden an approach referred to as "prudent
avoidance" of risks has been adopted (Raffensperger and Tickner, 1999).
Because the precautionary principle does not give explicit guidance as
to what constitutes adequate proof of safety, its application may require
an exploration of the appropriate balance between resources devoted to
establishing causation and resources devoted to reducing a risk.

Risk assessment may be thought of as promoting a fragmented approach
to environmental decision making. It applies a sometimes constraining
framework to a range of rich and complex questions, often simplifying
diverse impacts in a manner that appears to be unreasonable to some.

Assuming that risk assessment and risk management are kept completely separate, then it would seem that one's adherence to the precautionary principle might only have an impact on risk management. Still, considering the four steps in the risk assessment framework, it may be argued that there is a place for precautionary thinking at every step, especially during the initial hazard identification. In this case, the precautionary principle acts as an overall guiding philosophy. Its adoption colors choices about which problems to tackle, which assumptions and interpretations are made along the way, and finally which conclusions are drawn.

A precautionary approach, based on sound scientific practice and analysis, need not be incompatible with a risk-based (or, perhaps preferably, a "risk-*informed*"[4]) decision framework, but in both the need to identify hazards and prioritize them remains unsolved.

Tickner (1999) discusses issues that could assist in structuring a precautionary decision-making framework, acknowledging that there are necessary steps remaining for determinations of harm and acceptability under uncertainty. These elements encompass a range of challenges in adopting precaution as well as benefits, including this subset from Gee (1997):

- "Who/what gets the benefit of scientific doubt and who has the burden of proof?
- "What level of proof is appropriate?
- "What other benefits of risk reduction exist?
- "How should efforts to reduce risk be balanced with efforts to understand it better?
- "What is the likely size and distribution of false negatives and false positives?
- "What is the optimum mix of policy instruments, targets, and timetables that will maximize overall cost-effective public policy?"
- "How should innovation and risk be balanced?"

The precautionary principle has been suggested for guiding decisions about new developments, new activities, and new industries, assuming a "safe" status quo of "no action." With newly proposed activities, momentum may be relatively low since stakeholders may not be fully vested in the benefits of going ahead. Nevertheless, action to restrict or prohibit ongoing risky activities often carries economic consequences and risk, which ultimately may lead to decrements in personal and environmental health. These trade-offs should be considered with a full comparison of all alternatives, whether they are existing or newly proposed. The principle as strictly applied seems to overlook the importance of comparing the risk of new developments against alternatives.

In cases where the precautionary principle is to be used to support action in advance of scientific evidence of harm, the question remains (as it does within the risk-based decision framework), who decides when the

risk of harm is too high to accept? This question is analogous to that faced
in all risk management, for example in the process of selecting a "margin
of exposure" or a "bright line" for cancer-risk decision making. The answer cannot be determined independently of the risk of the alternatives.
Still, by sharpening the focus on costs and benefits of all alternatives (including no action), the precautionary principle could bring about a better
analysis and discussion of available options, including some currently left
unconsidered.

6. THE ROLE AND LIMITS OF QUANTITATIVE ASSESSMENT – THREE EXAMPLES

The previous sections describe basic tools for assessing uncertain risk and
suggest extensions to specialized approaches where warranted. There is no
shortage of examples of the application of quantitative tools to assessing
uncertain risk, but in the real world virtually all decisions are made on the
basis of multiple criteria of which a quantitative risk estimate is but one.
The three cases discussed here illustrate the role and limits of quantitative
assessment of risk in real decisions. The examples include waste combustor
permitting and siting, climate change assessment, and endangered species
determination. These examples are deliberately drawn from diverse decision contexts and scales of impact. All three grapple with inadequate information drawn from a range of sources and of varying degrees of relevance
to the problem at hand. All lie at the heart of interactions among passionate
stakeholders. And most importantly, all ultimately require the integration
of uncertain science and values in decision making, an enterprise to which
quantitative assessment can contribute, but one that requires a broad and
appropriate framework and process within which these contributions can
be effectively made (NRC, 1996).

6.1. Municipal Waste Combustors

Risk assessment performed in the context of municipal waste combustor permit applications illustrates the limitations of quantitative analysis
as few other cases can (Cullen and Eschenroeder, 1997). First, it must be
acknowledged that the management of waste in the United States is not
merely a practical or technical issue because for many it encompasses ethical and moral considerations. The argument runs, "our society is overly
consumptive and wasteful, and therefore generates far more garbage than
is necessary." Indeed, opposition to waste combustion is often motivated
as much by a desire to force pollution prevention and waste reduction as
it is by local health and safety concerns, though these issues are combined
with different emphases in different settings by national and local interests.

Waste management decisions raise clear issues of risk equity – the risks
generated by management of the wastes from product production and use

across a broad geographic area are often imposed upon a limited subset of those that benefit. Site selection, even if motivated by purely objective engineering cost considerations, can lead to the disproportionate selection of sites near poor or minority communities (U.S. CEQ, 1971; UCC Commission for Racial Justice, 1987). Certainly a broad spectrum of interested parties, from the regulated industry, to the public, to environmental groups, to all levels of government, vigorously debate waste management decisions. In the past, this debate has often singled out municipal waste combustors (MWCs) for special attention.

In 1986, the United States was home to 111 MWCs, which incinerated 5–10 percent of the municipal solid waste stream nationally. Historically, as landfills have filled up and open space has dwindled, MWCs have been proposed as an alternative, carrying along the attractive possibility of supplementary power generation. As many as 200 communities had plans to build MWCs with startup times in the mid-1990s. By 1995, only a small fraction of these facilities had been built. The rest were delayed or cancelled under heavy public opposition. Concern centered around health partly as a result of risk assessments; however, the quantitative estimates were less important than the acknowledgment that a nonzero risk existed. These assessments explored MWC in isolation and therefore never served as a basis for national policy making or as a means for weighing a range of alternatives in a common framework. Thus, it has been possible to block MWC sitings without ever considering the risks associated with alternative options such as landfilling (Cullen and Eschenroeder, 1997). MWC sitings in the United States have reached a virtual standstill. There is little evidence, however, that this has substantively contributed to pollution prevention or to any corresponding decrease in the rate of growth of municipal waste generation in the United States (World Watch Institute, 2000).

6.1.1. *Recent History of MWC risk assessment.* Since the early 1980s, state and local governments have required health risk assessments for MWCs. The discovery by researchers in the Netherlands, Canada, Japan, and Switzerland of dioxin in the ash released from MWCs raised public concern, since in toxicological experiments, dioxin acts as a potent carcinogen in animals, as well as being acutely toxic. Press coverage during the Agent Orange suit brought by Vietnam veterans and the evacuation of Times Beach, Missouri, led to extreme notoriety for dioxin. Scientists at the Dow Chemical Company (Bumb et al., 1980) imputed fire as a general cause for dioxin formation. Subsequent research refined the inventory of sources of dioxin and identified MWCs as high-profile emitters. Given the potent carcinogenicity associated with dioxins and furans, quantitative risk calculations for MWCs are dominated by this family of compounds. Decades of research have been devoted to developing a better understanding of the mechanistic behavior of dioxin; however, fundamental uncertainty still surrounds the vastly different responses observed in different species.

Approaches to assessing health risk from MWCs have evolved over time along with the field in general. Risk assessments have taken an increasingly comprehensive look at the contaminants released and the human receptors potentially exposed. For example, with better laboratory gas chromatography/mass spectrometry (GC/MS) techniques a more complete identification of the compounds present in stack gases has become possible. Also, routes of exposure to humans beyond inhalation, for example, ingestion of mother's milk and local foods, have been added. And rather than restricting consideration to hypothetical maximally exposed individuals, updated risk assessments identify multiple scenarios of exposure for potentially affected individuals and subpopulations using sophisticated models of contaminant fate and transport as well as information about the location of residences and workplaces. Despite the more extensive nature of these assessments over time, and the more inclusive set of health risks considered, estimates of individual risk have remained in the range of about 10^{-5} to 10^{-7}, with many results clustered around 10^{-6} or one in one million (Levin et al., 1991). There are at least two reasons why risk estimates have remained in the same quantitative range despite these changes. First, the estimates continue to be dominated by the risk attributable to dioxin exposure regardless of the inclusion of additional compounds. Second, a potential benefit of quantitative analysis is that facilities are now carefully designed to minimize the formation and release of dioxin in stack gases so that emission levels have been significantly reduced.

From the mid-1980s until the Clean Air Act Amendments of 1990, the main toxic emissions of MWCs went largely unaddressed by the federal government. In 1984, EPA was charged by Congress with producing a report of national scope covering the current state of resource recovery facilities burning solid waste and the presence of dioxin in their emissions, any health risks posed by these emissions, and appropriate operating practices for controlling these emissions. In 1987, a nine-volume report responding to these points was issued. The report focused on risk resulting from inhalation of dioxins and compared risk estimates between existing MWCs and those planned or proposed. This comparison showed significant decreases in potential health risks (as much as 90 percent overall) as a result of improvements in control technology for MWCs. This report provided a great deal of information about the state of the industry and the range and uncertainty of potential health risks. These results were later used to support regulation of MWCs under the Clean Air Act's new source performance standards (NSPSs) based on best demonstrated technology under Section 111, as opposed to a strict health basis under Section 112. In 1991, these NSPSs were introduced, but a suit brought by the National Resources Defense

Council (NRDC) and the Sierra Club led to their replacement with standards based on maximum achievable control technology (MACT). MACT standards rely on the performance of the top 12 percent of operating facilities to set a floor for pollution control that must be equaled or exceeded by new technology. Looking back, it was public protest and outcry driven by MWC risk assessments that led to the introduction of the improved control technology upon which these are based. As a top-ranked potential source category under Section 112 of the Clean Air Act, MWCs will be the subject of a national assessment to gauge residual health risk in the coming years.

6.1.2. The Role and Limits of Quantitative Assessment. During the last twenty years or more, community opposition has become close to 100 percent effective in blocking MWCs. Of the more than 200 MWCs proposed or planned in the mid-1980s only a handful have been built. Local governments require quantitative risk assessments on proposed MWCs; however comparative assessments of alternative disposal options, such as landfills, are largely lacking. Siting success appears to depend little on the quantitative estimates of risk reported during the application and permitting process, but rather *the admission of nonzero risks* for this particular option and the level of uncertainty and dread surrounding dioxin exposures. This result is consistent with the assertion that quantitative risk assessments are narrow and limited when viewed in the face of complex decisions involving community or individual values. This result is especially poignant given that the risks associated with landfill disposal are also nonzero and appear to be in the range of those associated with disposal by incineration (Cullen and Eschenroeder, 1997). Is this outcome a reflection of the emergence of a de facto precautionary approach in the United States? Are the potential risks from incineration more dreaded, less controllable, or less equitable than those from landfill disposal, or are we just averse to change in the type and nature of the risks that we face? To help assess these issues, it is interesting to compare the U.S. MWC experience to that of Europe, and in particular, France.

Municipal waste management in France is now in a similar situation to that faced by the United States twenty years ago in terms of impending capacity problems, but with a different baseline of management infrastructure and a different set of resource restrictions. While waste management in France has, in recent years, been dominated by incineration, many of the 300 MWCs currently operating in France are out of date. Construction of more than 100 replacement MWCs is planned for the next five to ten years (Boudet et al., 1999). Detailed risk analyses are being carried out for the replacement units that emit 10- to 1,000-fold lower concentrations of criteria pollutants in the stack gases than the original units. One of these units is the MWC in Grenoble (Boudet et al., 1999). Not surprisingly, expected levels of risk associated with a replacement facility are significantly lower than levels associated with the original facility, which is being used as a basis of comparison. Also interestingly, expected levels of dioxins and furans have not been established for the proposed facility in Grenoble owing to limits in the analytic capability or capacity of local laboratories. Still, the levels of human exposure to dioxins and furans from this source are expected to be low relative to background exposures from dietary sources. It is notable that exposure and risk comparisons to both existing MWCs and background levels are being conducted and presented. Finally, severe limitations in available open space color the feasibility of landfilling in France, so a full exploration of other options for disposal (or reduction of waste generation) might not be expected in this case either. Thus, even though it can be argued in the United States that precautionary approaches favor the restriction of MWC options, a different set of baseline conditions and available resources appears to ensure that municipal waste combustion, and the quantitative risk assessments that inform their siting and performance, remain central to waste management strategy selection in France.

6.2. Climate Change

One of the most contentious science policy issues at present is the management of human-induced climate change. The limitations of quantitative approaches are thrown into bold relief in this arena of analysis and decision making. In the past two decades, major investments in the natural sciences have resulted in significant advances in our ability to identify trends and disruptions in climate patterns. Research aimed at establishing the extent of anthropogenic influence on the stability of the Earth's climate relies on the application of quantitative tools to a natural system fraught with variability, and an often unclear level of model uncertainty (Casman, Morgan, and Dowlatabadi, 1999). Recently, the Intergovernmental Panel on Climate Change (IPCC), an international body that was spawned by the United Nations Environment Programme and the World Meteorological Organization in 1988 and charged with assessing the available scientific, technical, and socioeconomic information concerning greenhouse-gas-induced climate change, has reported the emergence of a scientific consensus that anthropogenic climate change is real and observable, despite the uncertainties (IPCC, 2001a). And yet as pointed out by Morgan and Dowlatabadi (1996), uncertainty surrounding value judgments about costs and benefits of alternatives to address climate change overshadows that about the science when it comes to decision making. Given our focus on assessing uncertain risk, we consider this example first from a relatively narrow viewpoint of hazard and risk assessment, and then return to a more holistic perspective.

6.2.1. Interactions of Atmosphere and Biosphere. All living things rely on the presence of a relatively consistent and limited range of climate conditions for survival, conditions mediated directly by the Earth's atmosphere. While some species and some individuals within species can withstand a wider range of temperatures, altitudes, and humidity than others, there is generally a preferred range sought out by any life form. Humans have survived in almost every location on the globe simply by heating or cooling their living space as necessary for comfort. But in general they avoid the locations with the most extreme conditions, as there are significant costs to maintaining livable conditions in these areas. Many species migrate seasonally to remain in their preferred climatic conditions. Other species cycle through active and dormant states with the seasons and have evolved to rely on these cycles to thrive over many years. In the short term, changes in Earth's atmosphere and climate can lead to the disruption of ecosystems, altering the patterns evolved by species for living, eating, and reproducing.

The similarity between the heat-trapping ability of the glass panes of a greenhouse and the absorption of infrared by gases in Earth's atmosphere was first proposed by Fourier, a French mathematician and physicist, in 1827. Before the end of the nineteenth-century Swedish chemist Svante Arrhenius (eventually a Nobel Prize winner in chemistry) predicted that the average temperature at the Earth's surface would rise by 4–6°C if atmospheric carbon dioxide levels doubled. The implications of this prediction did not receive much attention until about 1960 when CO_2 levels were observed to have risen to 315 parts per million (ppm), compared to a value of approximately 280 ppm in preindustrial times (IPCC, 1995, 2001a). CO_2 levels have now risen to approximately 365 ppm, a change attributed in part to the continued increase in fossil fuel burning to produce energy for human demands (IPCC, 1995; 2001a).

If one were to overlay a risk assessment framework on this system, what might it look like? It is first necessary to define the adverse outcomes of concern and to understand their causes. Do these include a rise in CO_2 levels, subsequent changes in temperature, the ultimate impacts of temperature changes, such as loss of species, habitat, productivity of farmland, and so on, or something beyond? Clearly the goal of such an exercise should not be to impose a single, rigid framework of analysis but rather to identify a full set of outcomes that people value, recognizing that the answers will depend upon the mix of stakeholders and decision makers included.

The IPCC synthesizes scientific information about climate change every five years and seeks to publish a consensus report. The last report was published in 2001, applying an international perspective to address the questions:

Have we detected climate change?
How much climate change can we expect under various assumptions about the emissions of greenhouse gases and about the emissions of CO_2 alone?

Climate change in this context refers to changes in the patterns of temperature and precipitation at the Earth's surface around the globe. National assessments are carried out by each country, as are regional assessments, such as in the northwest of the United States. These focus on potential risks to health, ecosystems, agricultural success, water storage, forestry, fisheries, recreation, and power generation resulting from climate change.

Over the past ten years, three main areas of climate change research have been pursued:

1. *Generating scenarios of greenhouse gas emissions in the past and future.* Human economic activities and associated greenhouse gas emissions have been monitored with accuracy only for the past century, and in many parts of the world estimates are still quite approximate. Predicting future emissions is fraught with great uncertainty, owing to unknown technological advances and possible structural changes in political and economic systems. Furthermore, these changes can interact with climate change in a synergistic or antagonistic manner. Changes in climate can cause

   - *increased* demand for energy use to maintain productivity, physical comfort, and lifestyle advances or
   - *decreased* demand through the adoption of more energy-efficient technology or lifestyles, and increased political support for treaties requiring emissions reduction in response to a growing, more apparent climate problem.

2. *Assessments of trends in the climate record.* At present scientists are on the edge of demonstrating the occurrence of climate change with statistical significance with the 100-year record of instrument data (leaving aside the paleo record). The record contains significant noise at low frequencies on the temporal scale. Successive years of the climate record are correlated, reducing the effective degrees of freedom afforded by climate datasets. Also, there is natural temporal variability such that apparent trends in the data could be the result of randomness. The question is, how do we separate the natural spectrum of climate variability from other trends and effects? Models for predicting climate change in response to greenhouse gas emissions play a critical role, both for assessing past trends and for predicting future outcomes. General atmospheric circulation models, now coupled with ocean and land-surface process models, can allow for historic reconstruction of recent climate, correcting for countervailing or compensatory effects, such as increases in atmospheric aerosol concentrations that have accompanied greenhouse gas emission and concentration trends. Research targeting climate variability, such as disruptions to the usual weather patterns, has also been pursued with great interest. A consideration of

potential changes in the frequency of extreme events introduces additional variability and uncertainty into the analysis. In addition, these changes have important implications for the need for precautionary response to avoid the associated disruption and damage they bring (Bier et al., 1999).

3. *Measurement/assessment of impacts*. Climate change is associated with a broad range of potential impacts to the Earth's geo- and biosphere. Assessments of effects such as sea-level rise, changes in plant productivity, species shifts in flora and fauna, and human health impacts are constrained by datasets often quite limited in their spatial and temporal coverage. Models for these processes are highly uncertain, both in their parameter values and their structure – often even the selection of system boundaries for models is unclear. This selection is important because complex, nonlinear interactions and feedback mechanisms can occur among many different components of the Earth's environmental and human system. For example, changing land use patterns in the tropics, such as by deforestation, can lead to changes in global and local climate that affect both the viability of agriculture and the potential for the regrowth of forests in these regions.

With an extensive and still growing research effort, we have gone from an inability to show anthropogenic effects ten years ago, to a near consensus that scientists are observing climate change now. Despite the difficulty of sorting out signal from noise in the variable state of atmosphere-ocean-land systems, the conclusion that humans have disrupted the usual patterns of climate variability has been embraced by the majority of the scientific community. This conclusion rests on evidence that CO_2 levels have risen and that global temperatures have risen in a corresponding manner, consistent with theory and models. But could this reflect natural variability rather than a real change? Proxy climate indicators suggest this is not the case, as the twentieth-century global mean temperature is at least as warm as any other century since A.D. 1400 or earlier (Goldfarb, 2000). Still, the assessment of the full range of effects from climate change, and the determination of the effect of different political and economic strategies on emissions growth and subsequent climate response, remain as uncertain as ever.

6.2.2. The Role and Limits of Quantitative Assessment. To date, despite a body of evidence and a cadre of scientists supporting the consensus that global climate change is real and anthropogenic, national governments seem reluctant to take steps to halt or reverse it.[5] Complicating the situation is the fact that governments are in a position to make only some of the relevant decisions. At every level from individuals to communities to states and provinces, countless decisions are made daily that have enormous cumulative impacts on climate (Morgan and Dowlatabadi, 1996). Furthermore, there is no international decision-making body empowered to specify or enforce mitigative actions in response to IPCC findings.

There are many factors to consider in decision making beyond the quantitative assessment of the probability that the climate is changing, including the projected cost, inconvenience, and disruption of taking actions to protect the stability of our planet's climate. Some would argue that inaction on the climate change issue is a result of a lack of scientific certainty. Indeed, those working on global climate have openly acknowledged uncertainty in their estimates of particular climate indicators, such as the magnitude of projected temperature changes. This recognition is apparent in recent IPCC reports, and has been necessary for obtaining consensus among the many scientists participating in the IPCC process. A related aspect of the problem may be that the lack of interlinked systems, which allow decision makers to access both detailed local-scale assessment information and existing global-scale information systems, is only beginning to be recognized and explored (Cash, 2000).

Another argument against action is based on the projected expense of changing behavior and practices that lead to emissions of CO_2 in the generation of energy. The assumption that a net social loss would occur with a major change in energy use rests on further assumptions about the balance between the inevitable winners and losers of such a shift. Those espousing a precautionary principle might choose to focus on such characteristic decision features as the potential for irreversible changes and the distribution of impacts between and across species when deciding whether to act to slow or halt climate change. In this example, it is interesting to consider how much and what type of scientific information (if any) would change opinions and preferences? All of these features challenge the adequacy of a strictly quantitative framework for illuminating the best decisions for the diverse populations and species sharing the planet.

6.3. Risk of Species Extinction

The final example concerns the application of a quantitative risk approach to ecosystem and species health explicitly. Under the U.S. Endangered Species Act (ESA) government agencies are responsible for evaluating the long-term viability of individual species in order to decide whether they warrant "listing" as critical or endangered. The goal of the ESA is to ensure that species are viable over the long term. The long-term survival of species is defined for this purpose by a 100-year period, consistent with extinction risk time frames for other species. The U.S. National Marine Fisheries Service (NMFS) is charged with these duties for marine species such as salmon. As described later, each salmon species encompasses multiple populations of fish, each with characteristic behaviors and all prized throughout history for symbolic and practical reasons. In 1999, Washington State's Puget Sound Chinook salmon were listed under the ESA. In addition to the role discussed previously, NMFS must participate, along with other groups, in the development of recovery plans for listed species to ensure that they will return to a viable status. Puget Sound Chinook became the focus of a NMFS Recovery Team study beginning in April 2000, tasked with identifying the measures necessary to bring the species back to health.

An interdisciplinary group of scientists convened by NMFS is currently tackling a pivotal step in this process, the establishment of the minimum number of individuals that constitutes a viable population. A salmon "species," the target of protection in this case, is considered to consist of an "evolutionarily significant unit" (ESU) made up of one or more populations of reproductively isolated salmon, which spawn in a specific location and season without significant exchange with other populations. The viable population varies with those characteristics described later. Yet this number serves an important role as it becomes an indicator for decision-making under the Endangered Species Act and also for assessing the impact of development or operations near crucial habitat for listed species.

Recently NMFS developed a detailed framework for identifying the biological requirements of viable populations of salmon. In this framework, the agency establishes four population parameters relating to viability (McElhany et al., 2000):

1. Abundance – possessing an adequate number of members;
2. Productivity – producing sufficient offspring per parent;
3. Diversity – representing adequate genetic, morphological, and life-history diversity; and
4. Spatial structure – distributing individuals at a density and in a manner that promotes species health.

A combination of quantitative and qualitative information is used to estimate the minimum viable population in light of these four parameters. As a result of severe limits in available data, in some cases the population requirements rely heavily on models describing the life cycle and robustness of salmon. Competing models lie at the heart of disputes about such habitat-altering developments as hydropower generating dams. Models have been developed by university researchers (e.g., the School of Fisheries at the University of Washington), government agency scientists (e.g., the Cumulative Risk Initiative (CRI) group at NMFS) and others. Some of the models rest on the assumption that simple proportions may adequately describe the rate of survival of fish faced with a barrier such as a dam, as they head to their spawning waters. Others use complex systems of equations to describe multiple life cycles and interactions in predicting survival under habitat challenges.

The absence of adequate quantitative information for determining the four critical population parameters leaves the agency in a position to discuss each in qualitative terms. Still, the number of fish constituting a viable population with respect to these parameters must be estimated for each population, and also in light of the characteristics of the overall ESU. This process is complicated in that the right number can never be known with certainty, and there is a small but finite probability of species extinction even with relatively large populations. In the face of this uncertainty, NMFS must estimate a minimum number of fish that will ensure viability while acknowledging that there is a small residual risk of extinction.

6.3.1. The Role and Limits of Quantitative Assessment. Implicit in the establishment of a viable population magnitude for salmonids is a decision about the magnitude of acceptable extinction risk. A working assumption is that a 5 percent risk of extinction within 100 years should correspond to the viable population estimate. This assumption lies at the margin between scientific assessment and risk management decision making. The development of a distribution reflecting uncertainty in the magnitude of the extinction risk versus population size would also present many challenges, but it would enable the decisions about the acceptable risk of extinction to be shifted from the shoulders of the scientists to those of policy makers. How should acceptable risk be established in this case? Is it more appropriately dictated by the values of society, or is it a purely scientific issue based on the history of species robustness in the face of challenge? This question will be the subject of public debate and discussion in the Northwest as the estimation of minimum viable populations continues beyond initial studies in April of 2000, led by the science-based Recovery Team for Puget Sound Chinook Salmon.

6.4. Summary of Examples

The limits of quantitative analysis are illustrated in the context of all three examples presented here. In these examples, uncertain risk to variable natural systems eludes definitive quantitative treatment. Profound knowledge gaps limit the role of quantitative analysis. Conflicts in values cloud the interpretation of quantitative results. Quantitative treatments across alternatives are not consistent. Decisions about risk acceptability fall both within and outside of the role of participating scientists.

Human health risk resulting from the operation of municipal waste combustors encompasses the smallest scale and possibly the best understood source of the three, though even here the health effects from the low-exposure concentrations associated with MWCs remain highly uncertain. At the opposite extreme, risk to the stability of earth's climate is arguably the least understood in terms of the complexity of the interactions between sources of hazard and receptor systems, and possible nonlinear or irreversible outcomes. Between these two lies the risk of species extinction with its dearth of relevant and applicable information.

In the case of municipal waste combustors, quantitative estimates of health risk, however uncertain, have in many locations (especially in the United States) become quite incidental to decisions about acceptability and siting. A risk estimate of any magnitude exceeds the zero risk assumed for the other options. This combined with the public reaction to dioxin sources of any description makes the decision about acceptability simple in the eyes of many, perhaps too much so. Meanwhile, concerns about distributional inequity in potential human impacts of MWCs are not compared with potential inequities associated with other forms of waste disposal. In the case of endangered salmonids, a science team has been put in the position of choosing an acceptable risk of extinction so that they can carry out their mandate to establish a minimum viable population size. The assessment struggles with profound uncertainty about the response of the populations to habitat interventions, but at its heart, the selection of an acceptable risk dictates the number of fish that will be defined as viable. Finally, in the case of climate change, a range of decision makers have been left to draw their own conclusions about the acceptability of the apparent risk to the planet of business as usual. These individuals and groups will make decisions that will be influenced by climatic, economic, political, and social outcomes, and perhaps even quantitative assessments of risk in a manner that is as uncertain as the risks themselves (Clark and Dickson, 1999). Much research is still needed to improve methods for quantitative risk and uncertainty analysis and to better understand its evolving contribution to individual, organizational, and societal decision making.

Notes

1. For example, in the September 1998 *Report of the Workshop on Selecting Input Distributions for Probabilistic Assessment* (U.S. EPA, 1998), Burmaster argues that: "It is wholly inconsistent for the Agency to proceed with policies that legitimize the use of probabilistic techniques for exposure factors while preventing the use of probabilistic techniques in dose-response assessment."
2. The late Senator Edmund Muskie is usually identified as the source of this request. During hearings on the design of the first U.S. Clean Water Act, he sought a scientist who would not end the answer to every question with the addendum, "but on the other hand."
3. With a linear, zero-threshold dose-response function, the population incidence is given by $I = \beta \mu_{\text{dose}} N$, where β is the slope of the dose-response function, μ_{dose} is the mean dose to the population, and N is the number of individuals in the target population. When other functional forms are used for the dose-response

function, the function must be integrated over the distribution of dose for the target population.

4. The U.S. Nuclear Regulatory Commission now identifies two approaches to its rules and regulation, one being "performance-based" and the other "risk-informed." See http://www.nrc.gov/NRC/COMMISSION/INITIATIVES/1999/index.html.

5. Significant progress toward implementing the Kyoto Protocol to the UN Framework Convention on Climate Change (UNFCCC) has been made since the initial drafting of this paper, following the November 10, 2001, meeting in Marrakech, Morocco, where over 170 nations completed nearly five years of negotiations to determine the rules for implementation of the greenhouse gas emission reductions determined by accords. The United States has as yet not participated in these agreements. (See Wiser, 2002.)

References

Adlassnig, K.-P. 1986. Fuzzy set theory in medical diagnosis. *IEEE Transactions on Systems, Man and Cybernetics*, SMC-16: 260–5.

Andrews, R. 1997. Risk-based decision making. In *Environmental Policy in the 1990s*, N. Vig and M. Kraft, eds., Congressional Quarterly, 208–230, Washington, DC.

Apostolakis, G. E. and S. E. Pickett. 1998. Deliberation: Integrating analytical results into environmental decision involving multiple stakeholders. *Risk Analysis*, 18 (5): 621–47.

Ashby, D., and Hutton, J. L. 1996. Bayesian epidemiology. In D. K. Berry, ed., *Bayesian Biostatistics*, pp. 109–38. Marcel Dekker, New York.

Bankes, S. C. 1993. Exploratory modeling for policy analysis. *Operations Research*, 41: 435–49.

Bankes, S. C. 1994. Computational experiments and exploratory modeling. *Chance*, 7 (1): 50–1, 57.

Bedford, T., and R. Cooke. 2001. *Probabilistic Risk Analysis: Foundations and Methods*, Cambridge University Press, Cambridge.

Berry, D. A. and Stangl, D. K., eds. 1996. *Bayesian Biostatistics*. Marcel Dekker, New York.

Bier, V. M., Haimes, Y. Y., Lambert, J. H., Ferson, S., and Small, M. J. 2003. Quantifying risk of extreme or rare events: Lessons from a selection of approaches. *This volume.*

Bier, V. M., Haimes, Y. Y., Lambert, J. H., Matalas, N. C., and Zimmerman, R. 1999. A survey of approaches for assessing and managing the risk of extremes. *Risk Analysis*, 19 (1): 83–94.

Bogen, K. T. 1990. *Uncertainty in Environmental Health Risk Assessment*, Garland Publishing, New York.

Bogen, K. T. 1995. Methods to approximate joint uncertainty and variability in risk. *Risk Analysis*, 15 (3): 411–19.

Bostrom, A., Fischhoff, B., and Morgan, M. G. 1992. Characterizing mental models of hazardous processes: A methodology with an application to radon. *J. Social Issues*, 48 (4): 85–100.

Boudet, C., Zmirou, D., Laffond, M., Balducci, F., and Benoit-Guyod, J-L. 1999. Health risk assessment of a modern municipal waste incinerator. *Risk Analysis*, 19: 1215–22.

Brasquet, C., Bourges, B., and Le Cloirec, P. 1999. Quantitative structure-property relationship (QSPR) for the adsorption of organic compounds onto activated carbon cloth: Comparison between multiple linear regression and neural network. *Environmental Science and Technology*, 33 (23): 4226–31.

Brattin, W. J., Barry, T. M., and Chiu, N. 1996. Monte Carlo modeling with uncertain probability density functions. *Human and Ecological Risk Assessment*, 2 (4): 820–40.

Brieman, L., Friedman, J. H., Olshen, R. A., and Stone, C. J. 1984. *Classification and regression trees*. Wadsworth&Brooks/Cole Advanced Books & Software, Pacific Grove, CA.

Browner, C. 1995. Policy for Risk Characterization at the US Environmental Protection Agency, Memorandum to assistant and regional administrators.

Bumb, R. R., Crummet, W. B., Cutie, S. S., Gledhill, J. R., Hummel, R. H., Kagel, R. O., Lamparski, L. L., Luoma, E. V., Miller, D. L., Nestrick, T. J., Shadoff, L. A., Stehl, R. H., and Woods, J. S. 1980. Trace chemistries of fire: A source of chlorinated dioxins. *Science* 210 (4468): 385–90.

Burmaster, D. E., and Thompson, K. M. 1997. Estimating exposure point concentrations for surface soils for use in deterministic and probabilistic risk assessments. *Human and Ecological Risk Assessment*, 3 (3): 363–84.

Burmaster, D. E., and Thompson, K. M. 1998. Fitting second-order parametric distributions to data using maximum likelihood estimation. *Human and Ecological Risk Assessment*, 4 (2): 319–39.

Burmaster, D. E., and Wilson, A. M. 1996. An introduction to second-order random variables in human health risk assessment. *Human and Ecological Risk Assessment*, 2 (4): 892–919.

Cao, H., and Chen, G. 1983. Some applications of fuzzy sets to meteorological forecasting. *Fuzzy Sets and Systems*, 9: 1–12.

Carrington, C. D. 1996. Logical probability and risk assessment. *Human and Ecological Risk Assessment*, 2: 62.

Carrington, C. D. 1997. An administrative view of model uncertainty in public health. Risk, 3. 273ff. Oct also http://www.tple.edu/risk/vol0/summer/ Carringt.htm.

Cash, D. W. 2000. Distributed assessment systems: An emerging paradigm of research, assessment, and decision-making for environmental change. *Global Environmental Change*, 10 (4): 241–4.

Casman, E. A., Morgan, M. G., and Dowlatabadi, H. 1999. Mixed levels of uncertainty in complex policy models. *Risk Analysis*, 19 (1): 33–42.

Chaloner, K. 1996. Elicitation of prior distributions. In D. K. Berry Stangl, ed., *Bayesian Biostatistics*, pp. 141–56. Marcel Dekker, New York.

Charniak, E. 1991. Bayesian networks without tears. *AI Magazine*, 12 (4): 50–63.

Clark, W. C., and Dickson, N. M. 1999. The global environmental assessment project: Learning from efforts to link science and policy in an interdependent world. *Acclimations*, 8: 6–7. http://www.nacc.usgcrp.gov/newsletter/.

Clarke, L. 1988. Politics and bias in risk assessment. *The Social Science Journal*, 15: 155–65.

Cooke, R. M. 1991. *Experts in Uncertainty: Opinion and Subjective Probability in Science.* Oxford Press, New York.

Costello, C. J., Adams, R. M., and Polasky, S. (1998). The value of El Niño forecasts in the management of salmon: A stochastic dynamic assessment. *American Journal of Agricultural Economics*, 80: 765–77.

Covello, V. T., and Mumpower, J. 1985. Risk analysis and risk management: A historical perspective. *Risk Analysis*, 5: 103–20.

Cox, D. C., and Baybutt, P. 1981. Methods for uncertainty analysis: A comparative study. *Risk Analysis*, 1: 251–8.

CRAM Presidential/Congressional Commission on Risk Assessment and Risk Management. 1997. *Risk Assessment and Risk Management in Regulatory Decision Making.* Final Report. Washington, DC.

Cullen, A. C. 1994. Measures of conservatism in probabilistic risk assessment. *Risk Analysis*, 14: 389–93.

Cullen, A. C. 1995. The sensitivity of Monte Carlo simulation results to model assumptions: The case of municipal solid waste combustor risk assessment. *Journal of Air Waste Management Association*, 45: 538–46.

Cullen, A. C., and Eschenroeder, A. Q. 1997. Coping with muncipal waste. In J. Graham and J. Hartwell eds., *The Greening of Industry*. Harvard University Press, Cambridge, MA.

Cullen, A. C., and Frey, H. C. 1999. *Probabilistic Techniques in Exposure Assessment: A Handbook for Dealing with Variability and Uncertainty in Models and Inputs*. Plenum Press, New York.

Dakins, M. E., Toll, J. E., Small, M. J., and Brand, K. P. 1996. Risk-based environmental remediation: Bayesian Monte Carlo analysis and the expected value of sample information. *Risk Analysis*, 16 (1): 67–79.

De Finetti, B. 1937. La Prevision: Ses Lois Logiques, Ses Sources Subjectives. *Annales de L'Institut Henri Poincaré*, 7 (1). pp. 1–68. English translation in H. Kyburg and H. Smokler, eds. 1964. *Studies in Subjective Probability*. Wiley, New York.

DeGroot, M. H. 1969. *Optimal Statistical Decisions*, McGraw-Hill Book Company, New York.

Dourson, M. L. and Stara, J. F. 1983. Regulatory history and experimental support of uncertainty (safety factors). *Regulatory Toxicology and Pharmacology*, 3: 224–38.

Draper, D. 1995. Assessment and propagation of model uncertainty. *Journal Royal Statistical Society, Series B*, 57 (1): 45–97.

Eddy, D. M., Hasselblad, V., and Shachter, R. 1990. An introduction to a Bayesian method for meta-analysis: The confidence profile method. *Medical Decision Making*, 10 (1): 15–23.

Eisenberg, J. N. S., and McKone, T. E. 1998. Decision tree method for the classification of chemical pollutants: Incorporation of across-chemical variability and within-chemical uncertainty. *Environmental Science and Technology*, 32 (21): 3396–404.

Evans, J. S. 1985. The value of improved exposure estimates: A decision analytic approach, *Proceedings of the 78th Annual Meeting of the Air Pollution Control Association* (June 16–21, 1985), Detroit, Michigan.

Evans, J. S., Graham, J. D., Gray, G. M., and Sielken, R. L. 1994a. A distributional approach to characterizing low-dose cancer risk. *Risk Analysis*, 14 (1): 25–34.

Evans, J. S., Gray, G. M., Sielken, R. L., Smith, A. E., Valdez-Flores, C., and Graham, J. D. 1994b. Use of probabilistic expert judgment in distributional analysis of carcinogenic potency, *Reg Tox and Pharm*, 20: 15–36.

Evans, J., Hawkins, N., and Graham, J. 1988. The value of monitoring for radon in the home: A decision analysis, *Journal of the Air Pollution Control Association*, 38 (11): 1380–5.

Evans, J. S. 1985. The value of improved exposure estimates: A decision analytic approach, *Proceedings of the 78th Annual Meeting of the Air Pollution Control Association*, (June 16–21, 1985), Detroit, Michigan.

Ferson, S. 1996. What Monte Carlo methods can not do. *Human and Ecological Risk Assessment*, 2: 990–1007.

Ferson, S., Ginzburg, L., and Akcakaya, R. Whereof one cannot speak: When input distributions are unknown. Unpublished (apparently) *Risk Analysis*.

Finkel, A. 1990. *Confronting Uncertainty in Risk Management: A Guide for Decision Makers*. Center for Risk Management, Resources for the Future, Washington, DC.

Finkel, A. M., and Evans, J. S. 1987. Evaluating the benefits of uncertainty reduction in environmental health risk management. *J Air Pollution Control Association*, 37: 1164–71.

Fischhoff, B., Riley, D., Kovacs, D. C., and Small, M. 1998. What information belongs in a warning? *Psychology and Marketing*, 15 (7): 663–86.

Fischhoff, B., Slovic, P., and Lichtenstein, S. 1982. Lay foibles and expert fables in judgments about risk. *American Statistician*, 36: 240–55.

Freudenburg, W. 1988. Perceived risk, real risk: Social science and the art of probabilistic risk assessment. *Science*, 241: 44–9.

Frey, H. C. 1992. *Quantitative Analysis of Uncertainty and Variability in Environmental Policy Making*. American Association for the Advancement of Science, Washington, DC.

Frey, H. C., and Burmaster, D. F. 1999. Methods for characterizing variability and uncertainty: Comparison of bootstrap simulation and likelihood-based approaches. *Risk Analysis*, 19 (1): 109–30.

Frey, H. C., and Rhodes, D. S. 1996. Characterization and simulation of uncertain frequency distributions: Effects of distribution choice, variability, uncertainty, and parameter dependence. *Human and Ecological Risk Assessment*, 1 (2): 423–68.

Gee, D. 1997. Criteria for Managing Uncertainty and Regulation in Public Policy. Draft manuscript.

Gelman, A., Carlin, J. B., Stern, H. S., and Rubin, D. B. 1995. *Bayesian Data Analysis*. Chapman & Hall, London.

Gentle, J. 1998. *Random Number Generation and Monte Carlo Methods*. Springer-Verlag, Heidelberg.

Giering, E. W., and Kandel, A. 1983. The application of fuzzy set theory to the modeling of competition in ecological systems. *Fuzzy Sets and Systems*, 9: 103–27.

Goldfarb, T. D. 2000. *Environmental Studies*. Dushkin McGraw-Hill, Guilford, CT.

Good, I. J. 1959. Kinds of probability. *Science*, 129: 443–7.

Gurian, P. L., Small, M. J., Lockwood III, J. R., and Schervish, M. J. 2001a. Benefit-cost estimation for alternative drinking water maximum contaminant levels. *Water Resources Research*, 37. (9): 2213–26.

Gurian, P. L., Small, M. J., Lockwood III, J. R., and Schervish, M. J. 2001b. Addressing uncertainty and conflicting cost estimates in revising the arsenic MCL. *Environmental Science & Technology*, 35 (22): 4414–20.

Haas, C. N. 1999. On modeling correlated random variables in risk assessment. *Risk Analysis*, 19 (6): 1205–14.

Habicht, H. 1992. *Guidance on Risk Characterization for Risk Managers and Risk Assessors*. USEPA memorandum to Assistant and Regional Administrators.

Haimes, Y. Y., and Lambert, J. 1994. When and how can you specify a probability distribution when you don't know much, *Risk Analysis*, 14 (5): 661–706.

Hammitt, J. K., and Shlyakhter, A. I. 1999. The expected value of information and the probability of surprise. *Risk Analysis*, 19 (1): 135–52.

Henrion, M. 1982. The value of knowing how little you know: The advantages of probabilistic treatment in policy analysis, Ph.D. dissertation, Carnegie-Mellon University, Pittsburgh.

Henrion, M., and Fischhoff, B. 1986. Assessing uncertainty in physical constants. *American Journal of Physics*, 54 (9): 791–8.

Hertwich, E. G., McKone, T. E., and Pease, W. S. 1999. Parameter uncertainty and variability in evaluative fate and exposure models. *Risk Analysis*, 19: 1193–1204.

Hilton, R. W. 1981. The determinants of information value: Synthesizing some general results. *Management Science*, 27 (1): 57–64.

Hoffman, F. O., and Hammonds, J. S. 1994. Propagation of uncertainty in risk assessments: The need to distinguish between uncertainty due to lack of knowledge and uncertainty due to variability. *Risk Analysis*, 14 (5): 707–12.

Hunt, S., Frewer, L. J., and Shepard, R. 1999. Public trust in sources of information about radiation risks in the UK. *Journal of Risk Research*, 2: 167–80.

Iman, R. L., and Conover, W. J. 1982. A distribution-free approach to inducing rank correlation among input variables. *Communications in Statistics*, B11 (3): 311–34.

Iman, R. L., and Hora, S. C. 1989. Bayesian methods for modeling recovery times with an application to the loss of off-site power at nuclear power plants. *Risk Analysis*, 9 (1): 25–36.

IPCC (Intergovernmental Panel on Climate Change). 1995. *Impacts, Adaptations and Mitigation of Climate Change: Scientific-Technical Analyses Contribution of Working Group II to the Second Assessment of the Intergovernmental Panel on Climate Change*, R. T. Watson, M. C. Zinyowera, and R. H. Moss, eds. Cambridge University Press, Cambridge.

IPCC (Intergovernmental Panel on Climate Change). 2001a. *Climate Change 2001: The Scientific Basis*. See http://www.ipcc.ch/pub/tar/.

IPCC (Intergovernmental Panel on Climate Change). 2001b. *Climate Change 2001: Impacts, Adaptation and Vulnerability*. See http://www.ipcc.ch/pub/tar/.

IPCC (Intergovernmental Panel on Climate Change). 2001c. *Climate Change 2001: Mitigation*. See http://www.ipcc.ch/pub/tar/.

Isukapalli, S. S., and Georgopoulos, P. G. 2001. *Computational Methods for Sensitivity and Uncertainty Analysis for Environmental and Biological Models*. Project Report to U.S. EPA Office of Research and Development (ORD) National Exposure Research Laboratory. EPA/600/R-01-068. Research Triangle Park, NC.

Jasanoff, S. 1987. Cultural aspects of risk assessment in Britain and the United States. In B. B. Johnson and V. T. Covello, eds., *The Social and Cultural Construction of Risk*, pp. 359–97. Reidel, Dordrecht, Netherlands.

Jasanoff, S. 1990. *The Fifth Branch, Science Advisers as Policymakers*. Harvard University Press, Cambridge, MA.

Jasanoff, S. 1993. Bridging the two cultures of risk analysis. *Risk Analysis*, 13: 123–9.

Johnson, B. B., and Slovic, P. 1995. Presenting uncertainty in health risk assessment: Initial studies of its effects on risk perception and trust. *Risk Analysis*, 15: 485–94.

Johnson, R., Pankow, J., Bender, D., Price, C., and Zogorski, J. 2000. MTBE: To what extent will past releases contaminate community water supply wells? *ES&T*, 34 (9): 210A–217A.

Julien, B., Fenves, S. J., and Small, M. J. 1992. Knowledge acquisition methods for environmental evaluation. *AI Applications*, 6: 1–20.

Kadane, J. B., and Wolfson, L. J. 1998. Experiences in elicitation. *Journal of the Royal Statistical Society, Series D*, 47: 3–19 (with discussion).

Kahneman, D., Slovic, P., and Tversky, A. 1982. *Judgment under Uncertainty: Heuristics and Biases*. Cambridge University Press, New York.

Kalagnanam, J., and Diwekar, U. 1997. An efficient sampling technique for off-line quality control, *Technometrics*, 38: 308.

Kammen, D. M., and Hassenzahl, D. M. 1999. *Should We Risk It? Exploring Environmental, Health, and Technological Problem Solving*, Princeton University Press, Princeton, NJ.

Kasperson, R. E., and Kasperson, J. X. 1996. The social amplification and attenuation of risk. *Annals of the American Academy of Political and Social Sciences*, 545: 95–105.

Kasperson, R., Renn, O., Slovic, P., Brown, H., Emel, J., Goble, R., Kasperson, J., and Ratick, S. 1988. The social amplification of risk: A conceptual framework. *Risk Analysis*, 8: 177–87.

Keeney, R. L. 1982. Decision analysis: An overview. *Operations Research*, 39: 803–38.

Klir, G. J., and Folger, T. A. 1988. *Fuzzy Sets, Uncertainty, and Information*, Prentice Hall, Englewood Cliffs, NJ.

Larkin, L. I. 1985. A fuzzy logic controller for aircraft flight control. In M. Sugeno, ed., *Industrial Applications of Fuzzy Control*. North Holland, New York.

Lave, L. 1978. "Eight Frameworks of Regulation: The Strategy of Social Regulation," Brookings Institute, pp. 8–28.

Lave, L. B., Resendiz-Carrillo, D., and McMichael, F. C. 1990. Safety goals for high-hazard dams: Are dams too safe? *Water Resources Research*, 26 (7): 1383–91.

Lee, P. M. 1989. *Bayesian Statistics: An Introduction*, Oxford University Press, Oxford.

Lempert, R., Schlesinger, M., and Bankes, S. 1996. When we don't know the costs or the benefits: Adaptive strategies for abating climate change. *Climactic Change*, 33: 235–74.

Levin, A., Fratt, D. B., Leonard, A., et al. 1991. Comparative analysis of health risk assessments for muncipal waste combustors. *Journal of the Air and Waste Management Association*, 41 (1): 20–31.

Lewis, H. W., Budnitz, R. J., Kouts, H. J. C., Lowenstein, W. B., Rowe, W. D., Von Hippel, F., and Zachariasen, F. 1979. *Risk Assessment Review Group Report to the U.S. Nuclear Regulatory Commission*. NUREG/CR-0400.

Lichtenstein, S., and Fischhoff, B. 1977. Do those who know more also know more about how much they know? *Organizational Behavior and Human Performance*, 20: 159–83.

Lockwood III, J. R., Schervish, M. J., Gurian, P., and Small, M. H. 2001. Characterization of arsenic occurrence in US drinking water treatment facility source waters. *Journal of American Statistical Association*, 96: 456, pp. 1184–1193.

MacKenzie, D. 1990. *Inventing Accuracy: A Historical Sociology of Nuclear Missile Guidance*. MIT Press, Cambridge, MA.

March, J. G., and Simon, H. A. 1958. *Organizations*. John Wiley and Sons, New York.

McElhany, P., Ruckelshaus, M., Ford, M. J., Wainwright, T., and Bjorkstedt, E. 2000. Viable Salmonid Populations and the Recovery of Evolutionarily Significant Units. National Marine Fisheries Service.

McKay, M. D., Conover, W. J., and Beckman, R. J. 1979. A comparison of three methods for selecting values of input variables in the analysis of output from a computer code. *Technometrics*, 21 (2): 239–45.

McKone, T. E., and Bogen, K. T. 1992. Uncertainties in health-risk assessment: An integrated case study based on tetrachloroethylene in California groundwater. *Regulatory Toxicology and Pharmacology*, 15: 86–103.

Merkhofer, M. W. 1987. Quantifying judgmental uncertainty: Methodology, experiences, and insights. *IEEE Transactions on Systems, Man, and Cybernetics*. 17 (5): 741–52.

Merz, J., Small, M. J., and Fischbeck, P. 1992. Measuring decision sensitivity: A combined Monte Carlo-logistic regression approach. *Medical Decision Making*, 12: 189–96.

Morgan, M. G., and Dowlatabadi, H. 1996. Learning from integrated assessment of climate change. *Climatic Change*, 34 (3–4): 337–68.

Morgan, M. G., and Henrion, M. 1990. *Uncertainty: A Guide for Dealing with Uncertainty in Quantitative Risk and Policy Analysis*. Cambridge University Press, Cambridge.

Morgan, M. G., Fischhoff, B., Bostrom, A., Lave, L., and Atman, C. 1992. Communicating risk to the public. *Environmental Science and Technology*, 26 (11): 2048–56.

Morgan, M. G., Henrion, M., and Morris, S. C. 1980, *Expert Judgment for Policy Analysis*. BNL 51358. Brookhaven National Laboratory. Brookhaven, NY.

Morokoff, W. J., and Caflisch, R. E. 1994. Quasi-random sequences and their discrepancies. *SIAM Journal on Scientific Computing*, 15 (6): 1251–79.

Nakatsuyama, M., Nagahashi, H., and Nishizuka, N. 1984. Fuzzy logic phase controller for traffic functions in the one-way arterial road. *Proceeding IFAC 9th Triennial World Congress*, pp. 2865–70, Pergamon Press, Oxford.

NAS. 1987. Report on the Delaney Paradox. National Academics Press. Washington, DC.

NCRP. 1996. *A Guide for Uncertainty Analysis in Dose and Risk Assessments Related to Environmental Contamination*. National Council on Radiation Protection and Measurements. NCRP Commentary No. 14. Bethesda, MD.

Norsys. 1998. *Netica*^{TM} *Application for Belief Networks and Influence Diagrams: User's Guide, Versions 1.03 for Macintosh and 1.05 for Windows*. Norsys Software Corporation, Vancouver, BC, Canada.

NRC. 1975. *Reactor Safety Study: An Assessment of Accident Risks in U.S. Commercial Nuclear Power Plants.* WASH-1400/NUREG 751014. U.S. Nuclear Regulatory Commission, Washington, DC.

NRC. 1983. *Risk Assessment in the Federal Government: Managing the Process* (also known as *"The Red Book"*). National Academy Press, Washington, DC.

NRC. 1994. *Science and Judgement in Risk Assessment.* National Academy Press, Washington, DC.

NRC. 1996. *Understanding Risk: Informing Decisions in a Democratic Society.* National Academy Press, Washington, DC.

NRC. 1999. *Our Common Journey: A Transition Toward Sustainability*, Board on Sustainable Development, National Research Council, National Academy Press, Washington, DC.

NRC. 2000. *Incorporating Science, Economics, and Sociology in Developing Sanitary and Phytosanitary Standards in International Trade, Proceedings of a Conference.* National Research Council, National Academy Press, Washington, DC.

Pate, E. 1983. Acceptable decision processes and acceptable risks in public sector regulations. *IEEE Transactions on Systems, Man, and Cybernetics*, SMC-13: 113–24.

Pate-Cornell, E. M. 1996. Uncertainties in risk analysis: Six levels of treatment. *Reliability Engineering and System Safety*, 54: 95–111.

Patt, A. 1999. Extreme outcomes: The strategic treatment of low probability events in scientific assessments. *Risk, Decision, and Policy*, 4 (1): 1–15.

Pearl, J. 1988. *Probabilistic Reasoning in Intelligent Systems.* Morgan Kaufman, San Mateo, CA.

Peters, R. G., Covello, V. T., and McCallum, D. B. 1997. The determinants of trust and credibility in environmental risk communication: An empirical study. *Risk Analysis*, 17: 43–54.

Petts, J. 1998. Trust and waste management information expectation versus observation. *Journal of Risk Research*, 1: 307–20.

Press, S. J. 1989. *Bayesian Statistics: Principles, Models, and Applications.* Wiley, New York.

Raffensperger, C., and Tickner, J. A., eds. 1999. *Protecting Public Health and the Environment – Implementing the Precautionary Principle.* Island Press, Washington, DC.

Raiffa, H. 1968. *Decision Analysis: Introductory Lectures on Choices Under Uncertainty.* Addison-Wesley Publishing, Reading, MA.

Raiffa, H., and Schlaifer, R.O. 1961. *Applied Statistical Decision Theory*, Harvard University Press, Cambridge, MA.

Ramsey, F. P. 1931. *The Foundation of Mathematics and Other Logical Essays.* Kegan Paul, London.

Renn, O., Burns, W. J., Kasperson, J. X., Kasperson, R. E., and Slovic, P. 1992. The social amplification of risk: Theroetical foundations and empirical applications. *Journal of Social Issues*, 48: 127–60.

Shlyakhter, A. I. 1994. "Improved Framework for Uncertainty Analysis: Accounting for Unsuspected Errors," *Risk Analysis*, 14: 441–7.

Simon, T. W. 1999. Two dimensional Monte Carlo simulation and beyond: A comparison of several probabilistic risk assessment methods applied to a Superfund site. *Human and Ecological Risk Assessment*, 5 (4): 823–43.

Slovic, P. 1993. Perceived risk, trust, and democracy. *Risk Analysis*, 13: 675–82.

Slovic, P., Fischoff, B., and Lichtenstein, S. 1979. Rating the risks. *Environment*, 21: 14–20, 36–9.

Small, M. J., and Fischbeck, P. F. 1999. False precision in Bayesian updating with incomplete models. *Human and Ecological Risk Assessment*, 5 (2): 291–304.

Smith, A. E., Ryan, P. B., and Evans, J. S. 1992. The effect of neglecting correlations when propagating uncertainty and estimating the population distribution of risk. *Risk Analysis*, 12: 467–74.

Solow, A. R., Adams, R. F., Bryant, K. J., Legler, D. M., O'Brien, J. J., McCarl, B. A., Nayda, W., and Weiher, R. 1998. The value of improved ENSO prediction to U.S. agriculture. *Climatic Change*, 39: 47–60.

Song, X.-H., and Hopke, P. K. 1996. Solving the chemical mass balance problem using an artificial neural network. *Environmental Science & Technology*, 30 (2): 531–5.

Spensley, J. W. 1995. National Environmental Policy Act. In T. F. P. Sullivan, ed., *Environmental Law Handbook*, 13th ed., Chapter 10. Government Institutes, Rockville, MD.

Spetzler, C. S., and von Holstein, S. 1975. Probability encoding in decision analysis. *Management Science*, 22 (3): 340–58.

Squillace, P. J., Zogorski, J. S., Wilber, W. G., and Price, C. V. 1996. Preliminary assessment of the occurrence and possible sources of MTBE in groundwater in the United States, 1993–1994. *Environmental Science and Technology*, 30 (5): 1721–30.

Stiber, N. A., Pantazidou, M., and Small, M. J. 1999. Expert system methodology for evaluating reductive dechlorination at TCE sites. *ES&T*, 33 (17): 3012–20.

Taylor, A. C. 1993. Using objective and subjective information to generate distributions for probabilistic exposure assessment. *Journal of Exposure Analysis and Environmental Epidemiology*, 3: 285–98.

Taylor, A., Evans, J., and McKone, T. 1993. The value of animal test information in environmental control decisions. *Risk Analysis*, 13: 403–12.

Tezuka, S. 1995. *Uniform Random Numbers: Theory and Practice*. Kluwer Academic Publishers, Dordrecht.

Thompson, K. M. 1999. Developing univariate distributions from data for risk analysis. *Human and Ecological Risk Assessment*, 5 (4): 755–83.

Thompson, K. M., and Evans, J. S. 1997. The value of improved exposure information for perchlorethylene (Perc): A case study for dry cleaners. *Risk Analysis*, 17 (2): 253–71.

Thompson, K. M., and Graham, J. D. 1996. Going beyond the single number: Using probabilistic risk assessment to improve risk management. *Human and Ecological Risk Assessment*, 2 (4): 1008–34.

Thompson, P. B., and Dean, W. 1996. Competing conceptions of risk. *Risk: Health, Safety & Environment*, 7: 361–84.

Tickner, J. A. 1999. A map toward precautionary decision making. In C. Raffensperger and J. A. Tickner, eds., *Protecting Public Health and the Environment – Implementing the Precautionary Principle*. Island Press, Washington, DC.

UCC Commission for Racial Justice. 1987. *Toxic Wastes and Race in the United States*. United Church of Christ and Public Data Access, New York.

UK-ILGRA. 1996. *Use of Risk Assessment within Government Departments.* UK Health and Safety Executive, HSE Books, Sudbury, Suffolk, UK. See also http://www.hse.gov.uk/dst/ilgra/minrpt1.htm.

UK-ILGRA. 1999. *Risk Assessment and Risk Management – Improving Policy and Practice within Government Departments.* UK Health and Safety Executive, HSE Books, Sudbury, Suffolk, UK.

U.S. CEQ. 1971. *Environmental Quality.* GPO, Washington, DC.

U.S. EPA Science Advisory Board. 1995. *Beyond the Horizon: Using Foresight to Protect the Environmental Future.* USEPA Science Advisory Board, Washington, DC.

U.S. EPA National Center for Environmental Assessment. 1997a. *Exposure Factors Handbook*, NTIS: PB98-124217, See also http://www.epa.gov/ncea/exposfac.htm, Washington, DC.

U.S. EPA Science Policy Council. 1997b. Proposed Policy for Use of Monte Carlo Analysis in Agency Risk Assessment. Memorandum of William P. Wood, Executive Director, Risk Assessment Forum, to Dorothy E. Patton, Executive Director of Science Policy Council (8104), January 29, 1997.

U.S. EPA Risk Assessment Forum. 1998. Report of the Workshop on Selecting Input Distributions for Probabilistic Assessments. Prepared by Eastern Research Group, Inc., for EPA under EPA Contract No. 68-D5-0028, Washington, DC.

U.S. EPA Risk Assessment Forum. 1999. Report of the Workshop on Selecting Input Distributions for Probabilistic Assessments USEPA 630/R-98/004. January 1, 1999. U.S. EPA Risk Assessment Forum, Washington, DC.

U.S. EPA Office of Research and Development. 2000. *Exploratory Research to Anticipate Future Environmental Issues.* FY 2000 STAR Program Request for Applications. U.S. EPA ORD National Center for Environmental Research, Washington, DC. http://es.epa.gov/ncerqa/rfa/explfuturefnl.html.

Winkler, R. L. 1967. The assessment of prior distributions in Bayesian analysis. *Journal of the American Statistical Association*, 62: 776–800.

Wiser, G. 2002. Analysis and perspective: Kyoto Protocol packs powerful implementation punch. *International Reporter Current Report*, 25 (2): 86. See http://www.ciel.org/Publications/INER_Compliance.pdf.

Wolfson, L. J. 1995. *Elicitation of Priors and Utilities for Bayesian Analysis.* PhD Thesis, Department of Statistics, Carnegie Mellon University, Pittsburgh.

Wolfson, L. J., Kadane, J. B., and Small, M. J. 1996. Bayesian environmental policy decisions: Two case studies. *Ecological Applications*, 6 (4): 1056–66.

Wolpert, R. L. 1989. Eliciting and combining subjective judgments about uncertainty. *International Journal of Technology Assessment in Health Care*, 5 (4): 537–57.

Wood, E. F., and Rodriguez-Iturbe, I. 1975. Bayesian inference and decision making for extreme hydrologic events. *Water Resources Research*, 11 (4): 533–42.

Word, C. J., Harding, A. K., Bilyard, G. R., and Weber, J. R. 1999. Basic science and risk communication: A dialogue-based study. *Risk: Health Safety and Environment*, 10 (2), 231–42.

World Watch Institute. 2000. *State of the World.* Norton, New York.

Wynne, B. 1980. Technology, risk and participation: The social treatment of uncertainty. In J. Conrad, ed., *Society, Technology and Risk*, pp. 83–107. Academic Press, New York.

Wynne, B. 1987. *Risk Management and Hazardous Wastes: Implementation and the Dialectics of Credibility*. Springer, Berlin.

For Further Reading

The following texts address a number of the methods and issues raised in this paper in more detail and are recommended for further reading. The books by Bedford and Cooke, Covello and Merkhofer, Cullen and Frey, Kammen and Hassenzahl, Morgan and Henrion, and Warren-Hicks and Moore are especially strong on methodologies for risk and uncertainty analysis. The publications by Crawford-Brown, Jaeger et al., Klapp, Porter, Slovic, and Sterling emphasize the social, cultural, and regulatory elements of risk and uncertainty.

Bedford, T., and Cooke, R. 2001. *Probabilistic Risk Analysis: Foundations and Methods*, Cambridge University Press, Cambridge.

Covello, V. T., and Merkhofer, M. W. 1993. *Risk Assessment Methods: Approaches for Assessing Health and Environmental Risks*. Plenum Press, New York.

Crawford-Brown, D. J. 1999. *Risk-Based Environmental Decisions: Methods and Culture*. Kluwer Academic Publishers, Boston.

Cullen, A. C., and Frey, H. C. 1999. *Probabilistic Techniques in Exposure Assessment: A Handbook for Dealing With Variability and Uncertainty in Models and Inputs*. Plenum Press, New York.

Jaeger, C. C., Renn, O., Rosa, E. A., and Webler, T. 2001. *Rational Action, Risk, and Uncertainty*. Earthscan, London.

Kadane, J. B., and Wolfson, L. J. 1998. Experiences in elicitation. *Journal of the Royal Statistical Society D*, 47: 3–19 (with discussion).

Kammen, D. M., and Hassenzahl, D. M. 1999. *Should We Risk It? Exploring Environmental, Health, and Technological Problem Solving*, Princeton University Press, Princeton, NJ.

Klapp, M. G. 1992. *Bargaining With Uncertainty: Decision-Making in Public Health, Technologial Safety, and Environmental Quality*. Auburn House, Dover, MA.

Morgan, M. G., and Henrion, M. 1990. *Uncertainty: A Guide for Dealing with Uncertainty in Quantitative Risk and Policy Analysis*, Cambridge University Press, Cambridge.

Porter, T. M. 1995. *Trust in Numbers: The Pursuit of Objectivity in Science and Public Life*, Princeton University Press, Princeton, NJ.

Slovic, P. 2000. *The Perception of Risk*. Earthscan, London.

Stirling, A. 1999. *On Science and Precaution in the Management of Technological Risk: Volume 1. A Synthesis Report of Case Studies*. Report EUR 19056/EN. European Commission for Prospective Technological Studies, Seville, Spain.

Stirling, A. 2002. *On Science and Precaution in the Management of Technological Risk: Volume II. Case Studies*. Report EUR 19056/EN/2. European Commission for Prospective Technological Studies, Seville, Spain.

Warren-Hicks, W. H., and Moore, D. R. J., eds. 1998. *Uncertainty Analysis in Ecological Risk Assessment*. Society of Environmental Toxicology and Chemistry (SETAC), Pensacola, FL.

7

Valuing Risk Management Choices

Robin S. Gregory

1. WHY VALUE RISKS?

Risks involve the possibility of damage or loss or injury to individuals or to groups. Whether a risk creates a high level of concern is established through judgments that are made about the definition of the problem, the nature and scope of its consequences, and the likelihood of different exposures. A risk that matters greatly to one person or one group may not be a concern to another because of differences in the context within which it is viewed. This context includes such considerations as the relative importance of other risks, the perception of accompanying benefits, the timing of the expected impacts, uncertainties associated with their receipt, the anticipated response of others, the understanding of realistic alternatives, and so forth. Risk is thus a multidimensional concept, defined differently by different people and in different cultures or at different times to help make sense of, and to create strategies for dealing with, a world that includes perceived dangers and hazards.

All decisions involve some weighing of risks, just as they involve some balancing of costs and benefits. Other things being equal, it is preferable to reduce risks in a given management context. Understanding the values and concerns that arise in the context of specific management options is important, however, because all other things rarely are equal; as a result, decision makers require information about the preferences and priorities of potentially affected individuals or groups toward the relevant set of risks, benefits, and costs. How we value risks thus provides a mirror of both the choices we make and the world we believe that we live in.

Many of the risks that we face are selected on the basis of the activities we choose to engage in and the products we choose to consume. For example, people may choose to live in a large urban center or ski or smoke cigarettes, knowing at least in a vague way that there are risks as well as benefits associated with these choices and that how these different components of

213

risk are valued provides a partial map of their preferences as well as a key to their future behavior. We also are presented with risks from the external world that we need to make sense of as part of our daily lives. For example, the health risks from ozone depletion or the environmental risks from global warming or species extinction affect many people but generally little change can be affected on the basis of an individual's actions. Similarly, many people living in less-wealthy countries, or in poverty in more industrialized nations, may have little choice about living in a contaminated area or working at a dangerous job. Although values may not directly link to exposures from identified risks, they may have a great effect on how we feel about a source of risk and the priority we place on its management. Over time, these affective and cognitive responses may result in actions that will change our exposure to the risks of immediate concern and, in turn, increase or decrease our exposure to other (identified and unidentified) risks.

In recent years, despite a great effort expended by governments and industry to make the world safer, many individuals have become more, rather than less, concerned about risk (Kunreuther and Slovic, 1996). Simultaneously, public views and perceptions about what constitutes important sources of risk have become more influential, playing an important and visible role in setting the agendas for many risk management agencies. Large differences in how the public and experts typically rate risks have led to a keen interest in the factors, or attributes, that lie behind people's conceptions of risk. This interest, in turn, has both led to, and been fed by, a growing sophistication in the methods used to identify and assess peoples' values in the context of risk management oppportunities.

2. VALUES AS AN INPUT TO RISK MANAGEMENT DECISIONS

Better methods for valuing the consequences of risks can lead to improvements in the choices that are made by individuals, groups, or societies concerning the selection and implementation of risk management actions. Because some risks are more worrisome than others, paying attention to values should help to focus attention and resources on policies or regulations or behaviors that address the more serious risks and on expenditures that are cost-effective in reducing risks, promoting safety, and mitigating losses. The assessment of a risk as high or serious does not necessarily mean that steps should be taken to reduce it (nor should a risk assessed as low be ignored) because risk policy decisions reflect a balancing of diverse benefits, costs, and risks in the context of management options. Valuations of risks are also often contentious because individuals differ widely in their understanding of risks, in their valuations of different risk attributes, and in their willingness to make sacrifices over time to deal effectively with sources of exposure to environmental or health risks.

2.1. Risk Objectives

The emerging science and art of risk valuation, which is the topic of this chapter, has been stimulated by a variety of diverse sources and interests. One major contributor is government, which at all levels has become concerned with managing risks and with preventing events that might lead to large human health or environmental or financial losses. Another obvious contributor is the research community, which has sought to understand how people make judgments about the importance of various risks and how these judgments might influence behavior. Industry also has been a major player, as it has sought to find approaches that will allow those activities it views as worthwhile to proceed in a safe, timely, and efficient fashion. And a large variety of public groups also play important roles in shaping approaches to risk valuation, including diverse international environmental and consumer groups as well as many community-based organizations.

In each case – whether government, researchers, industry, or public – the dominant concern has been a balancing of the anticipated risks, benefits, and costs associated with the impacts of a proposed activity or a particular decision. Thus, I may want to know more about the change in the probability that I will become ill if I engage in a particular activity or about the symptoms and duration of the illness. The reason why I seek this information typically is because I am faced with (or may soon be faced with) a choice, such as whether or not to purchase a product or to take on a new job that could lead to an increase in my exposure to that risk. What matters is not some abstract rating I place on the risks of the product or activity itself but, rather, the value I give to the risks of engaging in this action in a particular way, such as with or without safety equipment, and in a particular context, such as with friends or in the month of February or after eating supper. The value I assign to the risk in the context of a decision-making opportunity is therefore my translation and integration of a series of cognitive and emotional responses such that it can, or might, influence a choice that I will make, or that others will be making, across multiple risk and other objectives.

How a person narrows his or her alternatives in seeking to reduce the risks from an identified source depends on the identification of objectives and on how the risks of the activity (or, similarly, the risks of a product or technology) are defined. Objectives tell us what a person cares about in the context of the risk problem at hand (Keeney, 1992). If an individual cares about issues of personal safety, for example, then she may be very interested in information about the level of exposure she could face were she to participate in the action. If she cares about how she is treated relative to others, then fairness or equity considerations may be of interest. If she cares only about the short term, then longer-term considerations and most indirect consequences will not affect how she responds; if she cares only about the region, state, or province, she lives in then possible impacts on other states or countries will not be of interest.

As individuals seek to understand subjectively the key features of the risk situation, they will decide on what dimensions of the relevant risks, benefits, and costs they want to pay attention to and, on the basis of this reading of the situation as well as their other objectives, define a set of perceived options and how they feel about them. This establishes the background or underlying context within which a risk valuation will be made; Fischhoff and his colleagues refer to this problem-structuring and objectives-interpreting phase as a process of understanding how the decision task is construed (Fischhoff, Welch, and Frederick, 1999).

The link between risk objectives, risk consequences, and risk-based responses is dynamic. As an individual builds a strategy for dealing with the risk at hand, he is also deciding – implicitly or explicitly – on a context within which attributes or qualities are assigned to the risk. This information, in turn, shapes how he thinks and feels about the place or activity or product or technology in question. Values associated with risk management options, and in turn other preferences and related behavioral responses, change over time as the result of this process of preference construction. As a result, clarifying objectives for risk valuation has the ability to teach individuals (or government agencies or industries) something about their own values and, in turn, has implications for incorporating flexibility and learning over time into the definition of risk management options.

2.2. Risk Consequences

Risk consequences are what valuations are concerned about. In most cases, predictions of the impacts of actions on health or environmental risks are uncertain. This uncertainty may be easily subject to resolution, through the conduct of additional studies or discussions with experts, or it may (at least in the relevant time period) be irresolvable. Substantial advances have been made in the analysis of uncertainty as part of the evaluation of risk policy alternatives and in procedures for linking factual information about expected consequences to the assessment of management options. Studies comparing alternative risk communication techniques also have made important advances in the effective presentation of uncertain information to both technical and nontechnical audiences (Morgan and Henrion, 1990).

The connection between the probabilistic consequences of risk actions and the valuation of competing policies is fundamental to the construction of effective risk management options. Making choices about different actions that result in a wide variety of consequences can seem an overwhelming task until a link is made from the range of anticipated impacts back to objectives, so that priorities then can be assigned to denote which consequences are more or less important to the individual, group, or agency in a particular context. Numerous methods exist for accomplishing this task (as described in Section 2.3), and the insights provided by a strong values reference have the advantage of separating what is most important – to ourselves, to our community, to our employer or state or nation – from what is less important. Not only does this allow for a refocusing of attention on getting good information about the impacts and probabilities of the more important consequences, it also establishes a basis for defining areas of potential conflict and similarity across different individuals or parties involved in a risk management decision process.

Risk management decisions are often viewed as requiring the selection of a preferred option from among a set of alternatives. In contrast, placing the emphasis on objectives opens up the decision to include the current alternatives as well as a set of new alternatives, created expressly in response to the relevant concerns. For example, if a high value is placed on the intergenerational risk consequences of an action, then governments may be willing to spend more money now to add engineering improvements to a facility so as to increase the safety or health of a distant generation. Similarly, if it is considered important to reduce the uncertainty associated with a proposed action, then additional safeguards can be enacted or further studies conducted to increase the knowledge base or to achieve a higher anticipated probability of success. Commissioning further studies to decrease risks in the far future could mean that higher risks are faced in the short-term as a result of the requested delay in action. Thus, the risk consequences faced by society are shaped in part by the values of the decision makers, which argues for an explicit and conscious valuation process.

The value placed on risk consequences may depend to a great extent on whether personal or public risks are at issue. Personal risks involve identifiable consequences, including deaths or injuries, and identifiable costs. Decisions about public risks address statistical lives and incur statistical costs; the identify of those whose lives will be saved, or whose money will be used, is not known. A public program to reduce risks from avalanches by one chance in a million will save one statistical life (for every million people at risk), whereas an initiative to save one climber trapped on the mountainside may save one identified life. In general, much higher values are placed on saving identified lives or preventing identified damages (e.g., to animal or plant species) than on statistical lives and damages. Although the distinction sounds clear, great efforts often are made by proponents to make statistical losses more salient, for example through the use of photographs or narratives, in order to "personalize" the more abstract statistical risks and thereby promote a more highly valued risk consequence.

2.3. Risk Measures

Risk measures provide a link between qualitatively expressed risk objectives and the quantitative expression of risk consequences, based on an understanding of the risk dimensions, magnitudes, and probabilities. For many risk values, there exist familiar measures of possible consequences. Typically referred to as natural attributes, these include dollars for impacts that are easily monetizable (e.g., profits) and numbers for countable impacts (e.g., jobs, fish catch, temperature). Natural measures have the advantage of being easy to understand; therefore, they are assumed to be excellent for purposes of communicating the effects of risky activities. One problem with natural attributes, however, is that they may be more ambiguous than initially realized. For example, a risk cleanup project that promises to create jobs in a community will be received quite differently depending on whether those jobs are part- or full-time, are low-end or managerial, involve local or out-of-state workers, and so forth. Thus, there are ample opportunities for misunderstanding and miscommunication because of the unwarranted casualness with which the natural measure is interpreted.

For other risk values, even some for which natural measures exist, there may be good reasons to employ a proxy measure of impacts. This typically refers to the choice of a more easily, or less expensively, measured attribute as a proxy for the activity of concern. For example, sulfur dioxide concentrations might be used as a proxy for the risks of higher pollution levels on architectural monuments in a European city because they are more straightforward and less expensive to monitor on a regular basis. For similar reasons, spring run-off water levels in a downstream portion of a river may be used as a proxy for the risk of avalanches from unusually deep winter snowfalls.

A third type of risk attribute involves constructing a scale or index that is specific to the problem at hand. Such constructed indices are particularly useful in cases where the values in question are important to the decision but lack obvious measures. For example, biological diversity effects in the case of an estuary cleanup project may be measured by an index that includes the impacts on juvenile salmon, water temperatures, and the change in acres of protected wetland habitat. The components of the index could vary across sites, or be weighted differently, depending on the specific conditions at hand. Index scores would run from the best possible level of impacts, in light of the anticipated project alternatives, to the worst impact level. Several constructed scales were used, for example, in the influential mid-1980s analysis of the risks of alternative high-level nuclear waste repository sites (Merkhofer and Keeney, 1987), including indices of the "biological impacts" and "adverse archeological, historical and cultural" impacts from the repository and associated waste transportation.

The choice of a risk measure can greatly influence the identification of risk problems and the consequent selection of policy prescriptions. Wilson and Crouch (1982) demonstrate this point in the context of assessing changes in the risks of coal mining in the United States between 1950 and 1970. If the measure "number of deaths from accidents per ton of coal" is used, the statistics strongly support the view that coal mines became much less risky, whereas if the measure "deaths from accidents per employee" is adapted, the data show that coal mines became somewhat riskier over the same period. Other measures are also plausible: as previously noted (Slovic and Gregory, 1999), deaths per million tons of coal is an appropriate measure of risk from the national point of view, whereas a labor organization might prefer to rely on deaths per thousand persons employed. Each approach to measuring consequences brings with it a different expression of risk and can lead to different policy implications for reducing the identified losses.

2.4. Risk Management Alternatives

Valuations of risk consequences affect the choice of policies to address the different types of risks, benefits, and costs. Typical management alternatives include improved communications to encourage voluntary changes in behavior, setting standards, adjusting market prices, and regulating related activities. There exists a rich literature on each of these options; detailed discussions are included in several chapters of this volume. Information on the perspectives of different participants can facilitate the identification of preferred options from among a broad set of management possibilities.

In many cases, however, insufficient emphasis has been placed on relating the choice of a management option directly to the weighted objectives of the relevant stakeholders or decision makers. For example, if one purpose of the risk initiative is to encourage the identification of the best option among many alternatives, then adoption of a standard may contribute little new understanding because it will only distinguish those actions that pass from those that do not. On the other hand, a standard might be a helpful option if the management objective is to improve consistency or to reduce costs because the standard can be applied repeatedly to the set of all relevant proposals. Standards might also be useful in a situation where the objective is to make a statement that symbolically expresses the policy maker's concerns, such as the Delaney Amendment in the United States, or where the political will is lacking to identify differences in participants' weighting of risk attributes or to make tough decisions on a one-by-one basis (Fischhoff, 1984).

To make the link from objectives to risk policy alternatives successfully, those in charge of developing management options must be willing to adopt a process that expresses the full range of stakeholder concerns and has the analytical capability to identify trade-offs across these objectives. Over time, this input can be used to influence the development of new policies, as shown by the substantially tougher regulations on water and air emissions set by the U.S. EPA as the result of a perceived shift toward more environmentally protective values in society. Alternatively, market prices may be adjusted to pass on directly to consumers the higher costs associated with reducing risks. Voluntary changes in risk-inducing behaviors may be selected as a preferred management option owing to the lack of market prices or the absence of the monitoring or political resources needed to back up tougher regulations. The more that risk-management choices reflect values in ways that are inclusive and explicit, the better will be the design and selection of risk management alternatives.

3. GENERAL CONCERNS IN VALUING RISKS

This section discusses five fundamental concerns that typically arise during implementation of the more widely used approaches for valuing risk policies. Together, these five considerations form a common set of issues to consider as part of the application of, and choice among, alternative risk valuation methods. Although not typically included as part of any specific valuation approach, how each of these issues is handled can have a significant effect on risk valuation results and, in turn, on the selection of a preferred risk management approach.

3.1. Valuation Perspective

A first consideration is to identify the preferred perspective of the risk-valuation effort. One aspect of this question is to determine whether the intent is descriptive, for example leading to an improved understanding of community perspectives on a risk management proposal, or prescriptive, leading to improved actions and activities or decisions. Techniques such as risk perception surveys can yield important insights into individuals' relative ratings of risks or the desire of various groups for different levels of regulation for specified risky activities. Similarly, mental modeling techniques can yield helpful information about relationships among the different factors or attributes that contribute to how individuals think about a risk management scenario. Both methods, however, generally provide descriptive insights into people's cognitive understanding of a risk source or their likely support for various initiatives. Prescriptive techniques, on the other hand, are used in situations where the goal is to improve individuals' risk exposure decisions or the performance of risk managers. For example, multiattribute techniques might be used in small groups to design risk communication initiatives that will encourage safer behavior on the part of homeowners exposed to radon or teenagers exposed to sexually transmitted diseases.

A second issue fundamental to the valuation perspective concerns the expectations of the study's clients. Do they anticipate precise quantitative responses rather than values expressed as ranges? Do they want a dollar number to fit neatly into a cost-benefit analysis? Do they want stakeholders to agree on a single best management option? Do they want insights into the reasons why different stakeholders support or oppose a management initiative? Do they want suggestions for how to create a preferred alternative? Successful risk valuation efforts typically begin by clarifying what the client thinks will be the product of the study and adjusting this, as needed, by the realities of the valuation environment. Of course, in many cases, there is not a single client but instead multiple clients, each of whom may hold distinct objectives; a study completed for industry may also need to pass muster with community-based groups and with several levels of government regulatory agencies. As a general rule, however, the better valuation studies are those that start with an explicit acknowledgment of both endpoints – by considering what it is that policy makers are wanting and also what it is that public input and the available information can provide – and develop a realistic perspective on how appropriate measures of values can be built (Sunstein, Kahneman, and Schkade, 1998).

3.2. Participant Selection

If values determine what matters, then who should be listened to: who will have a voice in the risk valuation process? An easy answer is to say that this group should include all those individuals potentially affected by the decision. In very localized risk issues, such as rerouting a small stream to decrease property damage from seasonal floods, this inclusive answer works well. In the usual, more complicated situations, however, the approach may be unrealistic. One reason is that it is often not possible to include all potentially affected stakeholders. Many of the actions taken to address local risks are recognized as having national or international implications: the selection of a smokestack height in Ohio can affect acid rain levels in Quebec, and the selection of a policy to protect endangered species in Oregon can affect the vacation plans of German or Kuwaiti tourists. Who should be counted as a stakeholder, and should their votes all be counted equally in the risk deliberation process? In such cases, it may be useful to identify a small set of representative stakeholders, who then serve as a voice for, and conduit to, the larger group. Another reason why inclusiveness may falter is that some stakeholders may not know that they could be helped or harmed by the initiative under consideration. Others may be vaguely aware of the proposed change but not want to know more because of their own fears or because of constraints on their time or emotional and cognitive energies. In other cases, a large and important stakeholder group, as represented by the average taxpayer, may be left out of the picture entirely; this issue of who pays the bills is a particular concern in situations where similar locally important decisions are being made by numerous community-based stakeholder groups.

A similar set of questions arise in terms of the information needed to represent and to address stakeholder values, which can come from a variety of sources. What constitutes a comprehensive stakeholder involvement program? On what basis is it determined that a wider set of contacts should be initiated? What should be done when information is perceived to conflict? What will help data to be trusted by skeptical stakeholders? Who should be considered an expert? In the case of determining impacts on wildlife, for example, what weight should be given to years of hunting experience as compared to years of schooling or seniority within a management organization? These are important concerns, and for many the best guidance remains experiential and anecdotal because clear prescriptions from either the risk management or the judgment- and decision-making literatures are only now starting to emerge.

3.3. Problem Scope

A third valuation concern, often overlooked or taken for granted, is selection of the scope and bounds of the risk problem. In many circumstances, a government agency or private institution will select the issues to be valued from among a broader set of actions, often without reference to explicit objectives and often within constraints imposed by regulatory or statutory rules. They will then proceed with initial analyses; in other cases, a stakeholder committee will be convened to begin its assessment of the problem. The case-by-case nature of such risk reduction efforts can create a fundamental problem because it has the effect of legitimizing only particular types of actions. As a result, valued initiatives may be included implicitly (thereby distorting expressed risk values) or neglected entirely (thereby distorting risk-reduction expenditures) owing to the lack of a suitable forum for their expression.

Suppose, for example, that the issue under consideration is determining how much to spend to decrease the contribution of sewage overflows to water pollution in an urban river. More serious problems may be caused by other sources (e.g., leakage from contaminated sediments), but if these are not on the table, then funds may well be spent on the lesser problem of overflows: there is an understandable desire to do something rather than nothing, the opportunity costs may be considered irrelevant, and the issue of overflows may be seen as representative or symbolic of a larger class of water quality concerns (Kahneman and Knetsch, 1992). Thus, individuals may legitimately be concerned about a broad class of health or environmental risks, but their stated values will be inflated because the only opportunity for expressing these preferences is through a survey or referendum that is focused on a more narrowly defined project or policy initiative.

3.4. Gains Versus Losses

From an economist's point of view, resources have value to the extent that people are willing to make sacrifices of other things in order to acquire them or to prevent their loss. These values typically are measured by the maximum that people are willing to pay (WTP) to acquire or gain something considered desirable and, for losses, by the minimum amount that people would demand to accept them (WTA). Until recently, these two measures of value were considered to be equivalent in most relevant risk-policy situations (e.g., so long as there were no wealth limitations on the amount that individuals could pay). Empirical evidence from research by behavioral economists and psychologists, however, strongly supports the contrary view that people commonly value losses more highly than otherwise equivalent gains (Kahneman and Tversky, 1979; Kahneman, Knetsch, and Thaler, 1990). Although the magnitude of the disparity varies with context and the good in question, both laboratory experiments and studies of market behavior show that this "endowment effect" typically results in two-fold or higher disparities between WTP and WTA values.

The distinction between using WTP values for gains and WTA values for losses implies that the common practice of using WTP values for estimating risk initiatives resulting in either gains or losses is theoretically incorrect. The magnitude of the difference suggests that risk valuations of the same option may vary substantially depending on whether the expected change in status is perceived as a gain from the status quo or as a restoration of a loss (Gregory, Lichtenstein, and MacGregor, 1993). As a result, using measures of WTP to assess the value of risk initiatives designed to prevent a loss (e.g., an oil spill) will tend to bias risk assessments, discourage the use of mitigation as a remedy for damages, underestimate compensation and liability awards, and result in too few restrictions being placed on activities posing health or environmental risks (Knetsch, 1990).

3.5. Time

Risk management commonly involves decisions about how to take account of impacts occurring over time, including (in some situations) analysis of the consequences of current decisions that will accrue over decades or centuries or (as in the case of nuclear waste storage debates) even millennia. The current practice for comparing costs and benefits occurring in different periods is to weight the importance of future gains and losses using a single interest, or discount, rate, although there has been a debate for years (chiefly among economists) about exactly what this rate should be. Much of the more recent debate, however, focuses on suggestions for the use of varying rates depending on the particular circumstances or characteristics of a potential future event or outcome. For example, people appear to commonly discount future losses at a lower rate than future gains (Loewenstein, 1988) and to use much higher rates to discount outcomes in the near term relative to those accruing at more distant times (Benzion, Rapoport, and Yagil, 1989). Evidence also suggests that individuals may employ different discount rates for different types of goods, for example private versus public resources or financial versus health risks (Luckert and Adamowicz, 1993).

These results, to the extent that they accurately depict people's true time preferences, have important implications for valuations of risk policies. Incurring present costs to avoid potentially large losses in the distant future may not appear to be worthwhile using constant rates of discount (as in most current analyses) but may be worth undertaking when appropriate adjustments are made for the lower time preference rates on losses and for longer time horizons. Similarly, there may be less economic justification for actions providing short-term benefits (e.g., jobs, profits) at the expense of costs accruing over decades (e.g., loss of productivity in natural systems). Such reassessments of the use of discount rates to address time preferences may substantially affect valuations of risk programs concerned with issues such as global climate change, hazardous waste storage, or species protection.

4. APPROACHES TO VALUING RISKS

4.1. Risk Perceptions

Risk perception techniques depict the values associated with a risk initiative directly, in terms of the underlying cognitive and emotional factors, without translation to dollar terms or summary numerical scores. They share a common basis in psychology and survey methods, focusing on cognitive and contextual processes and rejecting the strict interpretation of expected value estimates as the sole basis for risk valuation or the choice of management options. Risk perception methods also reject the distinction between objective and subjective risk values, arguing instead that perceptions are real and matter to the extent that they influence individuals' behavior and judgments of their own utility or welfare.

4.1.1. Opinion Surveys. Opinion surveys have been used extensively to provide information about how individuals perceive a wide range of human health and environmental risks (Dunlap, Gallup, and Gallup, 1993). In recent years, important advances have been made that address problems relating to question form, wording, and order that can introduce inadvertent errors and misinformation into survey results by providing cues, biased framing, or other influences on respondents (Schuman and Presser, 1996). Generally, results of opinion surveys provide indications of relative values, for example the importance of difference types or sources of risk compared to each other or compared to other possible social initiatives. They provide a relatively inexpensive approach to eliciting citizen input, have the advantage of being user-friendly, and can be helpful in developing a broad-based understanding of public views or testing the acceptability of specific proposed policy actions.

In many cases, however, opinion surveys are not able to deliver the type of detailed evaluative information that is needed, either by respondents (to ensure an informed answer) or by decision makers (to ensure direct input to the risk question or policy action at hand). Information from opinion polls generally is limited to a single observation and, in the typical case, little time is provided for thinking through the many dimensions of what might be a very complicated risk problem. Extensive research suggests that views expressed in this shoot-from-the-hip manner are likely to reflect recent and highly salient incidences (e.g., availability) or the particular framing of a topic that is introduced by the interviewer (e.g., anchoring and adjustment). Further, the reliance on information from opinion polls, if considered on its own, also can convey a subtle message that the public's views are not really all that important to the solution of the risk problem; a quick and easy method was chosen for soliciting public views because they really do not count for all that much anyway.

4.1.2. Specialized Questionnaires. A variety of specialized survey and questionnaire approaches have been used with great success over the past twenty years to inform risk managers, researchers, and decision makers about individuals' perceptions and values. As described by Slovic (Slovic, 1987) and his colleagues (Slovic, Fischhoff, and Lichtenstein, 1985), an original impetus for the development of psychometric risk surveys was the recognition of a striking disparity between the statistical risks of many activities and individuals' estimated risks of these same items. This early research used extensive questionnaires to ask experts and laypersons for judgments of the relative contributions of selected attributes making up their assessments of risk levels. Factor analytic techniques revealed that many of these qualitative characteristics were highly correlated and led to the identification of a small number of characteristics that influence the publics' perceived risks of an activity, including its salience, the immediacy and voluntariness of exposure, how well scientists understand its risks, the distribution and longevity of the impacts, and whether its effects could prove catastrophic. Subsequent research showed that the location of a hazard within its factor space was an important predictor of peoples' desire to see its current risk levels reduced, for example through additional regulation of the risks.

Studies of public risk perceptions have helped to broaden the range of values that people legitimately bring to the assessment of risky options. By including psychological and procedural considerations that often are important determinants of risk values but may not influence mortality or morbidity estimates, research into the basis for risk perceptions has contributed to an increased appreciation of the multidimensionality and richness of risk-related attributes (Bradbury, 1989). Studies also have shown important differences in how different groups of people typically value risks. In research completed in the United States, for example, gender and race differences are especially striking: white men express less concern for nearly all risk sources, women show special concern for risks with adverse intergenerational effects, and nonwhite males and females are more similar in their perceptions of risk than are white males and females (Flynn, Slovic, and Mertz, 1994). Other research has shown how individuals' perceptions of hazards can be affected substantively by their underlying dispositions or worldviews (Dake, 1991) and by exposure to risk-related information (Smith and Johnson, 1988).

In addition, risk perception studies have served as the basis for an improved understanding of extreme risk events, in which impacts extend beyond the direct harm associated with an accident or other event (e.g., resulting in pollution or sabotage). A conceptual framework, emphasizing the importance of signal values in creating such ripple effects (Kasperson et al., 1988), has been developed for understanding how the interaction among psychological, political, and social factors results in this amplification of selected risks. The signal potential of an event is linked to the characteristics of the hazard so that even relatively small accidents occurring as part of an unfamiliar system or exhibiting characteristics associated with dread or catastrophic impacts could result in large economic and social consequences. More recent work has extended this concept of signals to understanding the processes by which a product, area, or technology may become stigmatized (Gregory, Flynn, and Slovic, 1995).

4.1.3. Cognitive and Affective Modeling. Research in behavioral decision making has led to several cognitive and affective modeling approaches to valuing perceptions of risks. These approaches assume that true values for risks do not exist prior to the elicitation but, instead, are constructed as part of an interview or survey process (Payne, Bettman, and Johnson, 1992). As a result, these techniques share a common interest in providing extensive supplementary help to participants in thinking through their concerns and priorities prior to making an evaluative judgment of risks.

Mental-models interviews are one important cognitive modeling approach that has been used to establish risk values in a variety of management contexts, principally as a tool in the design of risk communication surveys (Bostrom, Fischhoff, and Morgan, 1992). In contrast to a public opinion poll, which typically requires that participants choose from standardized responses to fixed questions, a mental-models interview protocol allows participants to express themselves in their own terms, discussing a common set of issues but with the ability to concentrate on those issues that matter most to them. The primary objective of a mental models interview is to identify, within the selected context, the major factors that account for how respondents think about the risk issue under consideration. In addition, differences in the expressed importance of these factors permit the mental model results to highlight differences across stakeholders in how issues are conceptualized and, within subjects, how the understanding of risk issues changes over time. The role of the interviewer, working from a script, is to ensure that an opportunity is provided for the respondent to address all major issues but also to add questions as needed so that the participant can provide additional detail, including relative valuations, for the more important risk sources and factors.

Decision-pathway surveys, another example of cognitive modeling approaches to perceptions of risk, attempt to draw out participants' reasoning by providing a set of linked, branching questions that encourage participants to self-select the response pathway that best reflects their thinking. In particular, decision-pathway surveys have been used to encourage respondents to consider trade-offs and potential value conflicts, thereby defining more clearly the relative benefits, costs, or risks associated with selected policies. For example, the province of Ontario used the results of a decision-pathway survey of the values of the general public, forest professionals, and residents of timber-dependent communities to develop new policies and risk-communication materials for addressing ecological and health risks in the context of managing the growth of unwanted forest vegetation (Gregory et al., 1997).

Several other valuation approaches also offer promising insights to understanding individuals' perceptions of risk. One technique uses conjoint analysis to build an understanding of an individual's values by asking the participant to make a series of paired comparisons. The results can be used to provide direct input to risk policy development or to estimate willingness-to-pay values for identified policy initiatives. For example, Opaluch et al. (1993) used a contingent choice survey based on paired comparisons between attributes of alternative landfill sites to develop overall scores for the sites and, in turn, estimate the percentage of votes that each site would receive in a hypothetical referendum.

Another approach, which emphasizes the affective and emotional effects of a risk initiative, explores the positive and negative images that are associated with risk policy options and their consequences (Slovic, 1992). Images are particularly useful when values are based strongly on fears, hopes, and perceptions that individuals associate with the anticipated effects of an action. In such cases, values may be poorly thought through and measures of affect, including people's dread or worry about a proposed management technique or initiative, may provide revealing indications of current risk values and future risk-related behaviors.

4.2. Risk Markets

An alternative approach to valuing risks is to establish monetary values for specific risk damages or policy initiatives within the familiar context of economic markets. Because dollar payments are often used as an indicator of value, these approaches have the advantage of being easily integrated into cost-benefit analyses of projects or programs and their results are readily understood by many policy makers.

4.2.1. Implicit Dollar-Risk Choices. Economists have devoted significant attention to the development of methods for estimating the implicit value of risks based on the decisions that individuals make in the course of their lives. The dominant approach uses labor market data on wages for risky jobs to infer attitudes toward risk, recognizing that workplace safety is costly to the firm but attractive to workers. Econometric analyses of wage-risk trade-offs have been conducted to estimate the (implicit) value of fatalities and the average (nonfatal) job accident risk. Economists also have analyzed the choices implied by peoples' decisions, such as whether to wear seat-belts (Blomquist, 1979), to infer a dollar-based value of life. Most estimates are in the range of $3 million to $7 million (Viscusi, 1993). Inventive techniques also have been developed to estimate risk-dollar values when morbidity effects (rather than fatalities) are of concern and when market data is largely missing. For example, Viscusi, Magat, and Huber (1991) developed an iterative computer-based methodology using risk-risk trade-offs to estimate the value of marginal reductions in the risks of chronic bronchitis.

   Substantial effort has been placed recently in moving from statistical value-of-life measures as a means for monetizing health benefits to measures that better reflect the quality of a person's life. Such measures of quality-adjusted life years typically take into account both years of life lost and the intensity of people's preferences for different states of health and illness. For example, Graham, Carrothers, and Evans (1999) used quality-adjusted life years to estimate the value of pollution control programs undertaken by EPA, resulting in considerably lower average values per life-year saved when compared to standard statistical-life analyses. Additional research focuses on estimating the indirect value of implicit dollar-risk choices by taking into account how exposure to air or water pollution, for example, might increase the probability that an individual will become ill prematurely and either incur a lower quality of life or die from other causes.

4.2.2. Hedonic Pricing. Hedonic techniques for valuing risks start from the recognition that many unpriced attributes associated with a hazard – such as the benefits of good visibility, incidence of floods, or proximity to hazardous waste sites – are reflected in the prices of marketed goods such as land or houses. The commodities in question are no different from other goods bought and sold by consumers except that, due to the absence of price information, some important characteristics cannot be unbundled. The economic value of these unpriced characteristics can be measured by the difference in prices between assets that are similar in all respects except that one is exposed to a higher level of risks. For example, the analysis of paired houses within a city, identical except that one is situated in an area with lower air quality, can provide a measure of the value associated with improvements in health owing to reductions in emissions from nearby factories (Freeman, 1993).

Although the concept behind hedonic studies is appealing, data requirements can be daunting. In particular, it is often difficult in risk management situations – where trust in managers, problem history, and other contextual concerns are often very important – to develop realistic comparison sets. It also can be difficult to specify fully the respective equations (e.g., the underlying production functions) so that it is problematic to separate out the effects of different attributes or to know whether some of the observed impacts might be due to unknown, and hence omitted, considerations.

4.2.3. Contingent Valuation. Among the most widely used of all economic techniques for valuing risks are contingent valuation methods (Mitchell and Carson, 1989). These approaches posit a hypothetical market for an unpriced good (e.g., an improvement in air or water quality or a reduction in risks to an endangered species) and ask people to state the maximum price they would be willing to pay to obtain more of the item or, alternatively, either their willingness to pay or their willingness to accept compensation in return for reducing its consumption or avoiding the item. Contingent valuation (CV) surveys typically use samples of as many as several thousand people, and the aggregated results are taken as indicators of the value placed by society on the risk-reduction measures in question. Contingent valuation methods have been used over the past twenty years to establish monetary values for a wide range of environmental and health risk initiatives, including such high-profile issues as valuing environmental and social damages caused by the *Exxon Valdez* oil spill (Mitchell and Carson, 1995). They also have been granted substantial authority by the popular press, by Congress, and by the courts (Kopp and Smith, 1993).

Contingent valuation approaches, at heart, are very simple – ask individuals about the monetary exchange they would make for an unpriced health or environmental good – so it is no surprise that a wide variety of CV methods have been used to value risks. Variations include the amount and form of information presented about the good, the use of visual or auditory stimuli, the format of the payment question (which can be elicited as an open-ended response or as a yes/no response to a referendum-type question), and the information provided regarding the social context for valuation (Fischhoff and Furby, 1988). Many more recent CV studies take extra care to help participants with some aspects of the valuation process that are known to be especially difficult, such as accurately processing probabilistic information.

Some of the more sophisticated CV approaches have been developed to establish dollar values for proposed health-risk changes and for developing value-of-life estimates. Gerking, DeHaan, and Schulze (1988), for example, used both willingness to pay and willingness to accept measures of value in a mail survey of possible changes in job risks. Jones-Lee (1989), in a CV survey completed in the United Kingdom, estimated respondents' willlingness to pay for reductions in risks related to motor vehicle accidents. In both these cases, the implicit estimated value of a statistical life was in the range of $3 million to $4 million (and closer to $9 million in the willingness to accept version). Several contingent valuation studies also have calculated economic values for nonfatal reductions in risks, including studies examining the value of morbidity risks from pesticides (Viscusi, Magat, and Huber, 1987) and the value of avoiding one day of various minor illnesses such as coughs, congestion, or itching eyes (Berger et al., 1987).

Despite this widespread support, serious questions have been raised concerning the legitimacy of the results of CV surveys as inputs to risk management decisions. In particular, their accuracy has been called into question by evidence demonstrating that minor variations in the information presented to survey participants or the way in which valuation questions are asked (e.g., their context, wording, and order) can have large effects on the magnitude of respondents' answers (Fischhoff et al., 1993). Many behavioral studies have questioned the ability of CV participants to make sense of the tough cognitive tasks they are given, including the integration across multiple dimensions of value and the translation of varied concerns into dollars; one result is the observed insensitivity of responses to changes in the problem scope (Kahneman and Knetsch, 1992; Ritov and Kahneman, 1997). As noted earlier, many CV studies also use a payment-based measure to estimate the value of losses, a conceptual error resulting in the underestimation of environmental or health damages (Knetsch, 1990) that can be avoided by using alternative, behaviorally more sophisticated approaches.

4.2.4. Damage Schedules. Damage schedules can provide scaled rankings of the relative importance of losses to various environmental and health concerns. The rankings reflect relative damages (of which people typically are more certain) rather than absolute values (which are more difficult for people to assess). Impacts included on a damage schedule can form the basis for regulatory and other controls in much the same way that schedules now are used to settle worker's compensation claims and establish workplace safety regulations (Rutherford, Knetsch, and Brown, 1998).

At this time, damage schedules for risks remain a promising experimental approach. They have the advantage of being straightforward to establish and of incorporating changes easily over time, in that new damages can be added to the list where appropriate or new information can be used to shift the placement of a given risk source. However, the greatest advantage of a schedule – its ease and consistency across a variety of contexts – may also prove to be a liability, in that individuals' perceptions of damages are quite sensitive to a variety of contextual effects that would need to be incorporated into the schedule and may result in changes across different situations, either in items or in their order.

4.3. Risk Trade-offs

Methods for valuing risks that focus on the acknowledgment of trade-offs typically look at people's risk values not as stable and coherent and well-defined, which is a view still common among economists, but instead as capable of evolving and changing dramatically in response to the properties of the decision task. These properties include both external inputs (e.g., selective media attention) and the internal information processing system of the individual (e.g., as people think more deeply about what risk attributes matter to them the most). This view raises fundamental issues in the context of risk values because it means that expressed trade-offs (e.g., across economic and ecological benefits or across expenditures and fatalities) will not be invariant to different elicitation procedures. Early evidence for this perspective came from studies of preference reversals, which demonstrated that expressed values systematically violated the fundamental principal of procedure invariance: object A may be clearly preferred over object B when using one method, whereas B is clearly preferred under a different (but equally plausible) procedure (Tversky, Sattath, and Slovic, 1988).

This view is consistent with recent research from behavioral decision theory emphasizing the constructed nature of how preferences and values are formed (Payne, Bettman, and Johnson, 1992). A constructive perspective is especially relevant to judgments and choices that are complex (e.g., multidimensional), important, and unfamiliar; this characterization closely describes many risk valuation contexts. In such cases, research suggests that constructive strategies will be used to make sense of the available information and build a preference order on the spot, based on a wide variety of heuristics or strategies such as relying on the more prominent dimensions, anchoring and adjustment, eliminating common elements, and ease of evaluability (Slovic, 1995).

Further, most of us wear many hats and confront risk choices differently depending on whether we see ourselves in the context of individuals, household and community members, or citizens. These hats change over time, as do the weights that we place on attributes associated with exposure to risks, in response to the context we encounter in making a trade-off across dimensions of the risk. In such cases, research in behavioral decision theory (Payne, Bettman, and Johnson, 1993; Gowda, 1999) shows that risk judgments tend to be highly contingent: a personal experience, time pressure and task complexity, or a biased media focus can dramatically alter a person's valuation or relative ranking of a risk, perhaps just for a short time and perhaps forever.

4.3.1. Rankings. Risk rankings have been the subject of substantial recent study as the result of a national initiative to develop improved data on comparative risk judgments. EPA's 1987 report *Unfinished Business* was an influential early risk-ranking exercise and helped to focus attention on problems of Superfund toxic waste sites. In the early 1990s, EPA initiated two further risk-ranking processes with state and local organizations; an important goal of these efforts was to enhance the agency's understanding of how to increase public support for proposed new risk management practices. Additional risk-ranking initiatives are currently under way within EPA and within many other federal and state agencies with responsibilities for managing health and environmental risk-reduction programs.

The general goal of these efforts is to develop relative rankings of risk so as to help decide which risks are most worthy of attention and, in turn, provide guidance for the allocation of society's resources. Although this is a worthwhile goal, the difficulties involved in ranking risks are often underappreciated. Risk sources are diverse, and so is the information needed to rank risks effectively – information on the magnitudes of risks (using which measures?), the costs of control strategies and their probability of success, or the effectiveness of restoration and monitoring. Even if all this information were of reasonable quality and available within time constraints, it would still be necessary to face difficult trade-offs across economic, ecological, health, cultural, and social considerations. Examples include trade-offs across the anticipated costs of risk reduction programs and the expected ecological damage, the expected number of deaths or injuries per year, the degree to which exposure to a risk is voluntary, the distribution of effects and whether there are unusual impacts on special populations, and so forth. The relative significance of these concerns could vary substantially across different risk management contexts, which makes it difficult to use a risk-ranking exercise to improve the consistency of funding decisions because the same risk attributes may be weighted diffferently in light of the other costs, benefits, and risks associated with management of the problem.

Over time, government agencies are beginning to acknowledge these problems and are moving to a more context-specific and interactive approach to ranking risks, making use of structured citizen input. For example, greater attention is being placed on presenting probabilistic information, establishing the quality and source of the scientific data, and getting information into a form that visibly addresses participants' concerns. More interactive ranking processes are being used, allowing for refinement of the factual information and ranking procedures, and there is a growing recognition that ranking exercises are more likely to be successful (because the required judgments are easier) to the extent that the risks in question have relatively comparable attributes and effects. Many of these risk-ranking processes use explicit multiattribute procedures, with risks within categories characterized on the basis of a small set of attributes that participants can then weight according to their values and perceptions. Recent experimental work along these lines (e.g., Morgan et al., 1999) provides evidence that participants in groups considering health and safety risks are able to conduct such multiattribute assessments relatively easily and, surprisingly, are able to reach a moderate level of agreement on ranking a diverse set of risks when involved as part of a well-moderated group process.

4.3.2. Paired comparisons. Judgment tasks are more accessible in valuation contexts when they require a less stringent level of measurement. For example, cardinal (or ratio) judgments are more demanding than ordinal, and ordinal (or interval) judgments are more demanding than pairwise comparisons. Mapping preferences onto a ratio scale requires knowing by how much one thing is preferred to another; choosing the preferred element in a paired comparison question requires only that the individual know which of two things is preferred. Keeping judgment tasks at a level appropriate to the time demands and computational abilities of participants helps to ensure that the instructions will be more easily understood and the tasks will be more meaningfully completed (Fischhoff and Cox, 1985).

Pairwise comparisons constitute a common psychometric method that is used to reveal an individual respondent's preference order among elements of the choice set. Applied to a set containing both goods and money, the approach elicits multiple binary choices that permit valuation in dollar terms through psychometric scaling or stochastic discrete choice analysis as well as by simple arithmetic (Peterson and Brown, 1998). Under certain assumptions, it is possible to use paired-comparison data to derive an interval preference scale; in some cases, the inclusion of monetary anchors (e.g., for the high and low ends of the scale) can facilitate translation of the relative values into dollar estimates. The method has been used in a variety of contexts to value risk reduction options, including both gains and losses from the status quo. For example, Irwin et al. (1993) used modified pairwise comparisons to study individuals' choices between higher air quality and improvements in selected consumer commodities. Viscusi et al. (1991) obtained choices between pairs of alternative geographic locations based on differences in the cost of living and in health risks.

The relative simplicity of making paired choices has the advantage of making it easier to extend valuation methodologies to resource users. For example, a paired comparison approach recently was used to compare the values of farmers and fishers to those of professional resource managers for a range of proposed environmental risk-reduction activities in Thailand (Chuenpagdee, Knetsch, and Brown, 2001). Pairwise methods for addressing trade-offs also share a common basis with conjoint methods for valuing risks and a variety of other approaches that collectively are often referred to as stated-preference methods. Johnson (1998), for example, used stated preference methods to estimate trade-offs between reducing risks to salmon and maintaining jobs in the Pacific Northwest, noting changes in the weights placed on attributes as information is processed in the course of the survey (e.g., the stated preference for jobs relative to salmon was lower at the end of the survey). Adamovicz et al. (1998) used a multiattribute approach to developing paired choices in asking respondents to choose from alternative bundles of attributes (instead of ranking or rating them) in a study focusing on the estimation of passive use values (e.g., maintaining wilderness habitat).

4.3.3. Decision Analysis. Decision analysis techniques share a conceptual basis with other trade-off-based valuation approaches in focusing on conflicting, multidimensional dimensions of risk problems but, in general, place relatively more attention on the initial phases of the value elicitation process – structuring the problem, defining objectives, and identifying measurable attributes (Keeney, 1992). These efforts, based in the theory of multiattribute utility analysis (MAUT), are intended to make explicit the diversity of values and opinions relevant to the risk management problem at hand (Keeney and Raiffa, 1993). These values can include both easily quantified concerns, such as dollar cost or the number of affected lives, and objectives such as community image, equity, or trust for which problem-specific scales can be constructed. Decision analysis techniques are therefore able to highlight the multiple values held by different stakeholder groups, elicit judgments about the quality of facts and relevance of underlying models, and provide decision makers with information on the implications of these judgments for their choice of a recommended policy (von Winterfeldt, 1992).

Decision analysis techniques have been used to elicit public value judgments in a variety of risk management contexts, many of which have been controversial (Edwards and von Winterfeldt, 1987). Examples include studies by Merkhofer and Keeney (1987), who evaluated the risks of alternative options for storage of high-level nuclear wastes; Keeney, von Winterfeldt, and Eppel (1990), who designed "public value forums" to assist the West German government in evaluating alternative energy policies; Gregory and Keeney (1994), who used decision analysis techniques to structure elicitations of stakeholder values in the context of searching for new policy alternatives in a controversial land-use debate in Malaysia; and McDaniels (1996a), who evaluated the environmental risks of electric utilities.

These examples demonstrate some of the strengths of a MAUT-based approach to valuing proposed risk-management initiatives: participants' values for actions or strategies can be structured carefully, multiple measures can be used to express these objectives, and a variety of alternatives can be placed on the table for discussion and evaluation. These strengths are one of the reasons why MAUT-based processes are often favored for interest-group negotiations (Raiffa, 1982) and why the use of decision analysis techniques is fundamental to a growing number of stakeholder negotiations concerned with finding solutions to problems characterized by controversial economic and environmental risk trade-offs. Tools such as value trees, which provide visual depictions of objectives hierarchies (Keeney and Raiffa, 1993), and influence diagrams (Schacter, 1986) have been used as aids in presenting the underlying logic of a decision analysis approach in ways that are relatively easy for individuals or small-group participants to understand. Means-ends networks (Keeney, 1992), which distinguish between values important in and of themselves (i.e., fundamental objectives) and those important because of their indirect contribution (i.e., means objectives), provide another useful tool for identifying values and choosing among alternatives in the context of risk management decisions.

One use of decision analysis techniques has been to develop estimates of the indirect impacts of risk reduction initiatives by examining the induced or indirect costs of regulatory and other risk reduction programs. Keeney and von Winterfeldt (1986), for example, examined some of the unintended side effects of health and safety regulations that can occur through pathways including accidents (e.g., as part of construction or equipment operation) and stress (e.g., from unemployment) as well as reduced household income. Their research, focusing on risk reduction regulatory initiatives proposed for power plants, concluded that these unintended side effects can in some cases be more severe than the risks the initiatives are intended to prevent. Keeney (1990) developed a general model that interprets the link between income and mortality risks as in induced relationship (i.e.,

lower incomes lead, on average, to higher risks) and examined the implications of cost-induced fatalities under specified conditions, concluding
that some expensive regulations intended to save lives may instead lead to
increased fatalities. The underlying philosophical question for risk managers (as raised by Wildavsky, 1979) is whether, on balance, people are
safer with the government buying safety (e.g., through air pollution regulations) or with individuals buying safety (e.g., through better nutrition or
the purchase of safer cars); the practical answer is that it depends on exactly
how the government and how individuals choose to spend the money in
question.

Another application of decision analysis approaches to risk valuation,
closely related to the pairwise comparisons discussed earlier, seeks to simplify complex valuation decisions by searching for trades or "even-swaps"
(Hammond, Keeney, and Raiffa, 1999) across decision objectives. The approach has respondents work through a series of choices across alternatives
that lead to expressions of one objective in terms of another: How many
additional dollars would be paid by a potential purchaser for a specified
improvement in the safety of an automobile? How many additional jobs
can be lost in return for decreasing the probability of extinction for a threatened species? This bartering mechanism, basic to a full multiattribute evaluation, can be used as a stand-alone technique for focusing the attention of
individuals on the key values that may be affected by a risk-management
decision. Although an even-swaps approach does not substitute for the
hard choices (and thinking) required of participants, it does help to simplify the decision problem (by eliminating some objectives) and to identify
dominant alternatives, thereby shifting attention to the most highly valued or controversial aspects of the risk management alternatives (Gregory,
2000).

Formal expert judgment elicitation procedures have been developed
by decision analysts to refine probability estimates of the impacts of risk-
inducing activities and to clarify the basis for technical judgments of consequences (Keeney and von Winterfeldt 1990). Elicitation of expert judgments usually is done as a one-on-one interview between the analyst and an
expert, with boundaries established for the event in question and probabilities assessed over key variables. An important aspect of this procedure is
understanding how the problem is thought about by different experts: how
it is decomposed, for example, highlights key considerations or stages for
the event in question and provides a starting point for understanding why
a risk source may be perceived differently by different individuals. Several
methods are available to aggregate judgments across experts, including
mechanical aggregation (e.g., averaging) and more behavioral techniques
(e.g., focusing on group interactions).

One concern with a decision analysis approach is that it will lead to excessive disaggregation of the risk problem, thereby omitting an integrative

element identified with the human or ecological system as a whole. Another concern is that a decision analysis approach will be perceived as excessively quantitative. Practitioners make the counterargument that, in contrast to the expert-centered focus of most formal risk analyses, the problem-structuring techniques of decision analysts can be used to assess the values of a wide range of potentially affected stakeholders. Furthermore, decision analysis approaches can comfortably employ multiple measures for assessing risk attributes. Most importantly, decision analysis approaches recognize that the level of acceptable risk is always decision driven: rather than seeking a magic number, as in many risk analyses and economic evaluations, a decision analysis approach seeks to define risk management options in terms that capture the problem context and reflect risk measures that make most sense to the affected parties (which, in turn, enhances the likelihood that a management plan will receive broad-based support).

4.4. Referenda

A variety of voting procedures provide a potentially attractive method for addressing risk trade-offs because of the familiarity with voting as a mechanism for expressing opinions about controversial policy options. Referenda processes, often following citizen-based initiatives (e.g., successful letter-writing or endorsement campaigns), have been used extensively in recent years to address a variety of environmental, health, and social risk initiatives. In many of these cases, however, the voting process likely suffered from the same problems of biased information and lack of careful thinking that are typical of elections for public officials or most public opinion polls – information is often overwhelming or biased, no explicit structure is given for addressing tough trade-off questions, and the objectives of the exercise are often unclear.

In an effort to overcome some of these drawbacks while maintaining the benefits of a referenda approach for valuing risks, several researchers have attempted to develop structured referendum processes based on the principles of value-focused thinking (Keeney, 1992) and structured decision processes. These approaches typically, in sequence, help the respondent to identify the decision to be made, provide a set of objectives for characterizing the problem, and identify several attractive alternatives whose impacts have been summarized using an "objectives-by-alternatives" matrix. Participants are then asked to vote for the one alternative that best satisfies their objectives; in some cases, participants instead have been asked to designate all those alternatives that they would support. This type of approach has been used by McDaniels (1996b), for example, in the context of a choice over three options for treating sewage from the city of Victoria, British Columbia. Based on the results of small-group discussions, all three options were described in terms of their anticipated impacts on environmental, health, aesthetic, and economic objectives. In the actual referendum, held in November 1992 with about 34,000 voters participating, over half (57 percent) identified the status quo (no treatment) option as the preferred risk management scheme over either of two new treatment facilities.

Value juries and science courts represent other forms of voting approaches that can be used to provide explicit or implicit valuations of competing risk policies. The members of a values jury might be asked to assess a specific environmental risk or to select a preferred course of action from a set of alternatives that imply different levels of risks and benefits (Brown, Peterson, and Tonn, 1995). Citizen members would vote in their capacity as direct representatives of the larger society, including future citizens who might be affected by an action or decision in the same way that jurors are asked to address other tough social problems such as the responsibility of an accused person or the compensation that should be paid to an injured party. Science courts are a similar idea, with members meeting over an extended period of time to discuss complex technical, social, and procedural questions and to arrive at a more in-depth understanding of the value trade-offs and factual information than would be possible in a short survey or two-hour group (Morgan et al., 1992). With both value juries and science courts, the underlying objective is to provide a forum in which citizen participants can make informed judgments, which requires not only access to critical scientific information but also assistance in structuring the problem, thinking through value conflicts, and learning over time.

5. IMPLEMENTING RISK POLICIES

After choosing an appropriate strategy for valuing risks, it is necessary to gain both political and public support so that the approach can be implemented swiftly and the results used to inform the development and finetuning of risk management policies. The connection between the analysis phase of risk valuation and implementation is subtle: too close a linkage and the analysis may be cut short or attenuated because of a fear that it might adversely influence implementation; too distant a linkage and the analysis may fail to address key issues that could influence public, government, or legal acceptance of the recommendations.

5.1. Public Acceptance and Support

A key lesson for risk managers interested in implementing the results of a valuation study is the overwhelming importance of process: no matter how carefully an analysis is conducted, acceptance of a study's results depends on key procedural elements. These include concerns such as the transparency of the valuation process, the identity of those making critical decisions (e.g., regarding the choice of models, interpretation and discussion of results, or the institutional affiliation of consultants), and the meaningfulness of stakeholder involvement (e.g., who is asked to the table, how early in the process, and what emphasis is likely to be given to their recommendations). Key process issues also focus on the visibility of citizens' participation in deliberating project options, in terms of the linkage from these alternatives back to the expressed values of stakeholders. For example, is only one best option provided for stakeholder review or are multiple alternatives suggested, and are these clearly shown in relation to the objectives of greatest concern to the different respondents?

One of the key topics in valuing risk management choices over the past decade has been the issue of trust. Clearly, there is a technical requirement for developing trust (related also to competency) that can influence valuations of risk management initiatives. Additional, less quantifiable factors include the extent of proponent cooperation in allowing close public scrutiny of analyses or operations, the opportunities that are afforded local representatives to take part in key aspects of decision making (e.g., related to engineering mitigation of a facility), and the willingness of operators to report unplanned incidents (e.g., cost overruns, accidental releases). Studies have shown an inverse relation between the perceived benefits and perceived risks of an undertaking (Alhakami and Slovic, 1994): as the perceived risks decline (because of engineering changes in a proposed facility or because of an increase in trust), the perceived benefits tend to increase. Factors such as trust, therefore, can fundamentally color the interpretation of factual information about a project's impacts and, in turn, participants' net (benefits minus risks) evaluation of a project's overall value. In a more general sense, studies of public acceptance for, and opposition to, risk management initiatives consistently have shown a tendency for blurring of the distinction between facts and values: values shape facts, by influencing what is looked at and in what way, and facts shape values by bringing certain issues to attention rather than others. Process considerations play a key role in entangling, or disentangling, these co-existing parts of the risk management puzzle.

Many risk management valuation initiatives conducted over the past decade have taken place within overall group-based decision-making structures that utilize dispute-resolution methods and seek consensus as an outcome of public involvement (Peelle, 1996). In many cases, the achievement of group consensus (e.g., regarding a decision whether to raise an additional $100 million through new taxes to pay for a specified water-quality improvement) is viewed as a goal that provides an indicator of the quality of the policy recommendation (Webler, 1997) and, therefore, a sign to decision makers (within a regulatory agency, for example, or elected officials) about how seriously a group's valuation should be taken. An alternative viewpoint looks at this emphasis on consensus as a costly error that can lead to distorted values and an inefficient allocation of resources because it gives an effective veto power to small but vocal minorities (Coglianese, 1999).

Another criticism of the search for group consensus, particularly in complex risk management decisions, is that it runs counter to the thoughtful exploration of participants' values and objectives. Issues may be selected in such a way that they offer a high potential for agreement, which has the result that less tractable issues may be ignored. Participants in community stakeholder forums may be selected more on the basis of their ability to get along and work well with others than on criteria relating to their understanding of the problem and their ability to articulate their interests clearly. Minority views within a group may be suppressed rather than explored, with conflict among group members being viewed as a problem to be overcome rather than as an opportunity for providing additional clarity regarding facts and values relevant to the decision at hand. Thus, this dissenting perspective suggests that differences in the expressed values and objectives of participants should be investigated carefully and fully documented, even at the risk of creating dissension within the group, because these same differences may serve as the building blocks for providing decision makers with sufficient insight about the differing viewpoints that they can craft a preferred and long-lasting agreement.

5.2. Heuristics, Affect, and Evaluability

Valuations of risk management options are not made solely on the basis of utility-maximizing cognitive judgments. This is one of the main messages of work over the past twenty-five years in behavioral decision making, which has explored the use (by experts as well as laypeople) of a variety of simplifying judgmental rules or heuristics that help individuals know what to do when faced with complex decisions. Individuals may rely on simplifying strategies to construct evaluative responses because the computational demands of the situation exceed their cognitive capabilities (March, 1978) or because they bring to the task a variety of conflicting goals that involve both deliberative and emotional responses: minimizing regret (Loomes and Sugden, 1982), maximizing their ability to justify a choice (Shafir, Simonson, and Tversky, 1993), and balancing desires to maximize accuracy yet minimize the effort required by the evaluation task (Payne, Bettman, and Johnson, 1993). Moral dimensions of the problem also are invoked in developing a contingent behavioral response, perhaps to the extent that participants are reluctant to make trades across different dimensions of the possible impacts (Baron and Spranca, 1997).

Recent research, building on the findings of neurologists such as Damasio (1994) and decision scientists such as Loewenstein (1996), also demonstrates a confounding of cognitive and affective assessments, as yet not well understood, that holds important implications for the evaluation of risk management alternatives. Research on the relation between affect and decision making shows that a key predictor of a person's (relative or absolute) valuation of an item often will be their general affective assessment of the good. The term *affect* is used here to describe the negative or positive reaction (i.e., the goodness or badness) associated with the stimuli (Finucane et al., 2000), rather than simply a more transitory emotional response (e.g., anger or mood swings). In general, affect may function as a cue for many important valuation choices, with evidence suggesting that items showing strong positive or negative evaluations on good-bad scales of affective intensity also tend to show a large disparity when comparing risk and benefit judgments.

Affective mapping has been shown to influence the importance placed on different aspects (dimensions) of a problem, which in turn can affect the overall evaluation of a risk management option. This is, in part, because affective considerations affect relative judgments of salience and precision which, in turn, influence how a stimulus enters into an overall, multidimensional valuation of an alternative. For example, Mellers, Richards, and Birnbaum (1992) have shown that weights in integrative valuation tasks are inversely proportional to their variance owing to the greater affective impression on judgments that are made by attributes with smaller variance.

Other relevant results on the topic of integrating values across the dimensions of complex risk judgments have been reported by Hsee (1996), who compared values assigned to the same two options when they were considered separately or together. In one of several insightful experiments, he asked subjects to assume they were music majors looking for a music dictionary. In a joint-evaluation condition, participants were shown two dictionaries (A and B) and asked how much they would be willing to pay for each. Respondents' mean willingness-to-pay was far higher for Dictionary B, presumably because of its greater number of entries. However, when one group of participants evaluated only A and another group evaluated only B, the mean willingness-to-pay was higher for Dictionary A. Hsee argues that this reversal provides evidence for the difficulty of making a choice based on the specified attribute for "number of entries" when the evaluator does not have a precise notion of how good or bad 10,000 (or 20,000) entries is. Thus, in the independent evaluation, more weight is given to the affective "defects" attribute, which translates easily into a good-bad response. Only under joint evaluation is the participant able to make an evaluative comparison and thereby see that option B is superior on the more important attribute.

This example calls into question the validity of valuation methods that ask participants to value only a single program or project: additional insight is provided through the display of multiple objectives across several alternatives because it enables respondents to make a more informed choice and to integrate more easily across cognitive and affective dimensions of the valuation problem. In addition, the concept of evaluability helps to explain why assessments showing multiple values, rather than a single or "true" value, for the same risk management alternative are often to be expected and, in general, underscores the significance of context in valuations of policy initiatives designed to reduce the risks of specified activities or technologies.

5.3. Flexibility and Learning

Learning, about both values and facts, is an important but neglected objective in many risk management settings. An emphasis on learning is consistent with the perspective of constructed preferences and a trust in the capability of both expert and public stakeholders to take in information critically, to think deeply, and then to consider revisions in the basis for their risk judgments, including weights on values, the expected magnitude of impacts, and the associated probabilities.

Sensitivity analyses, which typically are used as part of a formal risk analysis or as part of a cost-benefit study of risk management alternatives, reflect the influence of uncertainty on risk judgments and take account of learning that will take place over time. Decision analysts incorporate this same concept explicitly through reliance on value-of-information studies, which show how decisions can be made systematically about what sources of information to select and how much weight should be placed on them. Thus, if a value-of-information analysis on an uncertain event is low, there is little sense in spending substantial resources to reduce the uncertainty through collecting additional information. Concepts such as precautionary investment strategies, which (in situations where uncertainty is high and experts typically disagree) imply a commitment of current resources to safeguard against potentially adverse future outcomes, also reflect the importance of learning over time.

Adaptive management techniques provide another approach to management flexibility and learning that can influence society's valuation of proposed health and environmental risk options. Developed by the ecologists Holling (1978) and Walters (1986) in the context of coping with profound uncertainties in our understanding of ecological systems, an adaptive management perspective suggests that it is wise to develop strategies that will maximize learning over time and incorporate the inevitable surprises that are likely to come along. Thus, it is often advisable to incorporate adaptive management strategies into proposed plans as part of an ongoing, interactive valuation and selection process (rather than a one-time exercise), with close monitoring of results suggested in order to quickly learn from the results of management trials and to avoid costly failures.

Learning also has implications for the procedures followed as part of a risk valuation process. For example, a recognition of the importance of learning might affect the choice of a valuation process (e.g., favoring the participation of an ongoing small-group or values jury rather than a one-time survey) or the choice of participating stakeholders in an advisory group. Strategies for addressing a risk problem might be designed with objectives of flexibility and ease of adaptation in mind; for example, a more expensive engineering option might be selected because of the ability of machinery to adapt quickly to a variety of flow levels in a river or to a wide range of emission levels. Learning also can affect the design of participatory management structures for dealing with risk problems over time. One of the recommendations of a recent citizens stakeholder committee examining water-flow trade-offs between competing power, fish, recreation, and flooding objectives, for example, was for the provision of funding to be used by an ongoing Management Committee to oversee a program of ecological studies designed to reduce uncertainty in fisheries management results over time (McDaniels, Gregory, and Fields, 1999).

6. CONCLUSION: EMERGING APPROACHES TO VALUING RISKS

Risk management decisions are complex and difficult for many reasons, so perhaps it is no surprise that valuation studies are also complex and challenging. These challenges involve questions of social science, such as how issues are framed and public values are elicited; questions of natural science, such as how better information can be provided about the anticipated impacts of actions; and questions of politics, such as how expert and lay views will be integrated.

It would be reassuring if all the hard work of researchers, risk managers, and concerned stakeholders over the past several decades meant that valuation decisions for risk management initiatives were becoming easier. It is true that progress is being made on many fronts and that understanding of the many issues involved in soliciting citizens' values or representing consequences or integrating differing points of view is growing. However, it is equally true that the task of developing defensible valuation approaches remains daunting, in part because of what is being learned about the characteristics of a meaningful value. What is it, then, that might make any approach to risk valuation a success?

One place to look is to the psychological concept of validity. If an evaluative procedure measures what it is supposed to, then it should simultaneously include the most important features of the risk management problem under consideration (or face validity, a reference to objective-based structuring of the problem), reflect the same judgments and behaviors that would be made by the larger population that is being represented (predictive validity, such as could be achieved through a referendum among informed citizens), and be influenced by those concerns that should make a difference while being insensitive to irrelevant influences (construct validity, which seeks stability in responses over time and over equivalent measurement approaches). This is a tall order for even the simplest evaluation contexts, let alone ones as complex as to involve potential changes in quality-adjusted life years or biodiversity indices or risks to future generations.

A more comforting place to look is to the recent past. Not too long ago we tended to conduct valuations of risk management options, for example as part of analyses of regulatory alternatives or proposed standards, by looking at measures of expected fatalities and, if the analysts were unusually responsible, including associated estimates of morbidity. Environmental impacts were included only when they could easily be identified and, in many cases, were compared directly to expected income or employment or health benefits through simplistic monetary measures. Explicit objectives were seldom elicited or compared across participating stakeholders, attributes were rarely measured, and probabilistic judgments about impacts rarely were examined to understand their composition or assumptions. The diversity of psychological and affective factors contributing to perceptions of risk was largely ignored. Uncertainty was generally hidden from the view of decision makers by deterministic analyses and point estimates of impacts. Little attention was paid to public involvement because the views of technically well-informed experts were counted more heavily. No distinctions were made for quality-based adjustments in health or for changes to individuals' utility associated with supposedly equivalent gains and losses. Surveys of public views were rarely used to inform decision makers and, when conducted, paid little attention to questions of framing or context. In the unusual event that risk-management initiatives were subject to voting or open analysis, they were presented as single options and little capability existed for stakeholder groups to ask questions, to become better informed over time, or to define new and preferred alternatives.

Many of these what-are-now-obvious problems in risk assessment have been addressed, and, as a result, the standards for a defensible valuation study have been raised considerably. This is certainly comforting and is reason for celebration: the result is that relatively more lives are being saved at a lower cost and, when they are not, society is more aware of the psychological, epidemiological, political, or engineering reasons for why this is so.

The decided absence of jubilation among the community of risk analysts and managers, however, is a reflection of the discomforting thought that underlying the valuation process remain a host of unaddressed conflicts among methods and a rich storehouse of unanswered questions. Some of the methodological issues are being addressed by new valuation initiatives, such as stated-preference approaches that combine contingent valuation and multiattribute techniques or public participation approaches that combine input from representative groups of technical experts and laypersons with the results of community-based workbooks or broad-based surveys. Progress on other valuation issues is coming from basic research into how people form preferences, how the cognitive and affective dimensions of value combine to influence risk management choices, and how the diverse consequences of risk policies are integrated over periods of time.

The unanswered questions are more troubling: How do we know that expressed values for a complex, unfamiliar risk source have been constructed defensibly? What does it mean to have a well-structured or stable value? By what criteria should a valuation strategy be selected? How do we know that the questions we ask are the same ones that respondents answer? How do we know that the best attribute measure has been selected? How do we know that consequence uncertainty has been presented clearly or that enough has been done to make hidden value agendas explicit? How much information about the impacts of a project is needed to make an informed judgment? What is stakeholder participation intended to achieve? What is the role of discussion and affect in stakeholder deliberations? How do we know that participants are up to the judgmental and valuation tasks we have assigned them? How does a policy maker know that the values of stakeholders have been incorporated faithfully into policies?

Answers to these questions are as easy or as hard as we want to make them. The best that can be done, perhaps, is to forge ahead and to take some solace in the finding that, for valuation problems as tough as most of those facing risk managers, even a little insight and guidance can help a great deal.

References

Adamowicz, W., Boxall, P., Williams, M., and Louvière, J. 1998. Stated preference approaches for measuring passive use values: Choice experiments and contingent valuation. *American Journal of Agricultural Economics*, 80: 64–75.

Alhakami, A. S., and Slovic, P. 1994. A psychological study of the inverse relationship between perceived risk and perceived benefit. *Risk Analysis*, 14 (6): 1085–96.

Baron, J., and Spranca, M. 1997. Protected values. *Organizational Behavior and Human Decision Processes*, 70: 1–16.

Benzion, U., Rapoport, A., and Yagil, J. 1989. Discount rates inferred from decisions: An experimental study. *Management Science*, 35: 270–84.

Berger, M., et al. 1987. Valuing changes in health risks: A comparison of alternative measures. *Southern Economic Journal*, 53: 967–84.

Blomquist, G. 1979. Value of life saving: Implications of consumption activity. *Journal of Political Economy*, 87: 540–58.

Bostrom, A., Fischhoff, B., and Morgan, M. G. 1992. Characterizing mental models of hazardous processes: A methodology and an application to radon. *Journal of Social Issues*, 48 (4): 85–110.

Bradbury, J. 1989. The policy implications of differing conceptions of risk. *Science, Technology, and Human Values* 14: 380–99.

Brown, T., Peterson, G., and Tonn, B. 1995. The values jury to aid natural resource decisions. *Land Economics*, 71: 250–60.

Chuenpagdee, R., Knetsch, J., and Brown, T. (2001). Environmental damage schedules: Community judgments of importance and assessments of losses. *Land Economics*, 77 (1): 1–10.

Coglianese, C. 1999. The limits of consensus. *Environment*, 41: 28–33.

Dake, K. 1991. Orienting dispositions in the perception of risk: An analysis of contemporary worldviews and cultural biases. *Journal of Cross-Cultural Psychology*, 22: 61–82.

Damasio, A. R. 1994. *Descartes' error: Emotion, reason, and the human brain*. Avon, New York.

Dunlap, R. E., Gallup Jr., G. H., & Gallup, A. M. 1993. Of global concern: Results of the health of the planet survey. *Environment*, 35: 7–15, 33–9.

Edwards, W., and von Winterfeldt, D. (1987). Public values in risk debates. *Risk Analysis*, 7 (2): 141–58.

Finucane, M., Alhakami, A., Slovic, P., and Johnson, S. (2000). The affect heuristic in judgments of risks and benefits. *Journal of Behavioral Decision Making*, 13: 1–17.

Fischhoff, B. 1984. Setting standards: A systematic approach to managing public health and safety risks. *Management Science*, 30: 823–43.

Fischhoff, B. and Cox, L. A., Jr. (1985). Conceptual framework for benefits assessment. In J. Bentkover, V. Covello, and J. Mumpower, eds., *Benefits Assessment: The State of the Art*. D Reidel Publishing, Dordrecht, The Netherlands.

Fischhoff, B., and Furby, L. 1988. Measuring values: A conceptual framework for interpreting transactions with special reference to contingent valuation of visibility. *Journal of Risk and Uncertainty*, 1: 147–84.

Fischhoff, B., Quadrel, M., Kamlet, M. Loewenstein, G., Dawes, R., Fischbeck, P., Kleeper, S., Leland, J., and Stroh, P. 1993. Embedding effects: Stimulus representation and response mode. *Journal of Risk and Uncertainty*, 6: 211–34.

Fischhoff, B., Welch, N., and Frederick, S. 1999. Construal processes in preference assessment. *Journal of Risk and Uncertainty*, 19: 139–70.

Flynn, J., Slovic, P., and Mertz, C. K. 1994. Gender, race, and perception of environmental health risks. *Risk Analysis*, 14 (6): 1101–8.

Freeman, A. M. 1993. *The Measurement of Environmental and Resource Values: Theory and Method*. Resources for the Future, Washington, DC.

Gerking, S., DeHaan, M., and Schulze, W. 1988. The marginal value of job safety: A contingent valuation study. *Journal of Risk and Uncertainty*, 1: 185–99.

Gowda, R. 1999. Heuristics, biases, and the regulation of risk. *Policy Sciences*, 32: 59–78.

Graham, J. D., Carrothers, R., and Evans, J. 1999. Valuing the health effects of air pollution. *Risks in Perspective*, vol. 7 (July), Harvard Center for Risk Analysis, Cambridge, MA.

Gregory, R. 2000. Using stakeholder values to make smarter environmental decisions. *Environment*, 42: 34–44.

Gregory, R., and Keeney, R. L. 1994. Creating policy alternatives using stakeholder values. *Management Science*, 40: 1035–48.

Gregory, R., Flynn, J., and Slovic, P. (1995). Technological stigma. *American Scientist*, 83: 220–23.

Gregory, R., Flynn, J., Johnson, S. M., Satterfield, T. A., Slovic, P., and Wagner, R. 1997. Decision pathway surveys: A tool for resource managers. *Land Economics*, 73 (2): 240–54.

Gregory, R., Lichtenstein, S., and MacGregor, D. G. 1993. The role of past states in determining reference points for policy decisions. *Organizational Behavior and Human Decision Processes*, 55: 195–206.

Hammond, J., Keeney, R. and Raiffa, H. 1999. *Smart Choices: A Practical Guide to Making Better Decisions*. Harvard Business School Press, Boston.

Holling, C. S. 1978. *Adaptive environmental assessment and management*. Wiley, Chichester.

Hsee, C. K. 1996. Elastic justification: How unjustifiable factors influence judgments. *Organizational Behavior and Human Decision Processes*: 66: 122–9.

Irwin, J., Slovic, P., Lichtenstein, S., and McClelland, G. 1993. Preference reversals and the measurement of environmental values. *Journal of Risk and Uncertainty*, 6: 5–18.

Johnson, R. 1998. Task-specific information processing in multiple-response stated-preference surveys. Paper presented at EPA conference, *Alternatives to Traditional Contingent Valuation in Environmental Valuation* (October 1998), Nashville, Tennessee.

Jones-Lee, M. 1989. *The Economics of Safety and Physical Risk*. Basil Blackwell, Oxford.

Kahneman, D., and Knetsch, J. 1992. Valuing public goods: The purchase of moral satisfaction. *Journal of Environmental Economics and Management*, 22: 57–70.

Kahneman, D., and Tversky, A. 1979. Prospect theory: An analysis of decision under risk. *Econometrica*, 47 (2): 263–91.

Kahneman, D., Knetsch, J. L., and Thaler, R. H. 1990. Experimental tests of the endowment effect and the Coase theorem. *Journal of Political Economy*, 98 (6): 1325–48.

Kasperson, R. E., Renn, O., Slovic, P., Brown, H. S., Emel, J., Goble, R., Kasperson, J X , and Ratick, S. 1988. The social amplification of risk: A conceptual framework. *Risk Analysis*, 8: 177–87.

Keeney, R. 1990. Mortality risks due to economic expenditures. *Risk Analysis*, 10: 147–59.

Keeney, R. 1992. *Value-Focused Thinking: A Path to Creative Decisionmaking*. Harvard University, Cambridge, MA.

Keeney, R., and Raiffa, H. 1993. *Decisions with Multiple Objectives*. Cambridge University Press, New York.

Keeney, R., and von Winterfeldt, D. 1986. Why indirect health risks of regulations should be examined. *Interfaces*, 16: 13–27.

Keeney, R., and von Winterfeldt, D. 1990. Eliciting probabilities from experts in complex technical problems. *IEEE Transactions on Engineering Management*, 38: 191–201.

Keeney, R., von Winterfeldt, D., and Eppel, T. 1990. Eliciting public values for complex policy decisions. *Management Science*, 36: 1011–30.

Knetich, J. 1990. Environmental policy implications of disparities between willingness to pay and compensation demanded measures of value. *Journal of Environmental Economics and Management*, 18 (3): 227–37.

Kopp, R. J., and Smith, V. K. 1993. *Valuing Natural Assets: The Economics of Natural Resource Damage Assessment*. Resources for the Future, Washington, DC.

Kunreuther, H., and Slovic, P. 1996. Science, values, and risk. *Annals of the American Academy of Political and Social Science*, 545: 116–25.

Loewenstein, G. F. 1988. Frames of mind in intertemporal choice. *Management Science*, 34: 200–14.

Loewenstein, G. 1996. Out of control: Visceral influences on behavior. *Organizational Behavior and Human Decision Processes*, 65: 272–92.

Loomes, G., and Sugden, R. 1982. Regret theory: An alternative theory of rational choice under uncertainty. *Economic Journal*, 92: 805–24.

Luckert, M., and Adamowicz, W. 1993. Empirical measures of factors affecting social rates of discount. *Environmental and Resource Economics*, 2: 1–21.

March, J. 1978. Bounded rationality, ambiguity, and the engineering of choice. *Bell Journal of Economics*, 9: 587–608.

McDaniels, T. L. 1996a. A multiattribute framework for evaluating the environmental impacts of electric utilities. *Journal of Environmental Management*, 46: 57–66.

McDaniels, T. L. 1996b. The structured value referendum: Eliciting preferences for environmental policy alternatives. *Journal of Policy Analysis and Management*, 15 (2): 227–51.

McDaniels, T. L., Gregory, R. S., and Fields, D. 1999. Democratizing risk management: Successful public involvement in local water management decisions. *Risk Analysis*, 19 (3): 497–510.

Mellers, B., Richards, V., and Birnbaum, J. 1992. Distributional theories of impression formation. *Organizational Behavior and Human Decision Processes*, 51: 313–43.

Merkhofer, M., and Keeney, R. 1987. A multiattribute utility analysis of alternative sites for the disposal of nuclear wate. *Risk Analysis*, 7: 173–94.

Mitchell, R. C., and Carson, R. T. 1989. *Using Surveys to Value Public Goods: The Contingent Valuation Method*. Resources for the Future, Washington, DC.

Mitchell, R. C., and Carson, R. T. 1995. Current issues in the design, administration, and analysis of contingent valuation surveys. In P. Johansson, B. Kristrom & K. Maler, eds., *Current Issues in Environmental Economics*. Manchester University Press, New York.

Morgan, G., and Henrion, M. 1990. *Uncertainty*. Cambridge University Press, New York.

Morgan, G, Fischhoff, B., Bostrom, A., Lave, L., and Atman, C. J. 1992. Communicating risk to the public. *Environmental Science and Technology*, 26: 2048–56.

Morgan, K., DeKay, M., Fischbeck, P., Morgan, G., Florig, K., and Fischhoff, B. 1999. Development of a method for risk ranking. Unpublished manuscript, Dept. of Engineering and Public Policy, Carnegie Mellon University, Pittsburgh.

Opaluch, J., Swallow, S., Weaver, T., Wessells, C., and Wichelns, D. 1993. Evaluating impacts from noxious facilities: Including public preferences in current siting mechanisms. *Journal of Environmental Economics and Management*, 24: 41–59.

Payne, J. W., Bettman, J. R., and Johnson, E. J. 1992. Behavioral decision research: A constructive processing perspective. *Annual Review of Psychology*, 43: 87–131.

Payne, J. W., Bettman, J. R., and Johnson, E. J. 1993. *The adaptive decision maker*. Cambridge University Press, New York.

Peelle, E. 1996. Beyond the Nimby Impasse II: Public participation in an age of distrust. *Proceedings of the Spectrum 88 Conference*, Pasco, WA, pp. 575–82.

Peterson, G., and Brown, T. 1998. Economic valuation by the method of paired comparison, with emphasis on evaluation of the transitivity axiom. *Land Economics*, 74: 240–61.

Raiffa, H. 1982. *The Art and Science of Negotiation*. Harvard University Press, Cambridge, MA.

Ritov, I., and Kahneman, D. 1997. How people value the environment. In M. Bazerman, D. Messick, A. Tenbrunsel, and K. Wade-Benzoni, eds., *Environment, Ethics and Behavior: The Psychology of Environmental Valuation and Degradation*. New Lexington Press, San Francisco.

Rutherford, M., Knetsch, J., and Brown, T. (1998). Assessing Environmental Losses: Judgments of Importance and Damage Schedules. *Harvard Environmental Law Review*, 22: 51–101.

Schacter, R. 1986. Evaluating influence diagrams. *Operations Research*, 34: 871–82.

Schuman, H., and Presser, S. 1996. *Questions and Answers in Attitude Surveys: Experiments in Question Form, Wording, and Context*. Sage, Thousand Oaks, CA.

Shafir, E., Simonson, I., and Tversky, A. 1993. Reason-based choice. *Cognition*, 49: 11–36.

Slovic, P. 1987. Perception of risk. *Science*, 236: 280–5.

Slovic, P. 1995. The construction of preference. *American Psychologist*, 50: 364–71.

Slovic, P. 1992. Perception of risk: Reflections on the psychometric paradigm. In S. Krimsky and D. Golding, eds., *Social Theories of Risk*, pp. 117–52. Praeger, New York.

Slovic, P., and Gregory, R. 1999. Risk analysis, decision analysis, and the social context for risk decision making. In J. Shanteau, B. A. Mellers, and D. A. Schum, eds., *Decision Science and Technology: Reflections on the Contributions of Ward Edwards*, pp. 353–65. Kluwer Academic, Boston.

Slovic, P., Fischhoff, B., and Lichtenstein, S. 1985. Characterizing perceived risk. In R. Kates, C. Hohenemser, and J. Kasperson, eds., *Perilous Progress: Technology as Hazard*. Westview, Boulder, CO.

Smith, V. K., and Johnson, R. 1988. How do risk percpetions respond to information? The case of radon. *The Review of Economics and Statistics*, 70: 1–8.

Sunstein, C., Kahneman, D. and Schkade, D. 1998. Assessing punitive damages (with notes on cognition and valuation in law). *The Yale Law Journal*, 107: 2071–153.

Tversky, A., Sattath, S., and Slovic, P. 1988. Contingent weighting in judgment and choice. *Psychological Review*, 95 (3): 371–84.

Viscusi, K. 1993. The value of risks to life and health. *Journal of Economic Literature*, 31: 1912–46.

Viscusi, K., Magat, W., and Huber, G. 1987. An investigation of the rationality of consumer valuations of multiple health risks. *Rand Journal of Economics*, 18: 465–79.

Viscusi, K., Magat, W., and Huber, G. 1991. Worker learning and compensating differentials. *Industrial Labor Relations Review*, 45: 80–96.

von Winterfeldt, D. 1992. Expert knowledge and public values in risk management: The role of decision analysis. In S. Krimsky and D. Golding, eds., *Social Theories of Risk*, pp. 321–42. Praeger, Westport, CT.

Walters, C. 1986. *Adaptive Management of Renewable Resources*. Macmillan, New York.

Webler, T. 1997. Organizing public participation: A critical review of three handbooks. *Human Ecology Review*, 3: 245–54.

Wildavsky, A. 1979. No risk is the highest risk of all. *American Scientist*, 67: 32–7.

Wilson, R., and Crouch, E. A. C. 1982. *Risk/Benefit Analysis*. Ballinger, Cambridge, MA.

8

The Role of Efficiency in Risk Management

John D. Graham, Per-Olov Johansson,
and Junko Nakanishi

1. INTRODUCTION

Citizens throughout the world are demanding that their governments do a better job of investing resources in preventing, reducing, or mitigating risks to human health, safety, and the environment. If resources were limitless, nations could pursue a risk-free society. Yet citizen desires for risk reduction are constrained by the scarcity of resources of various forms (e.g., skilled and unskilled labor, capital, and even political energy). In fact, the resources devoted to protecting health, safety, and the environment could be used to serve other valuable public and private purposes, such as improving the quality of education, providing more day care services for parents and children, enhancing transportation networks, accelerating computer literacy, or buttressing criminal justice systems. The resources devoted to accomplishing ALL social objectives, including risk reduction, should be expended wisely and thus inefficient expenditures that accomplish little risk reduction should be questioned, unless policy makers have confidence that the diverted resources will be drawn from other sectors that are even less efficient than the proposed risk management measure. This comparison of alternative resource investments is exactly what analyses of economic efficiency do.

In recent decades, there has been significant progress, both intellectually and practically, in efforts to define and implement *efficient* approaches to managing risks. The purpose of this chapter is to review, for the nontechnical and noneconomist reader, the state of the art of the art of scientific approaches to efficiency in risk management. A particular goal of this chapter is to explain the technical differences between cost-benefit analysis and cost-effectiveness analysis, since these two tools of efficiency analysis are both widely used. In the process of providing this review, we reveal the assumptions that underpin economic studies of risk management and offer insight into the particular challenges posed by using quantitative risk information in economic studies.

We are aware that much work remains to be done, scientifically and politically, to advance the role of efficiency in risk management and that many obstacles stand in the way of efficient approaches to risk management. Indeed, some scholars and activists continue to question whether economic efficiency does or should play any significant role in public policy toward health, safety, and environmental risks. We do not consider this viewpoint to be on the rise (e.g., every U.S. President since Gerald Ford (1974–6) has, through executive order, subjected major federal regulations of risk to some type of efficiency analysis prior to adoption), and in fact we offer some evidence that international interest in efficiency is increasing. We explain the philosophical (utilitarian) foundation of economic efficiency, a foundation that is reflected (implicitly) in any society that permits a degree of free enterprise and capitalism. We also rely on key references to the philosophical literature for those seeking a more indepth philosophical case for efficiency in public policy.

Although our focus is economic efficiency, we respect the legitimacy of various "nonefficiency" considerations in public policy toward risk (e.g., various notions of fairness, justice, and equity) and thus advocate efficiency as an important but partial determinant of resource-allocation decisions (Lave, 1996). In this chapter, we do not address the difficult question of how to combine efficiency and equity in risk management, recognizing that this challenge is a vast research agenda that, lacking analytical resolution, currently remains in the province of accountable public officials.

The material in this chapter, though written for a diverse audience of life scientists, physical scientists and engineers, lawyers, social scientists, and practitioners, does not shy away from theoretical and technical issues, since an understanding of these issues is central to understanding the strengths and limitations of efficiency analysis. Although the material in this chapter will be quite familiar to economists, operations researchers, decision scientists, and policy analysts, it is our hope that even these audiences will find insight into our perspective about why cost-effectiveness analysis may have greater promise than cost-benefit analysis in many parts of the world and in international settings. Our theoretical and ethical discussions are accompanied by case studies in order to provide practical insight into how such studies are done, the nature of the typical input data, and the approach taken to selecting and comparing policy alternatives. An extensive list of references to classic articles and texts is provided for readers who are interested in digging more deeply into the subject.

2. TECHNICAL DEFINITION OF EFFICIENCY

The *American Heritage Dictionary* defines "efficient" as "acting or producing effectively with a minimum of waste, expense, or unnecessary effort" or "exhibiting a high ratio of output to input." This denotation captures much of the intuition behind the notion of efficiency, but, as we shall see, more precision is required to explain how the economic concept of efficiency applies to complex problems of risk management.

Imagine a steel plant that is polluting air and water. Suppose that there are ways to prevent or reduce pollution from the plant but that they are associated with some "cost" (expenditure of labor and capital). If pollution prevention is not costly to the firm (and in some cases firms have found that pollution prevention actually causes a net saving of labor and capital), then the typical economic assumption is that the firm will prevent pollution voluntarily, without the need for public policy. The more difficult issue arises when a firm has exhausted the cost-saving methods of pollution prevention and is still polluting air and water, though to a lesser extent. How should society decide how much pollution, if any, should be permitted by the firm? Stated differently, how much additional investment of labor and capital should society be prepared to invest in additional pollution prevention, where prevention could occur through changes in how much steel is consumed, how steel is made, or in how pollution is treated at the plant.

There is no obvious answer to this question because the benefits of improved environmental quality, which may be subjective as well as tangible, are not easily observed or measured. Even if the adverse effects of the pollution can be measured in physical units, such as additional cases of disease among residents near the plant, there is no obvious market where the price or value of a "unit" of environmental quality (or a unit of disease) is recorded. It would seem that we have to compare benefits and costs measured in different dimensions, which precludes a rigorous efficiency determination.

Economists have proposed the following solution to this conundrum (see, generally, Johansson, 1991). Assume (1) that improved environmental quality enhances citizen satisfaction ("utility") and (2) that higher costs of producing steel reduce producer and/or consumer utility (because producers/consumers will have fewer resources to spend on other goods and services that enhance their satisfaction). If each citizen in the community (including residents, steel consumers, and producers) is informed of a well-defined proposal to reduce pollution at a specified cost to each of them, it is assumed that each citizen is able to judge whether the change in pollution and cost, considered together, increase or decrease their individual level of utility. In this full-information case, it is assumed that citizens know as much as scientists do about the adverse effects of pollution on human health and ecosystems.

Even with these strict assumptions, we face the problem that we cannot observe or measure how a person's utility (welfare) is affected by the proposal. Yet economists propose that we transform the value of the improved environmental quality from unobservable units of utility to observable units of money. If this is feasible, we can compare the monetary value of the environmental improvement and the costs of the improvement and thus determine, for each individual, whether the proposal will make them better off or worse off.

One monetary measure is obtained by letting the individual make a hypothetical payment in exchange for experiencing an improvement in environmental quality. The payment is of such size that the individual neither gains nor loses utility (net) from the environmental improvement (after making the hypothetical payment). This particular payment is the individual's maximal willingness to pay (WTP) for the considered environmental improvement. It is a hypothetical payment because it may never be paid; analysts simply use it as a way to define the monetary value to the individual of the environmental improvement.

2.1. Single-Person Economy

Now suppose that there is just a single person in our economy and that all of the utility and cost consequences of the proposal will be experienced by that individual. If the individual's hypothetical payment exceeds the cost of the proposal, the proposal is said to be an improvement in efficiency. This is equivalent to saying that the individual's WTP for the improvement exceeds the cost of achieving the improvement. Transforming the benefits from units of utility to monetary units, we have arrived at a way to measure and compare the benefits and costs of the project.

In this single-person economy, we should invest in continuous improvements in environmental quality until the willingness to pay for an additional improvement is equal to the cost of achieving this improvement. If we proceed any further, we will find that the person will be made worse off by investing in additional reductions in pollution.

The efficiency rule just defined is a variant of the Samuelson condition for the optimal provision of a public good, referring to Nobel Prize-winning economist Paul Samuelson. Note that air quality is a "public good" (not because it may be required by the government) but because when it is provided to some citizens in a community it is virtually impossible to prevent other citizens in the same community from enjoying the benefits (i.e., benefits are "indivisible" and, once provided, are extended to additional "consumers" at virtually no cost).

2.2. Multiperson Economy

When the welfare of many individuals is at stake, the efficiency determination becomes more complicated. The people who experience the pollution may be different from the people who produce or consume the steel or steel-related projects. Moreover, individuals may differ in how much they care about pollution and consumption of steel-related products.

The original definition of economic efficiency, developed by European economists Nicholas Kaldor and John Hicks, called for the summation of the WTP of all citizens for the environmental improvement, and a comparison of that sum to the overall cost of the project. If the aggregate WTP in society exceeds the cost, the project is an efficiency improvement. In complex situations where projects induce changes in relative prices, gainers might not be able to compensate losers. For discussion of this point, see Broadway (1974) and Broadway and Bruce (1984).

In the case of a two-person economy, where citizen 1 is $10 better off and citizen 2 is $9 worse off, it is possible to imagine citizen 1 compensating citizen 2 for his loss while still leaving citizen 1 better off with the project. Unless actual compensation is carried out, an efficiency improvement may result in losers as well as gainers.

2.3. Jobs: Cost or Benefit?

One of the counterintuitive aspects of efficiency analysis is that a job created for someone who works to prevent pollution at a steel mill is treated as a cost (an investment of labor input), not a benefit. Doesn't the increased corporate compliance spending that results from environmental regulation of industry create employment "benefits" since jobs will be created at firms that supply pollution-control equipment to the steel mill or at the mill to monitor or operate the equipment or in the government to oversee the firm?

From an economic perspective, the human energy and skill that is devoted to environmental compliance/innovation has an "opportunity cost," the value of that same degree of human energy and skill devoted to an alternative, productive function in the economy. The monetary value of the additional labor devoted to these other functions (making cars, devising computer programs, or whatever) is measured (roughly) by the amount of wages firms in these other sectors are prepared to pay a worker who might otherwise be hired for pollution prevention. In a full-employment economy such as the one that the United States and parts of Europe were recently experiencing, environmental regulations induce shifts in where scarce labor in the economy is allocated, but they are unlikely to cause net changes in overall employment levels. Analyses of economic efficiency typically assume a full-employment economy, (Mishan, 1994). Thus, the social value of adding labor input to pollution prevention is measured by the social value of the diminished pollution, not by the wages provided to the worker engaged in pollution prevention.

If the economy is not fully employed, the analysis becomes more complicated, but it should be apparent that there is still an opportunity cost to allocating additional labor to pollution prevention (Mishan, 1994). It is measured by the depressed yet nonzero wages that are required to induce even unemployed workers to trade their leisure time for the pollution prevention work. Note, however, that the labor costs of pollution prevention will be greater when an economy is fully employed because wages will be higher than when unemployed labor is widely available. For this reason, analyses of economic efficiency, those taking a societal perspective, always treat jobs as a cost, not a benefit (Gold et al., 1996).

Another way to think about the unemployed case is to consider whether the government should hire the idle workers to prevent pollution or perform other productive public functions such as offer improved day care or educational services. The challenge, from an efficiency perspective, is to compare the social value of the additional pollution prevention to the social value of better day care or enhanced education of school children.

2.4. Accounting for Who Wins and Loses

Historically, many economists have regarded the "compensation question" (winners versus losers) as a distributional matter that is beyond the purview of economics. From a strictly utilitarian perspective, who wins and loses is irrelevant. The current practice of cost-benefit analysis rests on the premise that government should care (at least to some extent) about the overall efficiency of the project, regardless of how the gains and losses are distributed among citizens in society. This indifference toward winners and losers has provoked criticism that has caused efficiency determinations to be rejected as a sole determinant of governmental decision making. Politicians, for example, are deeply concerned about who loses and wins as well as about the overall efficiency calculation.

Theorists have sought to solve the distributional question through a construct called a social welfare function. Such a function yields someone's (possibly Parliament's) ranking of social states, taking into account both concern for efficiency and concern for a "fair" distribution of gains and losses. In effect, weights are assigned to gains and losses experienced by different people. This approach assumes that there is an accepted social welfare function, that we can measure changes in utility, and that we can meaningfully compare utility levels across individuals.

Another Nobel Prize-winning economist, Kenneth Arrow, is well known for a theorem proving that there is no social welfare function that satisfies several seemingly attractive criteria (e.g., transitivity in choice, no dictatorship, and several others). Thus, although public economists continue to search for social welfare functions to account for distributional concerns, no such approach commands broad acceptance in intellectual or political circles (Mueller, 1989; Johansson, 1991; Johannesson, 1996). Perhaps it should not surprise us that fairness is an elusive concept because it means very different things to different people.

2.5. Summary

A governmental proposal aimed at reducing risks to human health, safety, and the environment is said to be an efficiency improvement if those citizens who benefit from the proposal would be willing to pay the entire costs of the proposal (assuming they were required to do so). The ethical rationale is utilitarian in nature. Even if a proposal is not an efficiency improvement, it may still be worth adopting out of considerations of fairness and equity or other distributional considerations. If a proposal is an efficiency improvement, it may still be objectionable if it is judged to have unacceptable distributions of gains and losses. Thus, efficiency is not a necessary or sufficient condition for good risk management. It is best considered a contribution to resource-allocation decisions rather than a necessary or sufficient condition for public choice (Lave, 1996).

The informational requirements of efficiency analysis are substantial. In order to implement this criterion, analysts need to determine (1) what types and degrees of risks will occur under each alternative, (2) the preferences (e.g., WTP values) for those people who will experience the risk reduction or somehow be affected by the alternatives, and (3) what the additional costs of implementing each alternative would be. If multiple risk-management options are considered, the most efficient one has the largest degree of net benefits.

3. COMPLICATIONS IN MAKING EFFICIENCY DETERMINATIONS

The classical economic approach to making efficiency determinations was not designed for application to public decisions involving consequences for human health, safety, and the environment. The earliest applications, in the mid-nineteenth century, addressed the building of bridges and other social infrastructure. In the United States, applications began in the 1930s for investments in water resource projects such as building a new dam. Complications have arisen in health and environmental applications that have been the subject of continuing inquiry. We discuss here several major complications that arise repeatedly in such applications.

3.1. Willingness to Pay or Willingness to Accept?

The measure of benefit we have discussed is a citizen's maximal willingness to pay for an environmental improvement. However, an alternative measure of benefit is the minimum amount of compensation (willingness to accept or WTA) that a citizen would demand in exchange for a specified deterioration in environmental quality. This is the smallest amount of money that the individual would accept in order to be as well off as he or she would have been without the deterioration in environmental quality.

It is well demonstrated, theoretically and empirically, that these two measures of benefit, WTP and WTA, will diverge in some situations. For example, the residents living near a steel plant, those who are most adversely affected by air and water pollution from the plant, may report WTA values that are larger than WTP values, since WTP is constrained by a person's wealth, while there is no hard constraint on a WTA value.

Which is the "correct" measure of benefit, WTP or WTA? There is no clear conceptual answer to this question because it is not obvious who should be assigned the initial rights (the owner of the plant or the resident living near the plant). A variety of practical arguments have persuaded practitioners to favor the WTP over the WTA measure, but the normative argumentation for this preference is not entirely convincing (Mitchell and Carson, 1989).

3.2. Uncertainty and Risk

The adverse effects of the pollution from a steel plant are not typically known by science with any degree of certainty. Imagine a simple world with two health states, good health and bad health, and a project to reduce pollution from the plant that will increase the probability that people will be in good health. In turn, this will increase the expected utility of the residents living near the plant, since expected utility is the weighted average of the utility with good health and the utility with bad health (holding other factors, such as prices and income, constant), using the probabilities of good and bad health, respectively, as weights.

Several measures of the monetary value of a change in expected utility have been defined. The most popular measure, in practice, is the maximum amount of money to be paid such that expected utility with the higher probability of good health (and the payment) is equal to the expected utility of the smaller probability of good health. This is called the noncontingent compensation variation or simply an option price, which is attractive in situations were large numbers of people are exposed to relatively small probabilities of adverse events. Alternative measures of monetary value have been developed for settings where small numbers of people are exposed to large probabilities of adverse outcomes, or when low probabilities of catastrophic outcomes involving entire communities or countries are at stake. These alternative measures of monetary value permit a kind of risk premium for individuals who are risk averse with respect to actuarially fair gambles. In public-choice problems where risks can be diversified, the case for risk-averse behavior at the societal level is diminished (Arrow and Lind, 1970).

Early theorists drew a distinction between risky events, where probabilities are known by science, and uncertain events, where the probabilities of events are unknown. This distinction has been eroded owing to the emergence of subjective (judgmental) notions of probability (e.g., Bayesian statistics) combined with the development of practical tools to elicit subjective probabilities from scientists and engineers (Raiffa, 1968; Morgan and Henrion, 1990). Yet many analysts continue to distinguish different types of uncertainty in their analyses, such as not knowing the particular draw from a known probability distribution and not knowing the form (or parameters) of a probability distribution from which draws will be taken. The latter type of uncertainty is particularly common in risk assessment.

3.3. Treatment of Time Preference

If resources are consumed today to achieve health and environmental improvements in the future, then a procedure is needed to compare current costs to future benefits. A similar dilemma emerges if we wish to incur health risks today in order to achieve economic, health, environmental, or other benefits tomorrow.

There are two rationales for applying a discount rate to the future compared to today. One is the observation that consumers tend to prefer immediate as opposed to deferred gratification, leading to the concept of a "consumer rate of interest" or time preference. The second rationale is rooted in investment thinking. If resources are invested rather than consumed, they can accumulate interest, allowing more resources to be devoted to the welfare of people in the future. There is thus a "marginal rate of return on investment" that is the price we pay for expending resources now rather than investing them for future payoff.

In the discussion that follows, we distinguish consequences that occur within a single generation from consequences that might be imposed by the current generation on some future generation(s). In reality, generations are overlapping, and thus there will be investment linkages between now and the future. We suggest that it is useful, though, to consider differently the case where an investment is made now for several hundred years in the future.

4. INTRAGENERATIONAL DISCOUNTING

For governmental policies whose consequences (costs and benefits) will be experienced within a generation, there is a broad technical consensus that a real (inflation-adjusted) rate of discount should be used to convert future consequences into their present value (Gold et al., 1996). Even if the future consequences are health-related, there is a strong technical case for applying the same rate of real discount to money and health. If instead we were to apply a smaller rate of discount to future health than current economic consequences, then a perverse (yet logical) conclusion is that it always makes sense to delay adopting a health program that imposes immediate costs for deferred benefits. This Keeler-Cretin paradox, named for analysts at the Rand Corporation, underpins the convention that future health and economic consequences should be discounted at the same rate (Keeler and Cretin, 1983).

The choice of the appropriate discount rate is more controversial, in part because seemingly small changes in the discount rate can have a dramatic impact on the estimated efficiency of programs with consequences ten to fifty years in the future. Current analytical practice in the United States uses real annual discount rates that range from 0 to 7 percent, with base values of 3 or 5 percent being most typical (Gold et al., 1996). It is now customary for analysts to report whether efficiency estimates are robust with respect to plausible changes in the rate of discount.

5. INTERGENERATIONAL DISCOUNTING

The major consequences of some governmental policies, particularly those in the environmental arena, may extend beyond the current generation or even into multiple future generations. Examples of such policies include the efforts to protect global climate by reducing greenhouse gas emissions from the burning of fossil fuels and to regulate chemical exposures that may have developmental and reproductive effects on humans and other species.

There is far less consensus about how analysts should treat intergenerational time-preference issues (Portney and Weyant, 1999). The consumer time-preference rationale is inapplicable because the trade-off between current and future gratification is made between different consumers (those alive today and those who may not yet be alive). It is difficult to assess the preferences of people who have not yet been born. On the other hand, the investment rationale for some form of discounting is still applicable, though it is complicated by different viewpoints about whether future generations will be better off or worse off than current generations.

There is growing interest in determining whether the case for "sustainability" is compatible with the case for "efficiency." This interesting area is not yet worked out, in part because the concept of sustainability is variously defined and has not yet been framed with sufficient precision to permit a formal comparison with efficiency. It is apparent, however, that some notions of sustainability are incompatible with efficiency. For example, some have suggested that any policy proposal that entails consumption of an additional unit of nonrenewable resource (e.g., petroleum) should not be permitted, unless the proposal is accompanied by a proposal to create – at some cost – at least an equivalent amount of a renewable resource (e.g., solar energy) to substitute for the loss of the nonrenewable resource. The assumption is that society should be concerned now about whether future generations will have adequate nonrenewable resources available to meet their needs.

This version of sustainability will be inconsistent with efficiency in some applications because an argument can be made that future generations are likely to be wealthier and technologically more advanced than we are today. If a nonrenewable resource could be used now or possibly by future generations, the social value of use may be greater now, when we know the resource is needed, than when the combination of future ingenuity and wealth have created many viable substitutes for the scarce resource. If our desire is to offer a gift to future generations, a more efficient gift than saving a nonrenewable resource would be a gift of equivalent monetary value today but one that would grows at a secure interest rate. We offer these comments only to suggest that proponents of efficiency and sustainability could benefit from greater deliberation. Some of this mutual learning is beginning to occur in the field of environmental and natural resource economics.

6. THE ANALYTIC TOOLS

In order to support efficiency determinations, a variety of analytical tools have been developed and applied to a wide range of problems in medicine, public health, consumer protection, transport and occupational safety, and environmental protection. The tools share a grounding in the economic theory described earlier but differ in how strictly and completely they employ the tenets of social-welfare economics (Drummond et al., 1997).

6.1. Cost-Benefit Analysis

A risk-management policy is said to be an efficiency improvement when it brings about positive net benefits to the society. Cost-benefit analysis (CBA) is the tool employed to make this determination (Mishan, 1994). The distinctive feature of CBA is that an effort is made to express all the consequences of a policy change in monetary units, a direct implementation of the Kaldor-Hicks hypothetical compensation test (Leonard and Zeckhauser, 1986). As mentioned at the outset of this chapter, most people would reject a policy proposal to run public policy solely on the basis of the Kaldor-Hicks test, which is a variation on utilitarianism, precisely because people care about how wins and losses are distributed in society. The more limited claim that is made by most advocates of efficiency analysis is that policy makers should give at least some weight – how much is debatable – to the efficiency result in those cases where the distribution of wins and losses is considered undesirable or unfair.

For example, in a policy aimed at reducing human exposure to hazardous chemicals, costs may arise as a result of an increase in production costs initiated by an increase in the price of substitutes for the regulated hazardous chemicals or by additional capital or operating costs required to reduce emissions of hazardous chemicals into the environment. The benefits, on the other hand, might arise from reduced human mortality and morbidity (or ecological/natural resource damage) caused by diminished exposures to the chemicals.

The key practical challenges in CBA of risk-management policies are to (1) obtain appropriate quantitative estimates of risk levels associated with each policy and (2) obtain appropriate monetary values for the changes in health, safety, and environmental risks. During the last twenty years, significant scientific progress has been made in both of these areas, yet important challenges remain.

Informational Needs from the Risk Assessment Process. Analysts who practice CBA are not typically experts in the scientific and engineering aspects of risk management problems. Yet analysts will often need to draw information from risk assessment reports that are based on evidence from the physical and life sciences and engineering. Thus, high-quality CBAs often entail close collaboration between analysts and technical specialists. In health economics, for example, clinical specialists and economists often collaborate in the use of medical data from randomized controlled clinical trials. Even when such collaboration does not occur, it is often useful to have CBAs peer-reviewed by scientists/engineers/clinicians as well as economists to make sure that the most appropriate technical data have been identified and interpreted correctly.

During the last twenty years, it has become apparent that risk information is often generated in a manner that is not well suited for use in CBA. We discuss here some ways that risk assessment reports could be improved to enhance the quality of CBAs (Kopp, Krupnick, and Toman, 1997).

First, estimates of health, safety, and environmental risk may be highly uncertain, and this uncertainty needs to be conveyed to the author of the CBA. For example, it is well known that estimates of cancer risk from low levels of exposure to chemicals and ionizing radiation can vary by several orders of magnitude, depending upon what models are chosen to represent the dose-response function below the range of observation. Many risk assessment reports collapse this uncertainty into a single numerical estimate of risk. The properties of this single number may be unclear: it may be a worst-case estimate, it may be a plausible yet upper-bound estimate, it may be a "most-likely" estimate, it may be considered an "expected-value" estimate, it may be an optimistic or lower-bound estimate, or it may be a judgmental "best guess" that has no particular probabilistic interpretation (Pate-Cornell, 1996, 1998).

Point estimates of uncertain risks are problematic for CBA because the societal perspective employed in CBA usually imposes risk neutrality, which means that the only point estimate that is appropriate is the expected value of risk (taking into account all of the model (fundamental) as well as parameter (statistical) uncertainties in the risk estimation process). For the analyst to compute the expected value, the risk assessment report must contain a probability distribution on the true yet unknown risk that captures all the important uncertainties. If the societal perspective is adjusted to permit an appropriate degree of social risk aversion (e.g., in a problem involving low probabilities of potentially catastrophic losses) or "ambiguity aversion" (e.g., due to uneasiness about squishy probabilities), it is particularly critical that the CBA be informed by a risk assessment that contains a probability distribution on the true yet unknown risk. Fortunately, there has been significant progress in the subfield of formal uncertainty analysis in risk assessment (Morgan and Henrion, 1990; Cooke, 1991; Cullen and Frey, 1999).

Second, a risk assessment report may address multiple health and ecological consequences yet report quantitative information only on the most sensitive endpoint (e.g., a cancer or reproductive endpoint) or the most sensitive subpopulation or natural resource. The practice of reporting numbers only for the most sensitive endpoint diminishes the usefulness of a risk assessment report to practitioners of CBA. In order to make a comprehensive assessment of a policy's costs and benefits, quantitative information on all important health and ecological outcomes is required.

Third, risk assessment reports may report information on intermediate outcome measures (e.g., blood cholesterol levels, parameters of lung function, or indicators of ecosystem damage) that are not readily translated into health and ecological consequences that can be monetized. Wherever possible, it is useful for risk assessment reports to provide guidance to CBA practitioners on how intermediate outcome measures can be translated into outcomes (e.g., quality-adjusted life years saved) that can be translated by analysts into monetary units.

Finally, some risk assessment reports offer only a numerical determination as to whether exposures or stressors are above or below a "safe" level. The CBA practitioner may be given no information about how the probability or severity of outcomes will be affected by exposures or stressors that are progressively larger than the "safe" level. For purposes of CBA, a risk assessment report that reports probabilities of various health/ecological consequences is more useful than a report that simply determines whether a certain technology or exposure is safe.

Valuing Health and Ecology in Monetary Units. A challenging aspect of CBA is the exercise of translating human health or ecological outcomes into monetary units. Here we discuss some of the "valuation" methods that are now in use by CBA practitioners in the fields of health (Tolley, Kenkel, and Fabian, 1994) and environmental protection (Kopp, Krupnick, and Toman, 1997; Kopp and Smith 1993).

INDIRECT METHODS OF VALUATION. In some cases, it is feasible to use market prices to infer indirectly the value of an environmental good. Consider trips taken to a natural park, a case where the "travel cost method" has been employed to place a monetary value on recreational trips that might be jeopardized by pollution or poor health (Freeman, 1993; Johansson, 1991, 1995).

A trip is taken only if an individual perceives that the value of the visit exceeds (or possibly equals) the travel cost (money and time) of the visit. Thus, the travel cost yields a lower monetary bound for the unobserved utility of the visit. Because individuals visiting a particular site have different travel costs (e.g., depending upon how far away from the park they live), it is feasible to exploit this variation in travel cost to construct a demand function for trips to the site. The demand function reveals the maximum willingness to pay for each quantity of visits, not just the amount of travel cost incurred for each visit. The difference between a visitor's actual travel cost and their maximal willingness to pay for the visit is called the "consumer surplus" (Braden and Kolstad, 1991). Important assumptions often made are that visitors have homogeneous preferences and that people don't visit multiple sites, though both of these assumptions can be relaxed in more complex studies.

An alternative to the travel-cost method is the random utility model. This model assumes that the individual makes a choice between a fixed number of recreational sites based on what will give him or her the highest level of satisfaction (utility). The model can be used to assess the value of changes in the quality of a site or the complete elimination of a site.

Another indirect approach is the "averting behavior method," which is employed when individuals are observed taking defensive measures to reduce an unwanted hazard (Freeman, 1993). A homeowner might install a filter for the water tap to avoid ingestion of a hazardous substance or install an air conditioner to reduce the concentrations of contaminants in the air. In these circumstances, individuals are purchasing reductions in risk to human health, safety, and the environment. The costs they incur for defensive measures represent a lower bound on the monetary value to them of the perceived risk reduction that is accomplished by the defensive measures. If there is sufficient variation across people in the cost or effectiveness of defensive measures, analysts can estimate a model that supplies information similar to the demand function described previously.

When a market price reflects the riskiness of a good as well as other quality attributes of the good, a "hedonic model" can be employed to estimate monetary values for changes in riskiness or other quality attributes (Freeman, 1993). House prices, for example, are sometimes used to assess the value of changes in environmental quality or crime rates in a community because houses in cleaner and safer communities can command higher prices (holding constant other factors such as size and condition of house) than houses in polluted and dangerous communities. If houses differ with respect to a particular environmental quality attribute (e.g., air quality), the hedonic approach can be used to estimate the monetary value of changes in air quality.

The monetary value of changes in mortality and morbidity risk have been estimated using hedonic methods and labor market data. Workers are assumed to be willing to accept lower wages (higher risks) in exchange for a safer working environment (higher wages). These labor market models control for other determinants of wages such as labor unions, skill level of the job, region of the country, and seniority. There are now over twenty studies in different countries of the world that have estimated the relationship between wage compensation and annual mortality risk on the job. Although the results of these studies vary, workers in the United States appear to behave as if they would be willing to accept a $300 to $700 per year annual wage reduction in exchange for a reduction in annual mortality risk of 1 in 10,000 per year on the job. The so-called value of statistical life (VSL), $3 million to $7 million, represents the total wage loss in the group of 10,000 workers that would be considered equal in value to prevention of one statistical fatality in this group of workers. This does not mean that any one of the workers would be willing to pay $3 million to $7 million to avert certain death – most workers have asset positions far less than $1 million – but it means that small reductions in risk are valued as if each statistical fatality on the job is valued at $3 million to $7 million (Viscusi, 1992, 1993).

The distinguishing feature of indirect methods of valuation is that they draw inferences about people's values from the actual choices they make in daily life. The crucial assumptions in this approach are that (1) people have a range of options, making choice meaningful; (2) they are informed about the risks and understand the consequences of their options; and (3) they make choices in a manner that maximizes their expected utility. In the empirical studies, it is necessary to assume that analysts have controlled adequately for the skill levels of the jobs, the amount of effort required, and other aspects of the jobs. If these assumptions are not valid, then the standard inferences that are drawn from indirect methods of valuation can be erroneous. It is also often assumed (incorrectly we suggest) that the valuations of workplace risks by employees are relevant to valuations of other types of risks (such as risks of transport and/or risks of poor air quality). Because of controversy about whether these assumptions are valid and because of the absence of indirect methods in some settings, analysts have searched for alternatives to indirect methods of valuation.

DIRECT METHODS OF VALUATION. If a consumer makes daily purchases of berries or bird watching, it may seem plausible that value can be inferred from these purchasing decisions. Yet an individual may value a particular resource (e.g., a species or a lake) for reasons other than direct consumption. A person who does not personally value a national park might still value the possibility that others will visit the park. Likewise, an individual may never expend resources to see an endangered species yet may still value an assurance that a particular endangered species will be protected. These kinds of "non-use" or passive values are sometimes called altruistic or existence values. Because it is typically infeasible to measure non-use values through indirect methods, more direct methods (discussed later) have been developed (Mitchell and Carson, 1989; Kopp, Pommerehne, and Schwar 1997; Jones-Lee, 1989).

Even if a consumer does make choices in daily life that reflect value, it may be difficult for a researcher to identify, measure, and quantify the value in monetary units using indirect methods. For example, the value a consumer places on a safer (crime-free) home may be difficult to disentangle from the value the consumer places on a home that is in a clean neighborhood with good school systems. When indirect methods of valuation prove to be feasible or unreliable, more direct methods are employed.

The "contingent valuation method" (CVM) is the modern name for the use of survey methods to obtain monetary values for goods/services/resources that are contingent on a particular hypothetical market described to respondents. CVM collects preference information by asking individuals how much they are willing to pay for provision of a good or for the minimum compensation they would require if a good or service is not provided.

The CVM was probably first applied by Davis (1964) who used questionnaires to estimate the monetary benefits of outdoor recreation. Mitchell and Carson's (1989) classic textbook lists more than 100 U.S. studies based on the CVM, while Green et al. (1990) list 26 CVM studies in the United Kingdom. The literature in this field has exploded since 1990. Navrud (1992) surveys a variety of European studies while Carson et al. (1993) list over 1,400 studies related to the CVM. The CVM is now widely used to estimate the value of environmental consequences, and there is a large body of knowledge about the advantages and disadvantages of the CVM. The literature applying the CVM to human health and safety is newer and smaller but still quite substantial (Tolley et al., 1994).

The earliest applications of the CVM used what are known as open-ended valuation questions, where respondents are asked directly how much money they would be willing to spend to reduce (say) the number of asthmatic attacks they experience from four to two per month. Open-ended questions may be difficult for respondents to answer because they may never have considered the maximum amount they would pay for a particular improvement. Some have argued that only people who attend (art or antique) auctions on a regular basis are good at thinking about their maximum willingness to pay for particular goods.

In 1979, Bishop and Heberlein published a study employing what is now called the closed-ended valuation question. According to this approach, a respondent is offered an opportunity to receive (hypothetically) a particular good or environmental improvement at a specified (fixed) price (bid level). The bid has to be accepted or rejected (unless the respondent is offered a "no opinion" option). Instead of asking the respondent for the highest acceptable bid, the researcher varies the price (bid amount) across respondents (or subgroups of respondents). Econometric techniques can then be employed to estimate the demand function for the item of interest, building on classic choice theory (Hanemann, 1984; McFadden, 1973).

The advantage of the closed-ended approach is that it is easier for respondents to answer "yes" or "no" to a specified price than it is to locate one's maximal WTP, particularly since the familiar activity of shopping in a store is closer to a yes-no response to a price than it is to participating in an auction or negotiation. More recent studies have employed double-bounded (or even triple-bounded) approaches where the price for the good is increased or decreased depending upon whether the initial price was rejected or accepted. In this way, more information is obtained from each respondent, but it is not yet clear whether such approaches achieve realistic responses from subjects.

Does the CVM yield valid estimates of WTP for health, safety, and environmental improvements? Because the questions posed to subjects are hypothetical, validity is difficult to assess. There have been some efforts to validate the results of WTP surveys in settings where actual payments can be collected or where actual consumer behaviors can be compared to self-reported intentions about purchasing. Yet this literature is too limited to draw any confident, general conclusions about the CVM.

The validity debate becomes polarized when the CVM was used to assess the value of damages caused by the *Exxon Valdez* oil spill off the coast of Alaska in 1987. Researchers funded by the state of Alaska and Exxon came to some very different conclusions. The U.S. government responded by commissioning an expert panel of social scientists, co-chaired by Nobel Prize winners Kenneth Arrow and Robert Solow, to evaluate the CVM as applied to natural resource damages. The panel gave a cautious go-ahead to use of the CVM only if a variety of fairly rigorous guidelines are followed by researchers. The Arrow-Solow panel did not accomplish consensus in the technical community, but it has stimulated research into improvements in the CVM. Controversy remains about whether studies employing state-of-the-art CVM are biased and, if so, in what direction and in what magnitude. Until more is known about the properties of these studies, it is difficult to recommend their use to inform public policy and judicial proceedings.

Cost-Effectiveness Analysis. When there is a lack of consensus about how risk reduction should be valued in monetary units, the analyst may choose to employ cost-effectiveness analysis (CEA) instead of CBA (Drummond et al., 1997; Petitti, 1994). CEA can be used to identify the least-cost method of achieving a particular health or environmental goal, or it can be used to identify the policy that will achieve the most health or environmental protection for a given budget or fixed expenditure of resources. Algorithms for comparing the incremental cost-effectiveness ratios for multiple competing or noncompeting policies have been developed for use by practitioners (Weinstein and Zeckhauser, 1973). CEA offers no insight into what the goal should be or what the budget should be. The CEA produces a rank ordering of various policy options but does not recommend which ones should be adopted (Johannesson, 1996). The conditions under which decisions based on CEA will be equivalent to decisions based on CBA have only recently been defined rigorously (Garber and Phelps, 1997; Meltzer, 1997; Bleichrodt and Quiggin, 1999).

The crucial challenge in CEA is to express the health and/or environmental consequences of a policy in a common unit other than dollars, the so-called effectiveness metric. In medicine and public health, where CEA is more popular than CBA, there have been numerous effectiveness metrics developed over the past twenty years.

MORTALITY-BASED METRICS OF EFFECTIVENESS. The earliest approach was to compare policies according to the number of lives saved. Morrall (1986) used this metric to compare cost-effectiveness ratios for forty-four health, safety, and environmental regulations in the United States. Yet "lives saved" or "deaths averted" are now considered a misleading metric because everyone must die sooner or later. To account for prematurity of death, some interest was expressed in "early" deaths or "productive" lives lost (where death before age 65 or 70 might be employed as a cut-point). Yet this procedure ignored the benefits of reducing mortality risk among the elderly. It also seemed inappropriate to make no distinction between a death from accidents early in the life span and a death from chronic disease that occurs later in the life span (Ten Berge and Stallen, 1995).

The number of life years saved by policies becomes a popular metric because it accounts for premature deaths at all ages and because life-expectancy information can be obtained from standard life tables. When life years are used to measure effectiveness, it is typically assumed (at least implicitly) that all years of life are of equal value, regardless of who in society experiences them and when in the life span they are experienced. Building on early work by Schwing (1979) and Graham and Vaupel (1981), Tengs et al. (1995) and Tengs and Graham (1996) compared the estimated cost per life year saved of 500+ policies in the United States. Ramsberg and Sjoberg (1997) collected cost-per-life-year saved estimates for 165 programs in Sweden, while Kishimoto (1999) collected such information for 88 life-saving interventions in Japan. These studies document differences in the cost-effectiveness ratios for various policies that often exceed several orders of magnitude.

If applied strictly, this life-year approach will obviously give more weight to the lives of young people than old people. In the United States and Japan, more weight will be applied to the lives of women than men (because men live shorter lives than women); in the United States, more weight will be assigned to the lives of whites than blacks (since whites live longer lives than blacks). Conclusions from such analyses need to be tempered by equity concerns.

EFFECTIVENESS METRICS THAT COMBINE MORTALITY AND MORBIDITY INFORMATION. Policies that affect mortality rates are also likely to affect morbidity rates (rates of nonfatal injury or disease) and thus the field of CEA has developed measures of effectiveness that combine both types of health information. The literature on this topic has exploded in the last twenty years (Gold et al., 1996).

In developed countries, the effectiveness metric in most widespread use is the "quality adjusted life year" (QALY), where each year of life is weighted in subjective quality on a scale from 0 to 1.0, where 1.0 is a year in perfect health and 0 is equivalent to death (Zeckhauser and Shepard, 1976). The effectiveness of a procedure such as coronary artery bypass surgery is expressed as the net number of QALYs saved, where this quantity includes the expected number of life years gained, adjusted for quality, plus any improvement in the quality of lives that would have been lived at a diminished level of quality in the absence of surgery. Suppose a 60-year-old male with three-vessel coronary artery disease is estimated to have 5 years of life expectancy at an average quality level of 0.7. If surgery can double his life expectancy and achieve perfect health status for the entire period, the patient will have gained 6.5 QALYs, 5.0 from the longevity effect, and 1.5 (5×0.3) from the reduction in morbidity associated with heart disease.

Diseases vary in whether associated QALY losses are dominated by mortality or morbidity effects. For cancer, a disease that is often fatal, QALY losses tend to be dominated by life years lost, although the painful years of life while a tumor progresses provide some contribution to the overall loss in QALYs from cancer. Heart disease, on the other hand, has a larger burden of morbidity (nonfatal heart attacks, for example) relative to mortality, and thus it is more important to account for morbidity effects when heart disease is reduced than when cancer is reduced. Strictly speaking, all morbidity and mortality effects of a disease should be counted, regardless of their quantitative contribution to overall QALY loss.

Weinstein and Stason (1977) in the United States and Williams (1985) in the United Kingdom were pioneers in applying the cost-per-QALY method to medical advances. In a recent paper, Graham et al. (1998) collected standardized cost-per-QALY information on over forty interventions aimed at reducing cancer, heart disease, trauma, and infectious disease.

The procedures employed to elicit quality weights for health states from patients or community residents have become particularly sophisticated in the last decade. There are both general health-utility scales (where all adverse health conditions are rated according to specified dimensions) and disease-specific scales (that are tailored to discriminate among patients with a specific disease [e.g., Alzheimer's] that vary in severity). The most popular general tools are the Health Utilities Index, which was developed in Canada; the EuroQoL; and the Quality of Well-Being Index. The preference-elicitation methods employed by these tools include the rating scale, the time trade-off method, and the standard gamble method, the latter having the best grounding in expected utility theory (Torrance, 1986; Drummond et al., 1997). Questions have been raised about whether QALY measurements are more or less valid than self-reported WTP measurements (O'Brien and Viramontes, 1994).

Theorists have noted that the strict use of QALYs to make efficiency determinations is based on a variety of assumptions (Pliskin, Shepard, and Weinstein, 1993). For example, it is assumed that saving one QALY each for a community of forty people is equivalent in social value to saving forty QALYs for one person in the community. It is also typically assumed that saving ten QALYs for sure is equivalent in value to a 50–50 lottery at saving no or twenty QALYs. Researchers are currently exploring how implausible these assumptions may be in various settings, including development of modified methods to permit departures from these assumptions.

In developing countries, the "disability adjusted life year" (DALY) has proven to be a more popular effectiveness metric than the QALY (Murray, 1994). The two metrics are close cousins in the sense that they are both based on weighting schemes applied to life expectancy and morbidity data. Yet the weighting schemes for DALYs are based on expert judgement, while the weights for QALYs are based on the preferences of patients or lay citizens. Unlike the QALY approach, which gives equal weight to a year of perfect health (regardless of age), the DALY approach assigns the largest social value to healthy life years in the middle of the life span, with smaller weights assigned to healthy years at the beginning and end of the life span. Some commentators have questioned the ethical assumptions that underpin DALYs (e.g., the notion that the most important life years are those at the middle of the life span).

Value-of-Information Analysis. When risks are uncertain, it may not be feasible or advisable to execute a CEA or CBA. It may be argued that uncertainties are so great that it is preferable to gather additional scientific information prior to making a risk-management decision. Alternatively, it may be argued that "precautionary" action should be taken to protect people and the environment until scientific uncertainties about risk are reduced or resolved.

Value of information (VOI) analysis is a form of CBA that was designed for precisely these circumstances: situations where the costs and/or benefits of management alternatives are uncertain and where "waiting for the results of additional scientific research" can be subject to formal analysis as a decision alternative. From a VOI perspective, information denotes "uncertainty reduction," and because uncertainties are costly to decision makers and society, resource investments in collecting accurate information should be considered a viable policy alternative worthy of formal analysis. In the last twenty years, the analytical capacities to perform VOI analysis have improved, while the number of VOI applications in the peer-reviewed literature has expanded rapidly (Weinstein, 1983; North, 1983; Finkel and Evans, 1987; Evans, Hawkins, and Graham, 1988; Lave et al., 1988; Reichard and Evans, 1989; Hammitt and Cave, 1991; North, Selker, and Guardino, 1992; Taylor, Evans, and McKone, 1993; Dakins, Toll, and Small, 1994; Dakins et al., 1996; Thompson and Evans, 1997).

Under the VOI framework, the analyst compares the net benefits (benefits minus costs) of immediate protective action to the net benefits of a more informed decision that is made after a scientific investigation has been conducted. Note that the "costs" of additional study include the uncertain risks that may occur during the time period of study (which can be serious in the case of irreversible hazards) as well as the resource costs of performing additional studies (e.g., the labor time of the scientists performing the studies as well as any necessary equipment and materials). Since risks and other factors in the analysis may be uncertain, the net-benefit estimates of each risk management option are expressed in probabilistic rather than deterministic form. If the risk manager's attitude toward uncertainty is considered to be risk-neutral (typical for a societal perspective), then the efficiency rule under uncertainty is to conduct further study prior to taking protective measures only if the expected net benefits of further study are greater than the expected net benefits of immediate action. If the problem under study justifies some degree of societal risk aversion (e.g., from risk of global catastrophe), then an efficiency rule based on comparison of "certainty equivalents" – a measure that is more sensitive to severe downside losses – can be replaced by the typical rule maximizing expected net benefits.

One of the useful features of VOI analysis is that it can provide quantitative estimates of the expected societal benefits of conducting different kinds of scientific research, prior to taking protective action. For example, a risk manager faced with a toxic chemical issue might need to know which type of scientific study would be most useful: a definitive toxicology study, a comprehensive assessment of human and wildlife exposure to the chemical, an engineering-economics analysis of substitute chemicals, or a social science investigation of citizen preferences regarding reproductive versus carcinogenic effects. From a decision-making perspective, VOI analysis can provide insight into which types of additional studies will have the greatest expected societal net benefit.

The promise of VOI analysis is limited for a variety of reasons. In order to be implemented, VOI analysis must have, as inputs, quantitative characterizations of the uncertainties regarding all key inputs to the risk-management dilemma. Although significant progress has been made in the field of formal uncertainty analysis (Morgan and Henrion, 1990; Pate-Cornell, 1996; Claxton, 1999), particularly as it relates to risk, some of the most important uncertainties regarding risk tend to require judgmental quantification (Cooke, 1991). These judgments may have calibration difficulties and responsible experts may differ in their subjective probability distributions concerning the same quantity (e.g., the low-dose carcinogenic potency of ionizing radiation). In cases where uncertainty about risk is so chaotic that there are no experts willing and/or able to provide judgmental distributions, a VOI analysis may be impossible or may not be performed credibly. Even in these highly uncertain situations, there may be qualitative insights from the VOI framework that are useful to risk managers and stakeholders.

One of the most difficult challenges in VOI analysis is obtaining quantitative predictions of the likely outcomes of different research strategies, assuming that they are funded and executed. Some research questions are more tractable than others, and scientists in all fields of inquiry have exhibited a wide variety of errors when trying to predict the relative likelihood of different research outcomes. Nevertheless, VOI tools have been applied to research investment decisions such as the optimal allocation of federal funding for AIDS research to treatment, vaccine, and behavioral programs (Siegel, Graham, and Stoto, 1990). In addition to the decision-analytic literature on VOI, there is a related economic literature on quasi-option value (Arrow and Fisher, 1974; Henry, 1974) and value of information in financial investments (Dixit and Pindyck, 1994).

7. CASE STUDIES OF APPLICATION

In this section, we present the results of selected applications of formal efficiency analysis that appear in the peer-reviewed scientific literature. These applications are not necessarily representative because there are now thousands of applications published in the English literature. Our purpose here is to provide practical examples of formal efficiency analysis in diverse fields of application.

7.1. Medicine

Boyle et al. (1983) evaluated the efficiency of providing regionwide neonatal intensive care services to Canadian newborns in two weight categories: 500–999 grams and 1,000–1,499 grams. The comparator was conventional hospital care without long-term trachael intubation and assisted ventilation. Short-term estimates of mortality risk with and without neonatal care are based on data from clinical trials and epidemiology: Estimates of long-term longevity, quality of life, and cost are based on expert judgment and mathematical models.

The authors employ both CEA, using life years and QALYs as the measures of effectiveness, and CBA, using averted medical costs and enhanced productivity later in life as the monetary measures of benefit. Future health and economic consequences are discounted to present value at a real annual rate of 3 percent, with sensitivity analysis from 0 to 10 percent.

The analysts conclude, using CBA, that the net economic benefit is likely to be positive for infants weighing between 1,000 and 1,499 grams (at all discount rates less than 3.5 percent). Since the analysis excludes any parental willingness to pay to save the infants (beyond what is captured by medical costs and lifetime productivity), this conclusion is likely to be conservative. For infants weighing less than 1,000 grams, the net economic benefit of the services was estimated to be negative.

The estimated cost-effectiveness ratios were expressed in 1978 Canadian dollars (multiply by 0.877 to obtain U.S. dollars and by roughly 3.0 to obtain 1999 U.S. dollars). For the smaller infants (500–999 grams), the net cost-effectiveness ratios were $600 and $1,000 per life year and per QALY saved, respectively. These investment ratios are well within the range of the investments routinely made in developed countries for a variety of surgical and pharmaceutical treatments. For the larger infants, the numerator of the cost-effectiveness ratios were negative, indicating that neonatal care saves both life years (or QALYs) and resources. Even if savings in productivity and health care are excluded from costs, the gross cost-effectiveness ratios of neonatal care for the heavier infants (about $900 per life year or QALY saved) are modest compared to other well-accepted medical procedures.

The authors acknowledge that ethical issues beyond efficiency may influence whether governments decide to ration neonatal intensive care services. They also note that their economic estimates may not be appropriate outside Canada, particularly in countries that lack a well-developed health care system.

7.2. Radiation Protection

Ford et al. (1999) applied CEA to general and targeted strategies for residential radon screening and mitigation in the United States. The measures of effectiveness were lung cancer deaths prevented and life years saved, based on the risks of fatal lung cancer observed among workers exposed to radon in uranium mines. Savings in medical costs and productivity losses from lung cancer were accounted for in net-cost estimates. Costs are expressed in 1993 U.S. dollars (multiply by roughly 1.2 for 1999 U.S. dollars) and future costs and mortality/life years were discounted at a real annual rate of 4 percent.

The nationwide net cost of universal radon screening and mitigation, assuming a threshold of 4 pCi/L, was $3 million per death averted, or $480,000 per life year saved. If efforts are targeted at geographic areas at high risk for radon exposure, the cost-effectiveness ratios become somewhat more attractive ($2 million and $333,000 per life and life year saved, respectively). If mitigation is undertaken only after a second confirmatory radon test, the cost declines to $920,000 and $520,000 per life and life year saved, respectively, for the universal program and to $130,000 and $80,000 for the targeted program. Efforts to target houses occupied by smokers were found to be especially attractive because there is an interaction between smoking and radon in the radon epidemiology.

The study did not address nonfatal cases of lung cancer or the possibility that reduced radon exposure might delay the onset of lung cancer. Benefits to future generations from modifications of existing dwelling units were not assessed. A full cost assessment of efforts to ensure 100 percent citizen compliance with screening/mitigation recommendations was not conducted. Future studies may need to refine the effectiveness estimates now that new case-control studies of residential radon exposure and lung cancer are being published.

7.3. Transport Safety

Graham et al. (1997) examined the efficiency of installing driver and passenger airbags in new motor vehicles in the United States. Effectiveness was measured as net QALYs saved, considering both the safety effectiveness and risks of the technology in three age groups (0–9, 10–65, 65+ years). They employed CEA using real-world crash data as the basis for estimating the effectiveness, risk, and net costs of the present generation of airbag technology. Future health and costs (1995 U.S. dollars) were discounted at a real annual rate of 3 percent, with sensitivity analyses covering discount rates from 0 to 5 percent.

The authors conclude that the driver airbag costs about $24,000 per QALY saved. The incremental cost of adding the passenger airbag was estimated at about $60,000 per QALY saved. The inferior ratio for the passenger airbag reflects the lower rate of occupancy in the front-right seat plus the incremental risk of airbag technology to children, who have a large number of QALYs at risk. The authors note that both ratios are within the range of currently accepted clinical and public health investments, though the ratio for the passenger airbag could be lessened considerably if children under the age of 10 were seated in the rear – a practice that is commonplace in many continental European countries. The authors also note the cost-effectiveness ratios for airbags would be less attractive if rates of manual safety belt use in the United States could be increased from 50 percent (current levels) to 100 percent.

7.4. Food Safety

Buzby, Ready, and Skees (1995) examine a ban on the use of a specific postharvest pesticide (sodium ortho-phenylphenate or SOPP) used in grapefruit packaging houses in Florida (United States). It is assumed that an alternative, less toxic pesticide (thiabendazole, TBZ) can be used in place of SOPP, though at a slight increased cost to farmers and consumers. CBA is employed to determine whether the ban would be efficient.

The incremental lifetime cancer risk associated with consuming grapefruit treated with SOPP (1 in 10,000) is estimated to be 100 times larger than the risk associated with TBZ (1 in 1,000,000), given available toxicology data, average grapefruit consumption, and use of standard risk assessment methods recommended by the U.S. EPA. The authors assume that half of the estimated cancers will be fatal and consider only benefits attributable to the reduction in mortality risk. CVM is employed to estimate the dollar value of risk reduction, estimated to be about an extra $0.20 per grapefruit (+38 percent per grapefruit), or $4.1 million per life saved, or $80 million per year nationwide.

The costs of the ban, assumed to be worst case, are a 10 percent increase in postharvest loss from increased pathogen resistance and an increase in prices for fresh grapefruit. In total, the worst-case national cost estimate is $27.7 million per year. The long-run cost is likely to be smaller as grapefruit suppliers adjust to the assumed postharvest loss.

The authors conclude that the net benefits of the ban are positive, yet they stress several uncertainties. Benefit estimates are based on WTP estimates for risk reduction obtained using CVM. Respondent estimates of WTP may not be robust with respect to plausible changes in the way WTP questions are framed. Like all CVM-based benefit estimates, the stated WTP figures are hypothetical and cannot be verified as to their validity. The relative risks of the alternative pesticides are also uncertain because of the absence of direct human data, the lack of data on noncancer endpoints, and the absence of ecological information in the study. The costs of the ban are likely to be lower than the worst-case figures presented.

7.5. Air Quality Regulation

Wilson (1998) applied CBA to the air quality limit values for sulphur dioxide, nitrogen dioxide, particulate matter, and lead that were proposed by the European Union in October 1997. The estimated human health benefits of the limits were compared, in monetary units, to the estimated cost of compliance with the limits. The health effects from air pollution included some acute and chronic effects. WTP methods were employed to place monetary units on the reductions in mortality risk associated with improved air quality. The diminished costs of medications and enhanced productivity of workers (from less illness) were also counted as benefits.

Compared to current limits, the benefit-cost ratios for the proposed new limits were calculated to be 22.8 to 114, 9.6 to 23.4, 25.7 to 240, and 0.11 to 0.37 for sulfur dioxide, nitrogen dioxide, particulate matter, and lead, respectively. The analytical methods employed here were similar to the CBA techniques that have been employed previously by the U.S. Environmental Protection Agency (Morgenstern, 1997).

7.6. Toxic Metal Regulation

Nakanishi, Oka, and Gamo (1998) applied CEA retrospectively to the Japanese government's decision to prohibit the mercury electrode process in caustic soda production. A nonlethal endpoint from mercury exposure, paresthesia, was evaluated in terms of loss of life expectancy based on the results of epidemiological study of Minamata Disease (each case was associated with an expected loss of 1.85 life years). Taking into account the estimated cost of the prohibition and the number of cases prevented, the cost-per-life-year saved was estimated to be 570 million to 5700 million yen (divide by 100 to obtain U.S. dollars). The authors questioned the cost-effectiveness of this decision based on a comparison to the cost-effectiveness of other policies.

7.7. Pesticide Regulation

Gamo, Oka, and Nakanishi (1995) and Oka et al. (1997) applied CEA retrospectively to a 1986 decision of the Japanese government to ban chlordane, a pesticide used in termite control. The reduction in cancer risk associated with the ban of chlordane was compared to the acute neurotoxic effects of chloropyrifos, the substitute for chlordane. Life years was used as the effectiveness metric. The life years lost from cancer were calculated from Japanese statistics on cancer mortality and the 1990 life table for Japan. Each additional cancer death in a population of 100,000 people was calculated to be equivalent to an average loss of life expectancy of 65.8 minutes. The risk due to chloropyrifos was converted to lost life expectancy by expressing the severity of the adverse neurotoxic effects using the Cornell Medical Index (CMI) scores (based on a patient questionnaire) and by relating the scores to the score-mortality relationship among workers in the paper industry. The estimated cost-effectiveness ratio for this policy was 45 million yen (U.S.$0.45 million) per life year saved, which is within an accepted range for life-saving investments.

7.8. Natural Resource Protection

Bos and van den Bergh (1998) undertook an analysis of land/water use and nature conservation in a Dutch river plane composed of patches of lakes and polders. The area is dominated by the presence of wetland ecosystems. The study presents available information for a set of main wetland function and use categories. This includes estimates of the benefits and costs associated with particular activities. These are used in an illustrative cost-benefit analysis of a change to more nature conservation in the area under study. It is assumed that the nature conservation project reflects sustainable land use and cover in the area. For example, the costs of acquiring and restoring areas as well as sanitation of a river in the area are estimated. It is concluded that the nature conservation is socially profitable for discount rates up to 13 percent if the time horizon is ten years and up to 18 percent for an infinite time horizon.

8. TRENDS IN PRACTICAL USE BY GOVERNMENTS

Formal efficiency analyses are growing in importance throughout the developed world, though the governments of some countries place greater emphases on these analyses than other countries. Here is a survey of practical use of CBA and CEA in selected countries.

8.1. Australia

Pharmaceutical manufacturers must submit a CEA to the Australian Pharmaceutical Benefits Advisory Committee for decisions on whether drug products will be subsidized by the government. The guidelines call for use of "life years gained" as the measure of effectiveness for drugs that mainly influence mortality risk. The guidelines currently discourage use of QALYs owing to a lack of experience with this metric. The Australian approach to CEA of drugs is innovative, with other countries beginning to express interest in how the routine use of CEA is working.

8.2. Japan

The use of formal efficiency analysis in Japan is currently limited. The necessity of a CBA for governmental policy has been recognized by the ministries in charge of finance and economy. The public sectors of road and railway construction as well as flood control have conducted CBAs to help choose among alternative projects.

The guidelines for conducting CBA published by responsible authorities direct that only costs and benefits that can be evaluated in monetary units should be included in CBA. They also direct that a final decision should be made, giving careful consideration to the costs and benefits that could not be quantified in monetary units, such as costs and benefits related to human health and ecological risk.

The Environment Agency has never conducted a CBA or CEA on environmental programs or regulations. Except for the case of carbon emission control, impacts of individual environmental regulations on the national economy have not been assessed.

Environmental impact assessments (EIA) are required for major private or public development projects. Yet these analyses do not contain a CBA or CEA and the EIA is understood as a means of reducing adverse effects to the environment as much as possible.

If formal efficiency analysis is applied to Japanese health and environmental policy in the future, CEA may prove to be more feasible than CBA. Japanese data regarding WTP values are extremely limited, and there is no consensus in Japan, even among economists, on the validity and reliability of values obtained from the CVM. There is also suspicion that self-reported WTP values for ecological protection may exhibit a vulnerability to rumors, especially because the questions are hypothetical.

8.3. Sweden

The price of (prescribed) drugs is set in negotiations between pharmaceutical companies and a governmental body. In these negotiations, cost-effectiveness analysis is sometimes used. To the best of our knowledge, there is no formal requirement of or standard for such analysis. Recommendations to hospitals to turn to a particular medical treatment are often based on the cost-effectiveness of the recommended alternative.

The Swedish Board of Traffic uses CBA routinely to assess the social profitability of new roads. In these analyses, the Board uses a value of a statistical life equal to about $1.5 million. This value is derived from a contingent valuation study that was undertaken in the late 1980s. CBA is also used to assess investments in public transprotation. These analyses focus on travel time (costs), the environmental consequences, and sometimes on labor market issues. The latter is particularly true for investments in railways in rural areas of the country.

There is also ongoing work in Sweden to broaden conventional national income measures to cover both service flows from natural resources as well as changes in the stocks of (renewable and nonrenewable) natural resources. The difficult question of how to capture chemical substances in these accounts is still unresolved. It is also an open question whether the research and development aimed at developing the new national income measures really will be implemented on a regular (annual) basis.

The general impression, however, is that CBA and CEA are used occasionally rather than on a regular basis in public decision making in Sweden. The country also lacks a uniform standard for such analyses.

8.4. The United States of America

At the federal level, the U.S. government requires that all major federal regulations be accompanied by a CBA (a CEA is considered adequate in some circumstances). The regulatory decision is not necessarily dictated by the results of the analysis, but the analysis must be performed and be made available to the public. This process is monitored by the Office of Information and Regulatory Affairs of the Office of Management and Budget, an arm of the Executive Office of the President of the United States. Several dozen states in the United States have a similar process of analysis applied to state regulations that is monitored by the Governor's office.

In some regulatory arenas (e.g., consumer product safety and toxic substances control), the U.S. Congress has insisted by law that regulatory actions be supported by formal cost-benefit analysis. Yet in other regulatory arenas (e.g., pesticide regulation and air quality control), the United States has prohibited consideration of cost and benefits in favor a health-only approach to decision making. Even when U.S. agencies are prohibited or discouraged from considering costs and benefits, there is some evidence that decision makers implicitly consider costs and benefits.

Preference for analytic tools vary across U.S. agencies. The National Highway Traffic Safety Administration and Public Health Service employ CEA, while the Environmental Protection Agency and Consumer Product Safety Commission employ CBA. Some agencies, such as the Centers for Disease Control, apply CBA to some issues and CEA to other issues. No U.S. agencies have formally employed value-of-information analysis, although a recent Presidential/Congressional Commission on Risk Assessment and Management recommended that the U.S. Environmental Protection Agency experiment with VOI analysis as part of a precautionary approach to environmental protection.

The Food and Drug Administration has recently been drawn into the field of CEA by the efforts of pharmaceutical manufacturers to make promotional claims regarding whether their pharmaceutical products are "cost-effective." Although FDA approves drugs for use strictly on the basis of safety and efficacy, the U.S. Congress recently authorized FDA to oversee promotional claims made by manufacturers that address cost-effectiveness. FDA is thus in the process of formulating guidelines as to what constitutes a valid claim of cost-effectiveness.

9. RESEARCH NEEDS

Additional theoretical and empirical research is required to enhance the validity and influence of formal efficiency analyses. We identify two research needs here to stimulate workshop discussion of a larger menu of research needs.

9.1. Validity

When formal efficiency analyses are prepared prior to public risk-management decisions, analysts often lack precise knowledge of key input values (both physical quantities and preferences). In the face of uncertainty, analysts may make assumptions, or they may use imperfect or indirectly relevant data. For example, analysts may make assumptions about the long-run marginal cost of producing a new technology or they may project the real-world benefits of a new technology to humans based on limited experimental data with nonhuman subjects. Despite the uncertainties and imperfections inherent in such analyses, relatively little is known about the validity (accuracy) of the resulting point estimates of net benefits or cost-effectiveness ratios. In some cases, it is impractical or infeasible to perform a validity analysis; however, growing numbers of studies have found errors in the benefit and cost estimates made prior to implementation of a program or regulation.

On the cost side of the ledger, there is a limited literature suggesting that the actual (realized) costs of risk regulations tend to be less than they are projected ex ante. If true, we should not be surprised because firms subject to regulations may be more creative in finding ways to reduce risk that entail less resource (labor/capital) investment than regulatory analysts predict before a regulation is adopted. After all, we might expect that specialists in these firms know their business better than a regulatory analyst does. More research needs to be done to document and quantify the patterns of realized costs of risk regulations, laying the groundwork to assist regulatory analysts in making more realistic predictions of the long-term costs of risk regulation.

The benefit side of the ledger is even sparser with regard to literature on validation. It is very rare for a government agency or even an academic analyst to compare the projected benefits of a risk regulation to the amount of benefits actually experienced after the regulation is implemented. One recent study (reviewed later) found that the numbers of lives saved by a mandatory airbag regulation in the United States were actually a factor of 3 smaller than projected when airbags were mandated. In that case, the regulatory estimates of the costs of airbags proved to be roughly correct, though the costs of treating airbag-induced injuries were seriously underestimated.

A richer database on validation of formal efficiency analysis should be generated in the future. Such a database would provide useful feedback to analysts on systematic analytic errors while providing decision makers some insight into the historical track record of analysts in making accurate estimates. Attention to validity might also enhance the demand for greater use of deterministic and probabilistic tools of uncertainty analysis, encouraging analysts to be more transparent about the extent of uncertainty in their estimates of risk, benefit, and cost.

9.2. Influence

As formal efficiency analyses are produced in the public risk-management process, it would be useful to better understand what influence, if any, these analyses have on actual risk management decisions. Some commentators have suggested that analyses of efficiency are unlikely to have a significant impact on decisions because public decision makers are more interested in nonefficiency factors (e.g., politics, fairness, or equity) than efficiency (Lave, 1996). Other commentators have expressed fear that such analyses will have too much influence, allowing little room for equity or fairness considerations in risk management. Related issues that should be examined are the determinants of credibility of efficiency analysis, including factors such as the role of peer review, opportunity for stakeholder participation, clarity of analyses to nontechnical audiences, and inclusion of alternatives or benefit and cost considerations of interest to decision makers, stakeholders, and the public. Research on the influence of formal efficiency analysis will help lay the groundwork for more integrated research into how public risk-management processes can become more competent, credible, trustworthy, and responsive to the legitimate concerns of stakeholders and the public.

References

Arrow, K. J., and Fisher A. C. 1974. Environmental preservation, uncertainty, and irreversibility, *Quarterly Journal of Economics*, 88: 312–19.

Arrow, K., and Lind, R. C. 1970. Uncertainty and the evaluation of public investment decisions, *American Economic Review*, 60: 364–78.

Bishop, R. C., and Heberlein, J. A. 1979. Measuring values of extra market goods: Are indirect measures biased? *American Journal of Agricultural Economics*, 61: 926–30.

Bleichrodt, H., and Quiggin, J. 1999. Life-cycle preferences over consumption and health: When is cost-effectiveness analysis equivalent to cost-benefit analysis? *Journal of Health Economics*, 18 (6): 681–708.

Bos, E. J., and van den Bergh, C. J. M. 1998. *Economic Evaluation, Land/Water Use, and Sustainable Conservation of the "De Vechtstreek" Wetlands*. Department of Spatial Economics, Vrije Universiteit, Amsterdam.

Boyle, M. H., Torrance, G. W., Sinclair, J. C., Horwood, S. P. 1983. Economic evaluation of neonatal intensive care of very-low-birth-weight infants, *New England Journal of Medicine*, 308: 1330–7.

Braden, J. B., and Kolstad, C. D., eds., 1991. *Measuring the Demand for Environmental Quality*. North-Holland, Amsterdam.

Boadway, R. W. 1974. The welfare foundations of cost-benefit analysis. *Economic Journal*, 84: 926–39.

Boadway, R. W., and Bruce, N. 1984. *Welfare Economics*. Basil Blackwell, Oxford.

Buzby, J. C., Ready, R. C., and Skees, J. R. 1995. Contingent valuation in food policy analysis: A case study of a pesticide-residue risk reduction. *Journal of Agricultural and Applied Economics*, 27 (2): 613–25.

Carson, R. T., Wright, J., Alberini, A., and Flores, N. 1993. A bibliography of contingent valuation studies and papers. Natural Resource Damage Assessment, La Jolla, CA.

Claxton, K. 1999. Bayesian approaches to the value of information: Implications for the regulation of new pharmaceuticals. *Journal of Health Economics*. 8 (3): 341–64.

Cooke, R. M. 1991. *Experts in Uncertainty: Opinion and Subjective Probability in Science.* Oxford University Press, New York.

Cullen, A., and Frey, H. C. 1999. *Probabilistic Techniques in Exposure Assessment.* Plenum Press, New York.

Dakins, M. E., Toll, J. E., and Small, M. J. 1994. Risk-based environmental remediation: Decision framework and role of uncertainty. *Environmental Toxicology and Chemistry*, 13: 1907–15.

Dakins, M. E., Toll, J. E., Small, M. J., and Brand, K. P. 1996. Risk-based environmental remediation: Bayesian Monte Carlo analysis and the expected value of sample information. *Risk Analysis*, 16 (1): 67–80.

Davis, R. K. 1964. The value of big game hunting in a private forest, in Transactions of the twenty-ninth North American wildlife conference. Wildlife Management Institute, Washington, D.C.

Dixit, A. K., and Pindyck, R. S. 1994. *Investment Under Uncertainty.* Princeton University Press, Princeton, NJ.

Evans, J. S., Hawkins, N. C., and Graham, J. D. 1988. The value of monitoring for radon in the home: A decision analysis. *Journal of the Air Pollution Control Association*, 38: 1380–5.

Finkel, A., and Evans, J. S. 1987. Evaluating the benefits of uncertainty reduction in environmental risk management. *Journal of the Air Pollution Control Association*, 37: 1164–71.

Ford, E. S., Kelly, A. E., Teutsch, S. M., Thacker, S. B., and Garbe, P. L. 1999. Radon and lung cancer: A cost-effectiveness analysis. *American Journal of Public Health*, 89 (3): 351–7.

Gamo, M., Oka, T., and Nakanishi, J. 1995. A method evaluating population risks from chemical exposure: A case study concerning prohibition of chlordane use in japan. *Regulatory Toxicology and Pharmacology*, 21: 151–7.

Garber, A., and Phelps, C. 1997. Economic foundations of cost-effectiveness analysis. *Journal of Health Economics*, 15: 1–31.

Gold, M. R., Siegel, J. E., Russell, L. B., and Weinstein, M. C., eds. 1996. *Cost-Effectiveness in Health and Medicine.* Oxford University Press, New York.

Graham, J. D., Corso, P. S., Morris, J. M., Segui-Gomez, M., and Weinstein, M. C. 1998. Evaluating the cost-effectiveness of clinical and public health measures. *Annual Review of Public Health*, 19: 125–52.

Graham, J. D., and Vaupel, J. 1981. The value of a life: What difference does it make? *Risk Analysis*, 1 (1): 89–95.

Graham, J. D., Thompson, K. M., Goldie, S. J., Segui-Gomez, M., and Weinstein, M. C. 1997. The cost-effectiveness of air bags by seating position. *Journal of the American Medical Association*, 278 (17): 1418–25.

Green, C. H., Tunstall, S. M., N'Jai, A., and Rogers, A. 1990. Economic evaluation of environmental goods. *Project Appraisal*, 5: 70–82.

Hammitt, J. K., and Cave, J. A. K. 1991. *Research Planning for Food Safety: A Value of Information Approach.* Rand Corporation, Santa Monica, CA.

Hanemann, M. W. 1984. Welfare evaluations in contingent valuation experiments with discrete responses. *American Journal of Agricultural Economics*, 66: 332–41.

Henry, C. 1974. Option values in the economics of irreplaceable assets. *Review of Economic Studies: Symposium of Economics of Exhaustible Resources*, 41: 89–104.

Jones-Lee, M. W. 1989. *The Economics of Safety and Physical Risk*. Basil Blackwell, Oxford.

Keeler, E. B., and Cretin, S. 1983. Discounting of lifesaving and other non-monetary effects. *Management Science*, 29: 300–6.

Kishimoto, A. 1997. A comparative analysis of cost-effectiveness of risk reduction policies in Japan. *Japanese Journal of Risk Analysis*, 8 (2): 165–73.

Kishimoto, A. 1999. The cost-effectiveness of lifesaving interventions in Japan: Do the variations found suggest irrational resource allocation? *Proceedings of the 2nd International Workshop on Risk Evaluation and Management of Chemicals* (January 1999), Yokohama, Japan.

Kopp, R. J., and Smith, V. K., eds. 1993. *Valuing Natural Assets: The Economics of Natural Resources Damage Assessment*. Resources for the Future, Washington, DC.

Kopp, R. J., Krupnick, A., and Toman, M. 1997. Cost-benefit analysis and regulatory reform. *Human and Ecological Risk Assessment*, 3 (5): 787–852.

Lave, L. B. 1996. Benefit-cost analysis: Do the benefits exceed the cost. In R. W. Hahn, ed., *Risks, Costs, and Lives Saved*, pp. 104–34. Oxford University Press, New York.

Lave, L. B., Ennever, F. K., Rosenkrantz, H. S., and Omenn, G. S. 1988. Information value of the rodent bioassay. *Nature*, 336: 631–3.

Leonard, H. B., and Zeckhauser, R. J. 1986. Cost-benefit analysis applied to risks: Its philosophy and legitimacy, pp. 31–48. In D. MacLean, ed., *Values at Risk*. Rowman and Allanheld, Totowa, NJ.

McFadden, D. L. 1973. In P. Zacembka, ed. *Conditional Logit Analysis of Qualitative Choice Behavior*, Frontiers of Econometrics. Academic Press, New York.

Meltzer, D. 1997. Accounting for future costs in medical cost-effectiveness analysis. *Journal of Health Economics*, 16: 33–64.

Morgan, M. G., and Henrion, M. 1990. *Uncertainty: A Guide to Dealing with Uncertainty in Quantitative Risk and Policy Analysis*. Cambridge University Press, Cambridge.

Morgenstern, R., ed. 1997. *Economic Analyses at EPA: Assessing Regulatory Impact*, Resources for the Future, Washington, DC.

Morrall, J. F., III. 1986. A review of the record. *Regulation*, 10 (November/December): 25–34.

Mueller, D. C. 1989. *Public Choice II*. Cambridge University Press, Cambridge.

Murray, C. L. 1994. Quantifying the burden of disease: The technical basis for disability-adjusted life years. *Bulletin of the World Health Organization*, 72: 429–45.

Nakanishi, J., Oka, T., and Gamo, M. 1998. Risk/benefit analysis of prohibition of the mercury electrode process in caustic soda production. *Environmental Engineering and Policy*, 1: 3–8.

Navrud, S., ed. 1992. *Valuing the Environment. The European Experience*. Oxford University Press, Oxford.

North, D. W. 1983. Quantitative analysis as a basis for decision under TSCA. *American Chemical Society Symposium Series*, 213: 181–95.

North, D. W., Selker, F. K., and Guardino, T. 1992. The value of research on health effects of inorganic arsenic. In W. Campbell, ed., *Proceedings of the International Conference on Arsenic Exposure and Health Effects*, pp. 1–20. Society for Environmental Geochemistry and Health, Northwood.

O'Brien, B., and Viramontes, J. L. 1994. Willingness to pay: A valid and reliable measure of health state preference? *Medical Decision-Making*, 14: 289–97.

Oka, T., Gamo, M., and Nakanishi, J. 1997. Risk benefit analysis of the prohibition of chlordane in Japan. Japanese *Journal of Risk Analysis*, 8(2): 174–86.

Pate-Cornell, E. 1996. Uncertainties in risk analysis: Six levels of treatment. *Reliability Engineering and System Safety*, 54: 95–111.

Pate-Cornell, E. 1998. Risk comparison: Uncertainties and ranking. In A. Mosleh and R. A. Bari, eds., *Probabilistic Safety Assessment and Management*. PSAM 4 (3). Springer, New York.

Pliskin, J. S., Shepard, D. S., and Weinstein, M. C. 1993. Utility functions for life years and health status. *Journal of Health Economics*, 12: 325–39.

Portney, P. R., and Weyant, J. P. 1999. *Discounting and Intergenerational Equity*. Resources for the Future, Washington, DC.

Raiffa, H. 1968. *Decision Analysis: Introductory Lectures on Choices Under Uncertainty*, Addison-Wesley Publishing Company, Reading, MA.

Ramsberg, J. A. L., and Sjoberg, L. 1997. The cost-effectiveness of lifesaving interventions in Sweden. *Risk Analysis*, 17 (4): 467–78.

Reichard, E. G., and Evans, J. S. 1989. Assessing the value of hydrogeologic information for risk-based remedial action decisions. *Water Resources Research*, 25: 1451–60.

Schwing, R. C. 1979. Longevety benefits and costs of reducing various risks. *Technical Forecasting and Social Change*, 13: 333–45.

Siegel, J. E., Graham, J. D., and Stoto, M. A. 1990. Allocating resources among AIDS research strategies. *Policy Sciences*, 23: 1–23.

Taylor, A. C., Evans, J. S., and McKone, T. E. 1993. The value of animal test information in environmental control decisions. *Risk Analysis*, 13: 403–12.

Ten Berge, W. F., and Stallen, P. J. M. 1995. How to compare the risk assessments of accidental and chronic exposure. *Risk Analysis*, 15 (2): 111–13.

Tengs, T. T., and Graham, J. D. 1996. The opportunity costs of haphazard social investments in lifesaving. In R. Hahn, ed., *Risks, Costs and Lives Saved: Getting Better Results from Regulation*, pp. 167–82. Oxford University Press, New York.

Tengs, T. T., Adams, M. E., Pliskin, J. S., Safran, D. G., Siegel, J., Weinstein, M. C. M., and Graham, J. D. 1995. 500 lifesaving interventions and their cost-effectiveness. *Risk Analysis*, 15: 369–89.

Thompson, K. M., and Evans, J. S. 1997. The value of improved national exposure information for perc: A case study of dry cleaners. *Risk Analysis*, 17 (2): 253–98.

Torrance, G. W. 1986. Measurement of health state utilities for economic appraisal. *Journal of Health Economics*, 5: 1–30.

Viscusi, W. K. 1992. *Fatal tradeoffs: Public and private responsibilities for risk*. Oxford University Press, New York.

Viscusi, W. K. 1993. The value of risks to life and health. *Journal of Economic Literature*, 31: 1912–46.

Weinstein, M. C. 1983. Cost-effective priorities for cancer prevention. *Science*, 221: 17–23.

Weinstein, M. C., and Stason, W. B. 1977. Foundations of cost-effectiveness analysis for health and medical practices. *New England Journal of Medicine*, 296: 716–21.

Weinstein, M. C., and Zeckhauser, R. J. 1973. Critical ratios and efficient allocation. *Journal of Public Economics*, 2: 147–57.

Williams, A. 1985. Economics of coronary artery bypass grafting. *British Medical Journal*, 291: 326–9.

Wilson, D. 1998. A case study in cost-benefit analysis: Air quality standards for lead. *Proceedings of the OECD Workshop on the Integration of Social-Economic Analysis in Chemical Risk Management Decisions* (January 7–9, 1998), London.

Zeckhauser, R. J., and Shepard, D. 1976. Where now for saving lives? *Law and Contemporary Problems*, 40(4): 5–45.

Appendix A: Relevant Textbooks

Drummond, M. F., O'Brien, B. J., Stoddard, G. L., and Torrance, G. W. 1997. *Methods for Economic Evaluation of Health Care Programmes*. Oxford Medical Publications, Oxford.

Freeman, A. M. 1993. *The Measurement of Environmental and Resource Values*. Resources for the Future, Washington, DC.

Gold, M. R., Siegel, J. E., Russell, L. B., and Weinstein, M. C., eds. 1996. *Cost-Effectiveness in Health and Medicine*. Oxford University Press, New York.

Gramlich, E. M. 1984. *Benefit-Cost Analysis of Government Programs*. Prentice-Hall, Englewood Cliffs, NJ.

Johannesson, M. 1996. *Theory and Methods of Economic Evaluation of Health Care*. Kluwer Publishers, Norwell, MA.

Johansson, P-O. 1991. *An Introduction to Modern Welfare Economics*. Cambridge University Press, Cambridge.

Johansson, P-O. 1995. *Evaluating Health Risks: An Economic Approach*, Cambridge University Press, Cambridge.

Kopp, R. J., Pommerehne, W. W., and Schwarz, N. 1997. *Determining the Value of Non-Marketed Goods*. Kluwer Academic Publishers, Boston.

Mishan, E. J. 1994. *Cost-Benefit Analysis*, 5th ed. Praeger Publishers, New York.

Mitchell, R. C., and Carson, R. T. 1989. *Using Surveys to Value Public Goods. The Contingent Valuation Method*. Resources for the Future, Washington, DC.

Morgan, M. G., and Henrion, M. 1990. *Uncertainty: A Guide to Dealing with Uncertainty in Quantitative Risk and Policy Analysis*, Cambridge University Press, Cambridge.

Petitti, D. B. 1994. *Meta-Analysis, Decision Analysis, and Cost-Effectiveness Analysis: Methods for Quantitative Synthesis in Medicine*, Oxford University Press, New York.

Tolley, G., Kenkel, D., and Fabian, R., eds. 1994. *Valuing Health for Policy: An Economic Appraisal*. University of Chicago Press, Chicago.

PART THREE

NEW APPROACHES AND NEEDS FOR RISK MANAGEMENT

9

The Challenge of Integrating Deliberation and Expertise

Participation and Discourse in Risk Management

Ortwin Renn

1. INTRODUCTION

The State Department of Environmental Protection was holding a public hearing in Lancaster, a small town in central Massachusetts, concerning the siting of a municipal waste incinerator (Renn and Webler, 1992, p. 84f., similar accounts in Freudenburg, 1983; Elliot, 1984; Davis, 1986; Brion, 1988; Rosa, 1988; positive examples in Lynn, 1987). The town hall was still nearly filled at 10:20 P.M., almost an hour after the scheduled closing time. For two hours residents listened peacefully while regulatory officials explained why the plant was needed, how much trash it would burn, and how much pollution it could legally emit. Expert scientists supported the officials by supplying evidence and arguments about the "acceptability" of the level of risk posed by an incinerator. The regulators thought they were being sensitive to the citizens when they kept to the agenda and, at 9:00 P.M., opened the floor for half an hour of questions. Now it was late, but the citizens were in an uproar, and they showed no sign of letting up.

"What about our gardens?" one woman asked, "Will we be able to eat the vegetables we grow? I have an organic garden and I don't want it contaminated with dioxin." Another gentleman was interested in the contract. "Are you telling us that we have to sign a contract to supply 2,000 tons of trash every year? What if we want to start a recycling program?" And, after a lengthy discourse by one of the tired scientific experts to a question about respiratory illness and particulate emissions, one angry couple blurted out: "You're killing our children!" At 11:15 P.M., a respected town official suggested everyone go home for the evening. The state officials were tired but considered their job done. They had held the legally required public hearing. All they had to do was respond to comments in written form and continue with the permitting process. So they thought! During the next few weeks, some of the furious citizens organized an ad hoc citizen initiative and vowed to stop the incinerator. In their meetings, they expressed concerns over health and environmental damage from air pollution, over the huge refuse-hauling trucks speeding down their town roads, and over the fears that their town would be perceived as a dumping ground for the region. In the battle that ensued, the citizens finally won – to an extent. They were able to stop the incinerator. But the same regulatory officials who were so eager to help them solve their waste problem with an incinerator, were insulted and unwilling to help the town start a recycling program. The town made an attempt to initiate a voluntary recycling program, but eventually bought into an incinerator that was sited about 45 miles away. The problem was just too big for one small town to handle alone.

The Lancaster case is not an isolated incidence in risk management. Almost all industrial countries face severe acceptance problems with respect to siting facilities that pose health or environmental risks to residents (Morell and Margorian, 1982; O'Hare, Bacow, and Sanderson, 1983; O'Hare, 1990; Kunreuther, 1995; Petts, 1997; Löfstedt, 1997). Resolving conflicts about "unwanted" facilities has become one of the major activities of alternative conflict resolution in the United States and all over the world (English et al., 1993; U.S. EPA, 1983, 2001; President's Council, 1997; Wondolleck, Manring, and Crowfoot, 1996; Rowe and Frewer, 2000). But the problem is more than the usual conflict between local communities and regional or national planners. Deciding about the location of hazardous facilities, setting standards for chemicals, making decisions about cleanups of contaminated land, regulating food and drugs, designing and enforcing safety limits have all one element in common: these activities are collective attempts to reduce or control risks to human health and the environment. The term *risk* refers to possible effects of actions or events, which are assessed as unwelcome by the vast majority of human beings. Risks denote the possibility that human actions or events lead to consequences that harm aspects of what humans value.[1] This definition implies that the severity of experienced harm depends on the recognition of a causal relationship between a stimulus (human activity or event) and specific consequences. If we take a nonfatalistic viewpoint, consequences can be altered either by modifying the initiating activity or event or by mitigating the impacts. Therefore, risk is both an analytic and a normative concept (Rosa, 1997). If the vast majority of human beings assess potential consequences as unwelcome or undesirable, society is coerced to avoid, to reduce, or at least to control risks.

All attempts to reduce or control risks are based on two requirements. On the one hand, risk managers need sufficient knowledge about the potential impacts of the risk sources under investigation and the likely consequences of the different decision options to control these risks. On the other hand, they need criteria to judge the desirability or undesirability of these consequences for the people affected and the public at large (McDaniels, 1998). Criteria on desirability are reflections of social values such as good health, equity, or efficient use of scarce resources. Both components, knowledge and values, are necessary components of any decision-making process independent of the issue and the problem context. The literature contains countless procedural guidelines for combining knowledge and values in a rational manner (cf. brief overview in Edwards 1954 or Philips 1979; more elaborate review in Hyman and Stiftel 1988; related to risk management: von Winterfeldt and Edwards, 1986; Chen et al., 1979; Keeney, 1992). Yet decisions on risk pose additional difficulties. They need to address issues such as trade-offs between potential (i.e., uncertain) benefits and damages; the distribution of these potential benefits and risks among different regions, times, or social groups; and deeply rooted values on desirable lifestyles and social organizations (Freudenburg and Pastor, 1992; Fischhoff 1996; Metha, 1998). Risks are often associated with complex cause-effect relationships, uncertainty in the assessment of potential damages and probabilities, as well as far-reaching ambiguities when it comes to interpreting complex and uncertain results.

Dealing with complex, uncertain, and ambiguous outcomes often leads to the emergence of social conflict. Although everyone may agree on the overall goal of safety and environmental quality, precisely what that goal entails (How safe is safe enough?) and precisely how that goal will be obtained (Who bears the risks and who reaps the benefits?) may evoke substantial disagreement (Dietz, Stern, and Rycroft, 1989; Fiorino, 1989a; MacLean, 1986; Rayner and Cantor, 1987; Linnerooth-Bayer and Fitzgerald, 1996). Typical questions in this context are: What are the criteria for judging risks? What role should the assessment of uncertainty and ignorance play in dealing with risks? How should one balance a variety of options with different compositions of magnitude and probability of impacts but identical expected values? How should society regulate risks that benefit one party at the expense of a potential harm to another party?

These crucial questions of risk management demand procedures of decision making that go beyond the conventional agency routines. A variety of strategies to cope with this challenge have evolved over time. They include technocratic decision making through the explicit involvement of expert committees, muddling through in a pluralist society, negotiated rule making via stakeholder involvement, deliberative democracy, or ignoring probabilistic information altogether.[2] The combination of complexity, uncertainty, and ambiguity – typical characteristics of public risk-management decisions – requires new rationales for evaluating policy options on risks. The main thesis of this paper is that risk management agencies are in urgent need to revise their institutional routines and to design procedures that enable them to integrate professional assessments (systematic knowledge), adequate institutional process (political legitimacy), responsible handling of public resources (efficiency), and public knowledge and perceptions (social acceptance).

This chapter explores the possibilities and opportunities to develop risk management procedures for integrating knowledge, efficient handling of resources, political legitimization, and social acceptance. The core focus of this chapter is a critical review of the potential promises, merits, problems, and pitfalls of what analysts have coined an analytic-deliberative process (cf. Stern and Fineberg, 1996; Chess, Dietz, and Shannon, 1998; Renn, 1999a; Webler and Tuler, 1999; Tuler and Webler, 1999; critical reviews in Rossi, 1997; Coglianese, 1999). Such a process is designed to provide a synthesis of scientific expertise, a common interpretation of the analyzed relationships, and a balancing of pros and cons for regulatory actions based on insights and values. Analysis in this context means the use of systematic, rigorous, and replicable methods to formulate and evaluate knowledge claims (Stern and Fineberg, 1996, p. 98; cf. also Tuler and Webler, 1999, p. 67). These knowledge claims are normally produced by scientists (natural and social). In many instances, relevant knowledge also comes from stakeholders or members of the affected public. The main point is not where the knowledge is coming from but how adequately the knowledge serves the purpose of providing "useful" answers to the question it is supposed to address (for more detail on the role of scientific risk assessment for risk evaluation, see Shrader-Frechette 1991, pp. 29–30). *Deliberation* is the term we have adopted from the literature to highlight the style and nature of problem solving through communication and collective consideration of relevant issues (Stern and Fineberg, 1996, pp. 73 and 215ff.; original idea of discursive deliberation from Habermas 1970, 1978a). It combines different forms of argumentation and communication, such as exchanging observations and viewpoints, weighing and balancing arguments, offering reflections and associations, and putting facts into a contextual perspective. The word *deliberation* implies equality among the participants, the need to justify and argue for all types of (truth) claims, and an orientation toward mutual understanding and learning (in general: Habermas, 1987b, 1989, 1991; Dryzek, 1994; Sager, 1994; Sclove, 1995; Cohen, 1997; applied to risk management: Kemp, 1985; Warren, 1993; Tuler and Webler, 1995, Webler, 1995, 1999; Daniels and Walker, 1996).

Within this chapter, we will discuss the potential benefits and problems associated with the implementation of analytic-deliberative processes in the field of risk management. Section 2 is focused on the requirements of decision making in pluralistic and democratic societies and explicates the implications for risk management. It points to the potential contributions of economic, political, social, and scientific rationality for risk evaluation and management and highlights the need for effective, efficient, legitimate, and acceptable management decisions. Section 3 deals with the two elements of analytic-deliberative processes (i.e., scientific analysis and open deliberation). It defines the two components, presents arguments pro and con deliberation, explains the theoretical concepts behind deliberation, and articulates the requirements for a process to be called deliberative. Section 4 provides a brief review of different procedures and instruments that can be used for a hybrid model of analysis and deliberation. These instruments are ordered according to a proposed distinction between cognitive, reflective, and participatory discourse. One hybrid model, the model of cooperative discourse, is further explicated and described in Section 5. Section 6 discusses the problem of evaluation and testing. The conclusions in Section 7 summarize the major results and present some general insights.

2. EXPERTISE AND DELIBERATION IN RISK MANAGEMENT

2.1. Elements of Decision Making in Pluralist Societies

What attributes must a decision-making process have when a decision proposal implies a multiplicity of uncertain impacts on several affected groups? At the foundation of a society are the needs for *effectiveness, efficiency, legitimacy, and social cohesion*.[3] Effectiveness refers to the need of societies to have a certain degree of confidence that human activities and actions will actually lead to those consequences that the actors intended when performing these actions. Efficiency describes the degree to which scarce resources are utilized for reaching the intended goal. The more resources are invested to reach a given objective, the less efficient is the activity under question. Legitimacy is a composite term that denotes the degree of compatibility with the legal requirements, due process, and political culture. It includes an objective element such as legality and a subjective judgment such as the perception of acceptability (Luhmann, 1983; Scharpf, 1991). Lastly, social cohesion covers the need for social integration and collective identity in spite of plural values and lifestyles (Parsons, 1971; Renn and Webler, 1998, p. 9ff).

Within the institutional organization of modern societies, these four foundations are predominantly handled by different societal systems: economics, science (expertise), politics (including the legal systems), and social structure (Parsons, 1967; Münch, 1982; applied to technology assessment in Renn, 1999b). In the recent literature on governance, the political system is often associated with the rationale of hierarchical and bureaucratic reasoning; the economic system, with monetary incentives and individual rewards; and the social sphere, with the deregulated interactions of groups within the framework of a Civil Society[4] (overview in Etzioni, 1968; Ash, 1997; Shils, 1991; Seligman, 1993; Hirst, 1994). Scientific input in these models is seen as an integral part of either politics in the form of scientific advisory committees or the civil sector in the form of independent institutions of knowledge generation (Alexander, 1993; Held, 1995; specifically for handling environmental risks see Hajer, 1995). For characterizing risk management, I prefer to distinguish four separate subsystems since scientific expertise is of crucial importance to risk decisions and cannot be subsumed under the other three systems. Because science is a part of a society's cultural system, our proposed classification is also compatible with the classic division of society into four subsystems (politics, economics, culture, and social structure) as suggested by the functional school in sociology (Parsons, 1951; Münch, 1982). The picture of society is, of course, more complex than the division in four systems suggest. Many sociologists relate to the concept of "embeddedness" when describing the relationships between the four systems (Scott and Meyer, 1991; Granovetter, 1992). Each system is embedded in the other systems and mirrors the structure and functionality of the other systems in subsystems of their own. For making our argument, however, the simple version of four analytically distinct systems is sufficient.

Each of the four systems is characterized by governance processes and structures adapted to the system properties and functions in question. The four systems and their most important structural characteristics are shown as a scheme in Figure 9.1 (based on Downs, 1957; Münch, 1982; adapted from Renn, 1999b). What findings can be inferred from the comparison of these four systems?

1. In the market system, decisions are based on the cost-benefit balance established on the basis of individual preferences, property rights, and individual willingness to pay. The main goal here is to be efficient.

2. In politics, decisions are made on the basis of institutionalized procedures of decision making and norm control (within the framework of a given political culture and system of government). The target goal here is to seek legitimacy.

3. Science has at its disposal methodological rules for generating, challenging, and testing knowledge claims, with the help of which one can assess decision options according to their likely consequences and side effects. These insights help policy makers to be effective.

4. Finally, in the social system, there is a communicative exchange of interests and arguments, which helps the actors to come to a jointly agreed-upon solution. This creates cohesion and social acceptance.

Corresponding to their respective system logic, the four systems can be allocated certain methods or instruments, which are used basically or in hybrid forms when making decisions under uncertainty, setting trade-offs, or reconciling conflicting values (see Figure 9.1). Economic system logic has the instruments of (shadow) price setting, financial incentive systems, transfer of rights of ownership of public or nonrival goods and financial compensation (damages, insurance) to persons whose utilities have been reduced by the activities of others. The expert system uses a wide variety of

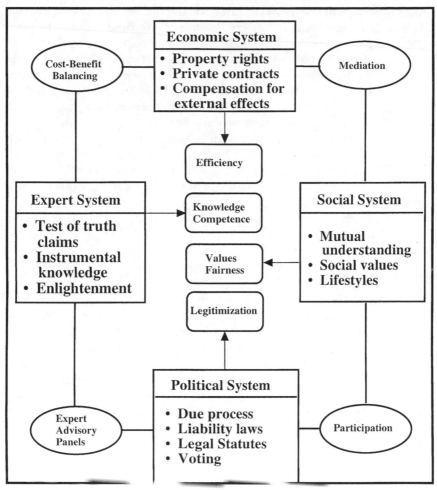

FIGURE 9.1. Four central systems of society.

knowledge-based decision methods (e.g., choice of appropriate methodology, peer review, Delphi, meta-analysis, (medical) consensus conferences, among others) to determine the likely and unlikely impacts of different decision options. In the social field, discursive forms of communication and joint problem solving are used within the framework of advisory committees, joint problem-solving groups, or citizen participation procedures. The conventional instruments of the political system comprise the processes of passing, implementing, and scrutinizing collectively binding decisions in executive, legislative, or judicial bodies. Votes in a parliament are as much part of this governance model as is the challenging of decisions before a court.

Socially relevant problems are rarely dealt with within the limits of one single system logic. Rather, they go through hybrid procedures, which are to be seen as combinations of the individual systems. The settlement of conflicts with the method of mediation or negotiated rule making can, for example, be interpreted as linkages between the economic and the social sector, while the cooperation between experts and political representatives in joint advisory committees (i.e., the experts provide background knowledge, while the politicians provide the preferences for making the appropriate choices) represents a combination of knowledge-oriented elements and political governance. Classical hearings are combinations of expert knowledge, political resolutions, and the inclusion of citizens in this process (for comparisons with these models see also the three political advisory models in Habermas, 1968, and the steering models of power, money, and knowledge in Willke, 1995, and more detailed descriptions in Renn and Webler, 1998, pp. 9ff).

2.2. Application to Risk Management

Risk decisions are routinely made within all four systems. Managers may decide on a risky strategy to market a product; consumers may decide to eat beef in spite of the BSE (bovine spongiform encephalopathy, in popular terms mad cow disease) warnings; scientists may find out that a specific concentration of a new substance may lead to cancer; and political bodies may require motorists to wear helmets when driving on public streets. However, many risk decisions require input from all four systems. This is particularly true for risk management decisions that impose restrictions on one part of the population to protect other parts or, vice versa, to allow some parts to impose risks on other parts (Kasperson and Kasperson, 1983; Vari, 1989; Kunreuther and Slovic, 1996; Linnerooth-Bayer and Fitzgerald, 1996). In these cases, legitimate decision making requires the proof that

- alternative actions are less cost-effective (*efficiency*),
- the required course of action (or nonaction) would result in the anticipated positive results (*effectiveness*),
- the actions are in line with due process and democratic procedures (*legitimacy*), and
- these actions are finding public approval (*social acceptance*).

When contemplating the acceptability of a risk, one needs to be informed about the likely consequences of each decision option and the opportunity cost for choosing one option over the other. One also needs to be cognizant of the potential violations of interests and values connected with each decision option (Vlek and Cvetkovich, 1989; Gregory, Lichtenstein, and Slovic, 1993; McDaniels, 1996). This is basically true for any far-reaching political decision. Decision making on risk, however, includes three additional elements that were already mentioned in the introduction: complexity, uncertainty, and ambiguity (Renn, 1999b; Klinke and Renn, 2002; based on March, 1978, and Elster, 1989; similar reasoning in Shrader-Frechette, 1985, 1991, pp. 182ff.).

Complexity refers to the difficulty of identifying and quantifying causal links between a multitude of potential candidates and specific adverse effects (Schellnhuber, 1999; WBGU, 2000, pp. 194ff.). The nature of this difficulty may be traced back to interactive effects among these candidates (synergisms and antagonisms), long delay periods between cause and effect, interindividual variation, intervening variables, and others (National Research Council, 1988, pp. 179ff.). It is precisely these complexities that make sophisticated scientific investigations necessary since the dose-effect relationship is neither obvious nor directly observable.

The second problem refers to uncertainty. Uncertainty is different from complexity. It comprises different and distinct components such as statistical variation, measurement errors, ignorance, and indeterminacy (Stirling, 1998; van Asselt, 2000, pp. 85ff; WBGU, 2000, pp. 52ff.), which all have one feature in common. Uncertainty reduces the strength of confidence in the estimated cause-and-effect chain (Priddat, 1996; Stirling, 1998). There is a clear connection between complexity and uncertainty. If complexity cannot be resolved by scientific methods, uncertainty increases. The opposite, however, is not necessarily true. Even simple relationships may be associated with high uncertainty if either the knowledge base is missing or the effect is stochastic by its own nature.

The last term in this context is ambiguity or ambivalence. This term denotes the variability of (legitimate) interpretations based on identical observations or data assessments (Renn, 1999b). Most of the scientific disputes in the fields of risk analysis and management do nor refer to differences in methodology, measurements, or dose-response functions, but to the question of what all this means for human health and environmental protection (Harrison and Hoberg, 1994, pp. 6 and 168ff.; Jasanoff, 1991; Margolis, 1996, pp. 99ff.). Emission data are hardly disputed. Most experts debate, however, whether an emission of x constitutes a serious threat to the environment or to human health. In risk management decisions, ambiguity plays an even stronger role. Questions often raised in this context are: Should regulation be confined to avoid significant health effects or should it be expanded to any measurable effect that could cause some still unknown damage (see the discussion on electromagnetic fields, EMF)? Should voluntary risks be regulated less stringently than risks that are imposed on people? Should natural risks receive identical attention and treatment compared to human-induced risks (afflotoxins versus pesticide residues, for example)? Again high complexity and uncertainty favor the emergence of ambiguity, but there are also quite a few simple and almost certain risks that can cause controversy and thus ambiguity. In Germany, for example, speed limits on highways are regarded as unacceptable by the majority of German citizens in spite of the fact that a considerable number of lives could be saved if they were introduced (Helten, 1998). Other countries find this policy absurd. Another classic example is smoking: Should the government regulate smoking because the link to cancer is clearly confirmed, or should it refrain from regulation since people voluntarily accept this risk (not including passive smoking here for the sake of the argument)? If one looks at the practice in different countries, one can observe a variety of approaches to deal with smoking ranging from strict to almost no regulatory actions. It is therefore important to distinguish between complexity, uncertainty, and ambiguity even if these three terms are correlated.

Different evaluation and management strategies follow from the analysis of these three challenges. If the problem is complexity, a risk manager is well advised to bring the best expertise together and regulate on the basis of the state-of–the-art knowledge in risk assessment. It does not make much sense to incorporate public concerns, perceptions, or any other social aspects into the function of resolving (cognitive) complexity unless specific knowledge of these groups help to untangle complexity (see also Charnley, 2000, pp. 16f.). Complex phenomena demand almost equally complex methods of assessments. And these methods can be offered by scientists and experts better than by anybody else. In terms of regulatory actions, quantitative safety goals and consistent use of cost-effectiveness methods would be the appropriate tools to deal with complex risk problems that show little uncertainty and no ambiguity (WBGU, 2000, pp. 288ff.). The major input from the four systems of society needed to manage complex risks is an assurance of effectiveness. As stated before, effective risk management relies on state-of-the-art expertise (WBGU, 2000, pp. 41ff.).

If the problem is uncertainty, however, knowledge is either not available or unattainable owing to the nature of the hazard. Under these circumstances, risk managers must rely on resilience as the guiding principle for action (Wynne, 1992a; Coolingridge, 1996; WBGU, 2000, pp. 176ff.) Most of the precautionary management tools would fall into this category (see Stirling, 1999; Bennet, 2000; Klinke and Renn, 2002). Knowledge acquisition may help to reduce uncertainty and thus move the risk back to the first stage of handling complexity. If uncertainty cannot be reduced by additional knowledge, however, or if action is demanded before the necessary knowledge can be obtained, the routine management strategies for resolving complexity would be incomplete because the objective here is to act prudently under the condition of uncertainty. Acting prudently means to design resilient measures that allow flexible responses to unexpected surprises (Stirling, 1998; WBGU, 2000, pp. 289ff.). Management tools that would fit the resilience approach include containment in space and time (to make exposure reversible), constant monitoring, development of equifunctional replacements, and investments in diversity and flexibility (Coolingridge, 1996; Klinke and Renn, 1999). Classic regulatory strategies such as the ALARA principle (as low as reasonably achievable), BACT (best available control technology), or state of technology are also elements of this approach (Pinkau and Renn, 1998; WBGU, 2000, pp. 217–18).

Decisions based on uncertainty management require, therefore, more than input from risk specialists. They need to include stakeholder concerns, economic budgeting, and social evaluations. The focal point is here to find the adequate and fair balance between the costs for being overcautious versus the costs of being not cautious enough (van den Daele, 2000, p. 215). Because both costs are almost impossible to quantify owing to the remaining uncertainties, subjective judgments are inevitable. These judgments should be performed on the basis of a careful analysis of resource allocation, but there is no way of avoiding painful value trade-offs. Assigning trade-offs belongs to the realm of the economic system as a means to ascertain efficiency and utility optimization (Graham and Wiener, 1995; Viscusi, 1998). It should be noted, however, that these economic tools of decision making under uncertainty prescribe only the procedure of making rational choices but do not specify the substantial ratios for the trade-offs (McDaniels, 1998). The ratios determine who will bear the costs – either in the form of additional damages by being not cautious enough or in the form of regulatory costs for being overcautious. It is obvious that those who bear either of the two costs are entitled to make the substantial judgments for the necessary trade-offs.

Setting trade-offs is even more complex when it comes to resolving ambiguity. Although scientific expertise is essential for understanding ambiguities better and dealing with them in an enlightened manner, it cannot prescribe the value trade-offs to resolve them (van Asselt, 2000, pp. 165ff; Charnley and Elliot, 2000). In addition, ambiguities cannot be resolved by increased efficiency because the outcome in itself is controversial not just the distribution of costs. Genetically modified organisms for agricultural purposes may serve as an example here. Our own surveys on the subject demonstrate that people associate high risks with the application of gene technology for social and moral reasons (Hampel and Renn, 2000). Whether the benefits to the economy balance the costs to society in terms of increased health risks was not a major concern of the polled public. People disagreed about the social need for genetically modified food in Western economies where abundance of conventional food is prevalent, about the loss of personal agency when selecting and preparing food, about the long-term impacts of industrialized agriculture, and about the moral implications of tampering with nature (Sjöberg, 1999b; cf. also Thompson, 1988).

These concerns cannot be addressed either by scientific risk assessments or by the right balance between over- and underprotection. The risk issues in this debate focus on the differences of visions about the future, basic

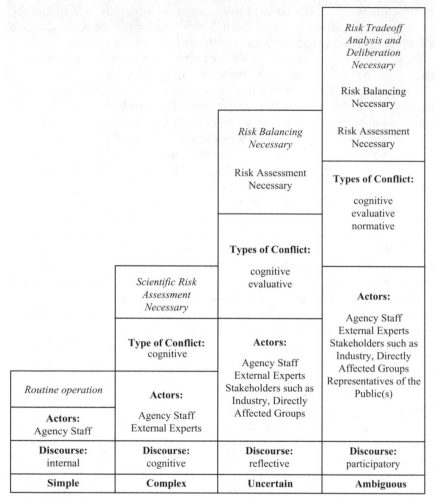

FIGURE 9.2. The risk management escalator (from simple via complex and uncertain to ambiguous phenomena).

values and convictions, and the degree of confidence in the human ability to control and direct its own technological destiny. These wider concerns require the inclusion of those who express or represent them (Krimsky, 1979; Barber, 1984; Dietz, 1987; Burns and Überhorst, 1988; Waller, 1995; Knight, 1998; Harter et al., 1998). Surveys may reveal some of these concerns, but they cannot resolve the underlying value conflicts or differences in vision. To make these concerns operational for risk management, some kind of discourse for conflict resolution and joint vision building is required. Coping with ambiguity necessarily leads to discursive management tools – communicative processes that promote rational value disputes

(Laksmanan, 1990; Laird, 1993; Rippe and Schaber, 1999, p. 77). This is the place where deliberative processes are required from a social-analytical as well as normative viewpoint (Bohman 1997, 1998; Cohen, 1997).

Most risks are characterized by a mixture of complexity, uncertainty, and ambiguity. Smoking may be a good example for low complexity and uncertainty but high ambiguity. Nuclear energy may be a good candidate for high complexity and high ambiguity but relatively little uncertainty. Endoctrine disrupters could be cited as examples for high complexity, uncertainty, and ambiguity. We could continue the list forever. If one looks at risks systematically, one can detect a gradual increase of sophistication starting from complexity over uncertainty to ambiguity as illustrated in Figure 9.2.

3. DELIBERATION IN RISK MANAGEMENT: POTENTIALS, PROBLEMS, AND REQUIREMENTS

3.1. The Concept of Deliberation

At the end of the last section, the conclusion was reached that complex risk issues need input from all four sectors of society and that the three challenges of complexity, uncertainty, and ambiguity demand expertise, cost trade-offs, due process, and the inclusion of public concerns. In theory, a benevolent dictator or an enlightened risk management agency could provide all this input without reliance on external feedback. In pluralist and democratic societies, however, differentiation of functional systems into semiautonomous entities (sociologists call them autopoietic) has produced a self-image of society, in which each system insists on generating and providing the functional input to collective decision making for which it is specialized (Luhmann, 1986; Scharpf, 1991; Bailey, 1994, pp. 285–322). In addition, the other systems usually expect that each system provides its specific contribution to collective decision making (Turino, 1989b). Experts expect to be heard and considered when risk managers make factual claims about potential outcomes of risky technologies or activities. Representatives of industry and the unions demand to be included when trade-offs with financial implications for them are made. Communities and affected citizens would like to be consulted if their backyard is being altered by locating a risky facility there. Not only do system representatives request to be included, it also seems wise from a purely functional viewpoint to use the collective pool of experience and problem-solving capacity of each of the systems rather than trying to reconstruct a mirror image of all societal systems in each decision-making agency (although some of this may also be necessary in order to process all the input from outside sources). Lastly, normative democratic theory would also suggest that those who are affected by public decisions should have a right to be involved or at least consulted (N. Rosenbaum, 1978; Kasperson, 1986; Sclove, 1995).

Based on these considerations, risk management agencies have established new routines for outreach as a means to tap into the experience of other systems. Numerous scientific advisory councils, stakeholder round tables, citizens advisory groups, and other forms of public outreach are examples of risk management agencies at work all over the world improving their basis for legitimizing decisions and practices.[5] The different forms of cooperation between risk management agencies and various constituencies of society have been described, analyzed, and reviewed under different conceptual headings:

- Within the tradition of the sociology of science, the interplay between experts and policy makers has been a popular topic of scholarly work (Primack and von Hippel, 1974; Nelkin, 1977; Knorr-Cetina, 1981; Jasanoff, 1982, 1991; Coppock, 1985; Rip, 1985, 1992; Majone, 1989; Functowicz and Ravetz, 1990; Shrader-Frechette, 1991; van den Daele, 1992; von Schomberg, 1992; De Marchi and Ravetz, 1999).
- Different traditions in the political sciences have focused on the role of stakeholders and organized groups in the risk arena (Kitschelt, 1980, 1986; Olson, 1984; Hillgartner and Bosk, 1988; Eder, 1992; Renn, 1992; Dahl, 1994; Schmidt, 2000).
- Other scholars have emphasized the relevance of trust and credibility in the relationships between governmental agencies, industry, stakeholder groups, and the general public (Renn and Levine, 1990; Slovic, 1993; Peters, 1995; Earle and Cvetkovich, 1995, 1999; Johnson, 1999; Löfstedt, 2001).
- Analysts of public participation and involvement have investigated different approaches of risk managers to have representatives of organized and unorganized interests in society participate in decision making (Guild, 1979; Kweit and Kweit, 1981, 1987; Fiorino, 1990; Laird, 1993; Renn et al., 1995; Creighton et al., 1998).
- The evolving concept of analytic-deliberative process is one of the few attempts to develop an integrative approach to cooperative risk management based on the inclusion of experts, stakeholders, and the general public (Stern and Fineberg, 1996; Chess, Dietz, and Shannon, 1998; Tuler and Webler, 1999; Renn, 1999a; Charnley, 2000).

In the following paragraphs and sections, I will explain the concept of deliberation, relate it to our analysis of the three challenges of risk management, and provide an example for one practical model that has been designed to put analytic-deliberative processes into action.

The term *deliberation* refers to the style and procedure of decision making without specifying which participants are invited to deliberate (Stern and

Fineberg, 1996; Rossi, 1997). For a discussion to be called deliberative, it must rely on mutual exchange of arguments and reflections rather than on decision making based on the status of the participants, sublime strategies of persuasion, or social-political pressure. Deliberative processes should include a debate about the relative weight of each argument and a transparent procedure for balancing pros and cons (Tuler and Webler, 1999). In addition, deliberative processes should be governed by the established rules of a rational discourse. The term *discourse* has different meanings in the social sciences (Renn, 1998). Discourse is often used to mean either whole language texts or narrative scripts in their context of use or the world-views that inform our understanding. In the theory of communicative action developed by the German philosopher Juergen Habermas, the term *discourse* denotes a special form of a dialogue in which all affected parties have equal rights and duties to present claims and test their validity in a context free of social or political domination (Habermas, 1970, 1987b). A discourse is called rational if it meets the following specific requirements (cf. McCarthy, 1975; Habermas, 1991; Kemp, 1985; Renn and Webler, 1998, pp. 48ff.; Webler, 1995, 1999). All participants are obliged to

- seek a consensus on the procedure that they want to employ in order to derive the final decision or compromise, such as voting, sorting positions, establishing consensual decision making, or involving a mediator or arbitrator;
- articulate and critique factual claims on the basis of the "state of the art" of scientific knowledge and other forms of problem-adequate knowledge (in the case of dissent, all relevant camps have the right to be represented);
- interpret factual evidence in accordance with the laws of formal logic and analytical reasoning;
- disclose their relevant values and preferences, thus avoiding hidden agendas and strategic game playing,
- process data, arguments, and evaluations in a structured format (e.g., a decision-analytic procedure) so that norms of procedural rationality are met and transparency can be created.

The rules of deliberation do not necessarily include the demand for stakeholder or public involvement. Deliberation can be organized in closed circles (such as conferences of Catholic bishops, where the term has indeed been used since the Council of Nicosea) as well as in public forums. Following our arguments, however, complex, uncertain, and ambiguous risk decisions require contributions from scientists, policy makers, stakeholders, and affected publics, a procedure is thus required that guarantees both the inclusion of different constituencies outside the risk management institutions and the assurance of a deliberative style within the process of decision making. I suggest using the term *deliberative democracy* when one refers to the combination of deliberation and third party involvement (see also Cohen, 1997; Rossi, 1997).

3.2. The First Component: The Role of Scientific Analysis in Risk Management

The first element of analytic-deliberative processes refers to the inclusion of systematic and reproducible knowledge in risk management. There is little debate in the literature that the inclusion of external expertise is essential as a major resource for obtaining and utilizing risk knowledge (Jasonoff, 1993; Stern and Fineberg, 1996; Bohnenblust and Slovic, 1998). A fierce debate has evolved, however, on the status of scientific expertise for representing all or most of the relevant knowledge. This debate includes two related controversies: the first controversy deals with the problem of objectivity and realism; the second one, with the role of anecdotal and experiential knowledge that nonexperts have accumulated over time.[6] This is not the place to review these two controversies in detail (see Bradbury, 1989; Shrader-Frechette, 1990, 1991; Clarke and Short, 1993, pp. 379ff.; Klinke and Renn, 2002). Depending on which side one takes in this debate, scientific evidence is either regarded as one input to fact finding among others or as the central or even only legitimate input for providing and resolving knowledge claims. There is agreement, however, among all camps in this debate that systematic knowledge is needed and that it should be generated and evaluated according to the established rules or conventions of the respective discipline. Methodological rigor aiming to accomplish a high degree of validity, reliability, and relevance remains the most important yardstick for judging the quality of scientific insights. Most constructivists do not question the need for methodological rules but are skeptical whether the results of scientific inquiries represent objective or – even more so – unambiguous descriptions of reality (Latour and Woolgar, 1979; Knorr-Cetina, 1981). They rather see scientific results as products of specific processes or routines that an elite group of knowledge producers has framed as "objective" and "real." In "reality," these products are determined by the availability of research routines and instruments, prior knowledge and judgments, and social interests (see also Beck, 1992, although he regards himself as a moderate realist).

For the analysis of analytic-deliberative processes, the divide between the constructivists and the realists (and all the ones in between) matters only in the degree to which scientific input is used as a knowledge base or as a final arbiter for reconciling knowledge conflicts. The analytic process in itself follows more or less identical rules independent of the philosophical stance on realism. A knowledge discourse deals with different, sometimes competing claims that obtain validity only through a compatibility check with acknowledged procedures of data collection and interpretation, a proof of theoretical compatibility and conclusiveness, and the provision of intersubjective opportunities for reproduction (Shrader-Frechette, 1991, pp. 46ff.). Obviously many research results do not reach the maturity of proven facts, but even intermediary products of knowledge, ranging from plain hypotheses via plausible deductions to empirically proven relationships, strive for further perfection (cf. the pedigree scheme of Functowicz and Ravetz, 1990). On the other hand, even the most ardent proponent of a realist perspective will admit that only intermediary types of knowledge are often available when assessing and evaluating risks (Starr and Whipple, 1980).

What does that mean for analytic-deliberative processes?

- First, scientific input is essential for risk decision making. The degree to which the results of scientific inquiry are taken as ultimate evidence to judge the appropriateness and validity of competing knowledge claims is contested in the literature and should therefore be one of the discussion points during deliberation. The answer to this question may depend on context and the maturity of scientific knowledge in the respective risk area (see also Horlick-Jones, 1998). For example, if the issue is the effect of a specific toxic substance on human health, anecdotal evidence may serve as heuristic tools for further inquiry, but there is hardly any reason to replace toxicological and epidemiological investigations with intuitions from the general public. If the issue is siting of an incinerator, anecdotal and local knowledge about sensitive ecosystems or traffic flows may be more relevant than systematic knowledge about these impacts in general (a good example for the relevance of personal knowledge can be found in Wynne, 1989).
- Second, the resolution of competing claims of scientific knowledge should be governed by the established rules within the respective discipline. These rules may not be perfect and may even be contested within the community. Yet they are usually superior to any other alternative (Shrader-Frechette, 1991, pp. 190ff.; Harrison and Hoberg, 1994, pp. 49ff.).
- Third, many problems and decision options require systematic knowledge that is either not available or still in its infancy or in an intermediary status. Analytic procedures are demanded as a means to assess the relative validity of each of the intermediary knowledge claims, show their underlying assumptions and problems, and demarcate the limits of reasonable knowledge (i.e., identify the range of those claims that are still compatible with the state of the art in this knowledge domain; Renn, 1995).
- Fourth, knowledge claims can be systematic and scientific as well as idiosyncratic and anecdotal. Both forms of knowledge have a legitimate

place in analytic-deliberative processes. How they are used depends on the context and the type of knowledge required for the issue under question (Wynne, 1989; Webler, 1999).

All four points show the importance of knowledge for risk management but also make clear that choosing the right management options requires more than looking at the scientific evidence alone.

3.3. The Second Component: Deliberation in Risk Management

What about the second part of the process: deliberation? What needs to be deliberated? First, as I argued earlier, deliberative processes are needed to define the role and relevance of systematic and anecdotal knowledge for making risk choices. Second, deliberation is needed to find the most appropriate way to deal with uncertainty and to set efficient and fair trade-offs between potential over- and underprotection. Third, deliberation needs to address the wider concerns of the affected groups and the public at large. These three elements of deliberation are related to resolving complexity, uncertainty, and ambiguity.

Why do I expect deliberative processes to be better suited for dealing with these three challenges than using data from surveys among the relevant constituents or organizing systematic feedback from them? To respond to this question, it is necessary to introduce the different theoretical concepts underlying deliberative democracy. The potential of deliberation has been discussed primarily in three schools of thought.

* *The utility-based theory of rational action* (basics in Fisher and Ury, 1981; Bacow and Wheeler, 1984; Raiffa, 1994; Susskind and Fields, 1996; Knight, 1998; critical review in Nicholson, 1991; review of pros and cons in Friedman, 1995; Jaeger et al., 2001). In this concept, deliberation is framed as a process of finding one or more option(s) that optimizes the payoffs to each participating stakeholder. The objective is to convert positions into statements of underlying interests. If all participants articulate their interest, it is possible either to find a new win-win option that is in the interest of all or at least does not violate anybody's interest (Pareto optimal solution) or to find a compensation that the winner pays to the losers to the effect that both sides are at least indifferent between the situation without the risk and no compensation and the risk plus compensation (Kaldor-Hicks solution). Deliberation helps to find either one of the two solutions and provides acceptable trade-offs between over- and underprotection for all participants.
* *Theory of communicative action* (Habermas, 1984, 1987a; Apel, 1992; Brulle, 1992; Benhabib, 1992; Webler, 1995; Cohen, 1997; Renn and Webler, 1998, pp. 48–57). This concept focuses on the communicative process

of generating preferences, values, and normative standards. Normative standards are those prescriptions that do not apply only to the participants of the discourse but also to society as a whole or at least a large segment of the external population. Normative standards in risk management are, for example, exposure limits or performance standards for technologies. They apply to all potential emitters or users regardless of whether they were represented at the discourse table or not. The objective here is to find consensus among moral agents (not just utility maximizers) about shared meaning of actions based on the knowledge about consequences and an agreement on basic human values and moral standards.

- *Theory of social systems* (Luhmann, 1989, 1990, 1993; Markowitz, 1990; Eder, 1992; Japp, 1996). The (neo)functional school of sociology pursues a different approach to deliberation. It is based on the assumption that each stakeholder group has a separate reservoir of knowledge claims, values, and interpretative frames. Each group-specific reservoir is incompatible with the reservoir of the other groups.[7] This implies that deliberative actions do not resolve anything. They represent autistic self-expressions of stakeholders. In its cynical and deconstructivist version, deliberation serves as an empty but important ritual to give all actors the illusion of taking part in the decision process. In its constructive version deliberation leads to the enlightenment of decision makers and participants (Jaeger et al., 2001). Far from resolving or even reconciling conflicts, deliberation in this viewpoint has the potential to decrease the pressure of conflict, provide a platform for making and challenging claims, and assist policy makers (Luhmann, 1989). Deliberations help to reframe the decision context, to make policy makers aware of public demands, and to enhance legitimacy of collective decisions through reliance on formal procedures (Luhmann, 1983; Freudenburg, 1983; Skillington, 1997). In this understanding of deliberation, reaching a consensual conclusion is neither necessary nor desirable. The process of talking to each other, exchanging arguments, and widening one's horizon is all that deliberation is able to accomplish. It is an experience of mutual learning without a substantive message.

These three understandings of deliberation are not mutually exclusive although many proponents of each school advocate their line of reasoning as if there were no room for alternative explanations or prescriptions. Based on the previous arguments, it is quite obvious that the rational actor approach is well suited to deal with questions of uncertainty management and the search for a balance between over- and underprotection. The communicative action approach provides a theoretical structure for understanding and organizing discourses on ambiguities. In particular, this concept can highlight those elements of deliberation that help participants to deal competently with moral and normative issues beyond personal interests. The system-analytic school introduces some skepticism toward the claim of the other schools with respect to the outcomes of deliberation. Instead, it emphasizes the importance of procedures, routines, and learning experiences for creating links or networks between the major systems of society (see also Stern, 1991). Deliberation is the lubricant that helps each system to move mostly independently in society without bumping into the domains of the other systems.

Turning back to the question of suitability, one can deduce from these three theoretical concepts the following inferences:

- Deliberation can produce common understanding of the issues or the problems based on the joint learning experience of the participants with respect to systematic and anecdotal knowledge (Webler et al., 1995b; Pidgeon, 1997).
- Deliberation can produce a common understanding of each party's position and argumentation and thus assist in a mental reconstruction of each actor's argumentation (Warren, 1993; Tuler, 1996). The main driver for gaining mutual understanding is empathy. The theory of communicative action provides further insights in how to mobilize empathy and how to use the mechanisms of empathy and normative reasoning to explore and generate common moral grounds (Webler, 1995).
- Deliberation can produce new options and novel solutions to a problem. This creative process can either be mobilized by finding win-win solutions or by discovering identical moral grounds on which new options can grow (Renn and Webler, 1998, pp. 64ff.). Even seen from a system-analytical perspective, joint statements can be generated in deliberative processes that may not resolve the issue but, instead, define the conditions under which each system can handle the problem using its own resources. A good example is the conflict about genetically modified food. Proponents and opponents may not find a common solution to the management of food risks, they may not even find a common problem definition, but they might agree to label all food items and leave it to consumers to decide whether they want to accept the risk or reject it.
- Deliberation has the potential to show and document the full scope of ambiguity associated with a risk problem. Even if one shares the skeptical view of system-analysts, deliberations helps to make a society aware of the options, interpretations, and potential actions that are connected with the issue under investigation (Wynne, 1992b; De Marchi and Ravetz, 1999). Each position within a deliberative discourse can only survive the crossfire of arguments and counterarguments if it demonstrates internal consistency, compatibility with the legitimate range of knowledge claims, and correspondence with the widely accepted norms and values of society. Deliberation clarifies the problem, makes people

aware of framing effects, and determines the limits of what could be called reasonable within the plurality of interpretations (Skillington, 1997).

Deliberations can also produce agreements. The minimal agreement may a consensus about dissent (Raiffa, 1994; Renn and Webler, 1998, p. 64). If all arguments are exchanged, participants know why they disagree. They may not be convinced that the arguments of the other side are true or morally strong enough to change their own position; but they understand the reasons why the opponents came to their conclusion. At the end the deliberative process produces several consistent and – in their own domain – optimized positions that can be offered as package options to legal decision makers or the public. After these options have been subjected to public discourse and debate, political bodies such as agencies or parliaments can make the final selection in accordance with the legitimate rules and institutional arrangements such as a majority vote or executive order. Final selections could also be performed by popular vote or referendum (Hartman, 1983; Wehrli-Schindler, 1987).

Deliberation may result in consensus. Often deliberative processes are used synonymously with consensus-seeking activities (Coglianese, 1997). This is a major misunderstanding. Consensus is a possible outcome of deliberation but not a mandatory requirement. If all participants find a new option that they all value more than the one option that they preferred when entering the deliberation, a "true" consensus is reached (Renn and Webler, 1998, p. 69). It is clear that finding such a consensus is the exception rather than the rule. Consensus is either based on a win-win solution (examples in Waldo, 1987) or a solution that serves the "common good" and each participant's interests and values better than any other solution (examples in Renn, 1999a). Less stringent is the requirement of a tolerated consensus. Such a consensus rests on the recognition that the selected decision option might serve the "common good" best but at the expense of some interest violations or additional costs. In a tolerated consensus, some participants voluntarily accept personal or group-specific losses in exchange for providing benefits to all of society. In our own empirical work, deliberation has often led to tolerated consensus solution, particularly in siting conflicts (one example in Schneider, Oppermann, and Renn, 1998). Tolerated consensus is difficult to explain in terms of the rational actor paradigm but is quite compatible with the theory of communicative action. Consensus and tolerated consensus should be distinguished from compromise. A compromise is a product of bargaining where each side gradually reduces its claim to the opposing party until they reach an agreement (Raiffa, 1994). All parties involved would rather choose the option that they preferred before starting deliberations, but because they cannot find a win-win situation or a morally superior alternative, they look for a solution that they can "live with" knowing that it is the second or third best solution for them. Compromising on an issue relies on full representation of all vested interests. If parties were not involved that are obliged by the compromise to take over costs or risks, the compromise would be a decision on the expense of a third party.

In summary, many desirable products and accomplishments are associated with deliberation. Depending on the structure of the discourse and the underlying rationale, deliberative processes can

- enhance understanding;
- generate new options;
- decrease hostility and aggressive attitudes among the participants;
- explore new problem framing;
- enlighten legal policy makers;
- produce competent, fair, and optimized solution packages; and
- facilitate consensus, tolerated consensus, and compromise.

Given the many potentials for deliberative actions, one should take a closer look at the potential risks, pitfalls, and problems of deliberation as well. This topic will be covered in the next section.

3.4. Problems of Deliberation

In recent publications, the value and potential contributions of deliberative democracy have been questioned, in particular with respect to risk management (Breyer, 1993; Coglianese, 1997, 1999; to some degree also Rossi, 1997). Cary Coglianese mentions six potential "pathologies" of consensus-seeking deliberation (Coglianese, N.D., p. 22):

- tractability having priority over public importance;
- regulatory imprecision;
- the lowest common denominator problem;
- increased time and expense;
- unrealistic expectations;
- new sources of conflict.

Let us examine these pathologies in more detail:

1. One of the major arguments has been that public preferences do not match people's best interests. The keywords are ignorance and incompetence (Okrent, 1996; Cross, 1998). This argument is not directed against deliberation as a method of making decisions within competent bodies but as a means to incorporate stakeholders and representatives of the

public in the decision-making process. Does it make sense to replace the best institutional knowledge with intuition and personal interest? Indeed, research shows that public perception of probabilities and risks differs considerably from professional analysis (Covello, 1983; Fischhoff, 1985; Borcherding, Rohrmann, and Eppel, 1986; Slovic, 1987, 1992; Boholm, 1998; Sjöberg, 1999a; Rohrmann, 1999; Rohrmann and Renn, 2000). Whereas experts usually give equal weight to probabilities and magnitudes of a given risk, the intuitive risk perception of lay people reflects higher concern for low-probability, high-consequence risks (Covello, 1983; Margolis, 1996, pp. 21ff.). The preferences that guide social groups and individuals to assign risk-benefit trade-offs may therefore be distorted by misconceptions about the factual probabilities and magnitudes of the respective risk as well as the appropriate model of their integration. Without educating the potential participants, the outcomes may likely be based on prejudices and ignorance (Aron, 1979). Most critics of deliberative democracy doubt that stakeholders or representatives of the affected public have either the capability or the desire to become educated about the subject (Cross, 1998; Coglianese, 1999). All they want, so the argument goes, is to make sure that their personal interest is served. Agencies should therefore convince the representatives of the other systems that it is in their best interest to delegate the decision-making power to democratically legitimized and scientifically competent agencies.

2. The second argument goes even further. Based on the insights from the system-analytic school of sociology, several scholars claim that regulators, stakeholders, and representatives of the public are unable to communicate with each other since they pursue different rationalities with respect to knowledge and evaluations (Rayner, 1990; Markowitz, 1990). Risk bearers expect and demand sufficient, if not absolute, protection of health, environment, and economic interests (Slovic, Fischhoff, and Lichtenstein, 1981; Drottz-Sjöberg, 1991). They may underestimate the impact of random occurrences and will blame the originator of a risk or the regulator if something goes wrong. Regulators, however, use the randomness of events as an excuse for management mistakes or bad performance. By referring to "Acts of God" or "Unpredictable Human Error" they blame the victims for not understanding probabilities and randomness (Luhmann, 1990, 1993). The demand for absolute safety on the side of the risk bearers and the attribution of responsibility to random events or fate on the side of the risk originators or regulators have created a paradox that neither of the parties can overcome through means of communication. Conflicts are not resolved but further fueled (Huntington, 1970). Deliberation in this sense is only a platform for mutual blame and a common search for the scapegoat. Furthermore, deliberation may aggravate the erosion of trust and credibility since the rationalities of the participating groups have no common ground (Earle and Cvetkovich, 1999). Deliberative processes

cannot resolve this confidence gap. All parties feel betrayed by the process and even more alienated from politics than before. The recent debate in Germany on BSE is a good example for this spiral of blame and frustration: the consumers blame the farmers, the farmers blame the regulators, and the regulators blame the consumers for the obvious management mistakes. The conclusion here is to refrain from deliberative democracy because it cannot meet its promises. It is better to tell people in advance what such a process can or cannot accomplish than to disappoint them later.

3. A similar argument refers to the coexistence of scientific (analytic) and communicative (deliberative) rationality (Habermas, 1971; Evers and Nowotny, 1987; von Schomberg, 1995). The claim is that risk managers cannot bridge the gap caused by the intrinsic conflict between the perspectives of the scientific community and the public in general. This conflict has manifested itself in many political arenas. Nuclear energy, genetic engineering, pesticides, transmitters for mobile phones, and other modern technologies face a severe crisis of public acceptance (see juxtaposition of expert and lay risk judgments in U.S. EPA, 1987; review in Rohrmann and Renn, 2000). Most experts think that those technologies pose only minor risks to the population, while most laypersons believe that these risks are a major threat to society. One of the reasons for this discrepancy is the fact that laypersons associate a multitude of attributes with modern technologies when they use the term *risk*, while experts have a narrow definition of risk focusing on probabilities and extent of harm (Freudenburg, 1983; Slovic, 1992; Jasanoff, 1998). In deliberations between scientists and affected people, these different concepts of risk cannot be reconciled. In addition, risk managers need to address a multitude of risk-related issues for which they have neither the jurisdiction nor the institutional expertise. Such discourses often end in mutual frustrations (Rossi, 1997).

4. The fourth argument stems from the economic analysis of risk. Many critics claim that people are either unable or unwilling to accept trade-offs and to search for efficient or cost-effective solutions (Zeckhauser and Viscusi, 1996). The argument is that deliberative processes tend to favor solutions that increase the overall risk level to society for any given amount allocated to be spent on safety (Cupps, 1977; N. Rosenbaum, 1978; Viscusi, 1998). Deliberation may even aggravate environmental damage or impacts on human health (Perry 6, 2000). By pursuing priorities that the public demands, regulators are likely to spend time and effort on those environmental threats that are relatively benign but highly visible in the public eye and neglect those threats that are publicly unknown but very potent in their consequences (Coglianese, 1999, N.D.). In the long run, more people will suffer from future damages than necessary since the funds for safety and risk reduction are spent inefficiently.

5. The fifth argument refers to the problem of scale and representativeness. Inviting stakeholders to take part of the process implies giving their specific, often self-centered interest special weight in the deliberation (Cupps, 1977; Reagan and Fedor-Thurmon, 1987; Lijphart, 1997). This would not be a problem if all social interests were represented and if the representation would be proportional to the potential benefits or risks that are at stake. Analysts of pluralist societies have demonstrated, however, that the relative power of interest groups do not match the relative importance of the issue for society but depend on factors such as exclusiveness of representation, availability of power and resources, and potential for social mobilization (Olson, 1965, 1984; Breyer, 1993). Particular interests usually have a better chance to dominate the decision-making process and will use the opportunity of deliberation to influence the opinion-forming process and impose their specific interest on the agenda. Under these premises, deliberative bodies constitute mere mirror images of the power distribution in society rather than a correction of an agency's perspectives (Waller, 1995). This argument refers particularly to deliberative procedures such as negotiated rule-making or mediation where stakeholder groups are asked to feed their interest into the decision-making process without further public scrutiny (cf. Schoenbrod, 1983; Edwards, 1986; Renn et al., 1995).

6. This criticism leads to a related argument dealing with the selection of participants (Gusman, 1983). Who is allowed to take part in the deliberation? Deliberative processes restrict the number of participants to small groups because only in direct face-to-face situations can one expect a mutual and fair exchange of ideas and arguments. If more people want to attend the discourse than can be reasonably accommodated, some means of selection must be used. There are three possibilities to assign or select people to a deliberative process (Webler and Renn, 1995):

- selecting representatives of groups or organizations that have shown interest in the issue (stakeholders);
- asking for volunteers within the affected population (e.g., through public announcements or advertisements);
- using random selection or an equivalent method to accomplish a statistical representation of a given population.

Selection of any sort poses serious legitimization problems. Representatives of organized groups do not reflect all the relative concerns of the people affected by risk management decisions, and their relative power is not directly linked to their degree of support within the relevant population. This was already mentioned in the last paragraph. The second solution, using volunteers, is likely to result in an overrepresentation of activists, thus giving hardly any voice to the noncommitted part of the population.

Empirical studies show that deliberate selection of group representatives as well as self-recruitment by volunteers lead to serious distortions of public values and interests (Conner, 1993; Vari, 1995).

Ideally, random selection assures that all values and preferences of the affected population are given an equal opportunity to be brought into the process (Dienel, 1989; Stewart, Kendall, and Coote, 1994). In theory, people who are not selected should be satisfied that their interests will be protected because there essentially is a guarantee that another person with similar interests will be selected. In practice, of course, such satisfaction is not forthcoming. People who are immediately affected and not selected in the random sampling feel deprived of a fundamental democratic right to protect their own interests. As a consequence, they may decide to seek other avenues to make their voices heard.

Another problem with random selection is the acceptance rate of those who have been selected. The rate of acceptance varies between 5 and 20 percent of those polled (Renn and Webler, 1998, pp. 82f.). As a problem is brought to public attention, the attitudes of the general publics may shift during the time that elapses between the selection and the end of the process. Although few drop out of the process, there is always the risk that the motivation to attend decreases over time (Renn et al., 1993). The less people feel compelled to accept the invitation, the less do randomly selected participants represent the greater community.

Different political cultures favor different types of recruitment (see Appendix 9.1 for a review of the historic development of deliberative processes in Europe and the United States). If policy making is characterized by an adversarial style as in the United States, selection of participants through organized groups combined with volunteering is regarded as most legitimate, whereas random selection is often associated with an allegedly hidden agenda to keep the interested and knowledgeable persons out of the process (Brickman et al., 1985; Vogel, 1986; Renn, 1995). In countries that do not share the U.S. policy of granting their citizens freedom of information, nonpublic discourses (e.g., negotiated rule making among stakeholders) are hardly discussed as instruments of participation. They are rather regarded as additional elements of a streamlined decision process that enhances cognitive competence but excludes most people from the decision process (Eder, 1992).

7. Another argument challenges the assumption of the connection between deliberation and participation. The claim is that deliberation relies on an intense exchange of arguments in a setting free of strategic maneuvering (Lindner, 1990; Rossi, 1997). Advocates of this line of criticism do not necessarily argue that participation is bad for democracy, but there is a deep concern that too much participation may disrupt the social system operation (Burke, 1968; Aron, 1979; Cross, 1998). It might result in increased immobility and stalemate. Some European analysts have claimed that the

European style of closed-shop negotiation has been much more effective in regulating environmental risks than the adversarial and open style of the United States (Coppock, 1985; Weidner, 1993). As more people take part, it will take more time to come to any conclusion. Effective government, so the argument goes, rests on a limited opportunity to participate. Too much deliberation immobilizes the regulatory system.

8. The last argument in this line of reasoning deals with the outcome of deliberative processes. The main criticism is that deliberative processes lead to trivial results (Coglianese, 1999). The more public input is allowed to enter the process, the more window-dressing is going to occur. If all have to find a common agreement, the language will remain vague, and the outcomes will lack specificity and clear direction. This argument is directed, of course, against deliberative procedures that require consensus or tolerated consensus.

3.5. Pro and Cons: The Two Sides of Deliberation

How valid are these arguments against the use of deliberative methods in risk management? The critical remarks against analytic-deliberative processes have provoked a naive and a sophisticated response within the wider risk community. The naive response has been to emphasize the democratic ideals of participation and to label all critics as conservatives and supporters of elitist societies. This line of argument is neither convincing nor fair. Advocates of representative democracy or proponents of a science-dominated risk management process may have a different interpretation of what democracy means (compared to the advocates of direct democracy or participatory democracy), but they are certainly supporters of democracy in general and not involved in an antidemocratic conspiracy. Normative claims such as who represents the "better" vision of democracy do not lead anywhere. Instead, we would like to turn to the more sophisticated and substantive responses to the critical points mentioned in the last subsection.

One of the major issues that critics of analytic-deliberative processes have voiced is the legitimacy of public perceptions for providing input to decision making on risk issues (Cross, 1998). The ordinary view here is that public knowledge is always inferior to the systematic knowledge of the experts and that the experts should not place their values into the decision process (more politely phrased in Breyer, 1993; see also Breyer et al., 2001). Several decades of participation research and its critical evaluation have demonstrated that such a simple division neither works nor does justice to either perceptions or expertise (Hyman and Siftel, 1988; Lynn, 1986; Wynne, 1992a; Slovic, 1992; McDaniels, 1998). In many decision-making contexts, anecdotal knowledge has often been proven as equally important to the systematic knowledge of experts (Wynne, 1989; Jasanoff, 1998). At the same time, the value reflections of experts constitute a welcomed input to the evaluation of options and have been usually cherished by the legal decision makers (Grima, 1983). Selection and evaluation of knowledge and values should be based on methods and procedures, not on persons (Shrader-Frechette, 1991, pp. 190ff.; Renn, 1995). All systematic knowledge claims need to be tested against the accepted rules of methodology, and all value judgments need to reflect the distribution of the potentially applicable values within the affected population (Kunreuther and Slovic, 1996). The two criteria "truth" (as fuzzy as it may even appear in many scientific contexts) and "representativeness" are neither interchangeable nor replaceable. All collectively binding decisions need to meet both criteria.

From this normative position, it is obvious that decision makers should not use risk perceptions prima facie as normative guidelines for managing risks. On this point, I am in total agreement with many skeptics of deliberative actions. Perceptions are partially based on false knowledge claims, cognitive biases, distortions, and nongeneralizable anecdotal evidence (Breyer, 1993; Okrent, 1998). Having said this, one should also acknowledge that neither experts nor regulators represent the full scope of personal experiences, values, and interpretations that shape the horizon of legitimate positions on risk issues within the affected population (Kasperson, 1974; Jasanoff, 1993; Keeney, 1996). Any decision on the acceptability of a given risk implies crucial value judgments on at least three levels (Renn, 1998). The first set of value judgments refers to the selection of criteria on which acceptability or tolerability ought to be judged. The second set of value judgments determine the trade-offs between these criteria, and the third set of values should assist risk managers in finding optimal strategies for coping with remaining uncertainties and ambiguities. Using methods of public participation and deliberation on all three value inputs does not place any doubt on the validity and necessity of applying the best of technical expertise for defining and calculating the performance of each option on each criterion (see also Charnley, 2000; van Asselt, 2000, pp. 172ff.).

Public input is an essential contribution for determining the objectives of risk policies and for weighing the various criteria that ought to be applied when evaluating different options. To know more about perceptions can help to create a more comprehensive set of decision options and to provide additional anecdotal knowledge and normative criteria to evaluate them (Gregory et al., 1993; McDaniels, 1998). In essence, a deliberative process can handle the problem of ignorance, incompetence, and distorted perspectives if it succeeds in feeding the relevant knowledge and the full diversity of values and interest into the deliberation procedure. How this will be done is the topic of the next section.

The second line of arguments refers to the role of representative versus direct democracy. As far as we know, there is hardly any voice in the contemporary risk debate that would recommend the abolition of representative government.[8] All proposals to increase participation or to make sure that expert judgments and stakeholder values will get more weight in the decision-making process are embedded in a firm commitment toward the institutions of representative government as the final decision maker on risk management options. Representatives of legitimate decision-making bodies need, however, to prepare themselves for making prudent judgments. This was the major insight from the sociological analysis in the second section. Hence, policy makers in pluralist and democratic societies are well advised to consult with experts and the public as a means to meet the requirement of serving the common good (original argument by Mill, 1873; modern version in Dahl, 1989). Any legitimate decision maker is most likely interested in what consequences each decision option might produce and how these consequences are evaluated by those who have to live with them (Chess et al., 1998). It is therefore a constitutive element of representative democracy to ensure a continuous consulting process by which the decision makers become familiar with the state of the art in assessing the likely consequences (knowledge) as well as with the preferences of those whom they are supposed to serve (values).

The highly cherished solution of the past has been to have expert panels feed in the facts and have democratically elected representatives reflect these facts on the basis of public values and make informed decisions (Renn, 1995). Unfortunately, this concept, based on the decisionistic model of decision making, has several major flaws. The selection of facts relies largely on the choice of concerns, and the value preferences of the elected representatives are at least partially dependent on the knowledge about the likely consequences of each decision option. Separating facts from values by division of labor leads to a vicious cycle (von Schomberg, 1995; Tuler and Webler, 1999; Webler, 1999). In addition, uncertainty about consequences, ambiguity of the knowledge base, and dissent among experts motivate decision makers to interact directly with experts and get an impression of the present state of the art (von Schomberg, 1992). For analytical reasons, it is helpful to separate facts from values and design different processes for predicting the likely impacts and evaluating the desirability of each of these consequences. In practical discourse, however, it is counterproductive to run the two processes in isolation and to assign these tasks to different agents because the answers of the first task codetermines the answers to the second task and vice versa (Primack and von Hippel, 1974; Shrader-Frechette, 1990; Renn, 1995). At the same time, as mentioned earlier, those groups and individuals who are exposed to the risk usually demand that their values and preferences be taken into account directly by risk managers without the detour of activating the often only remotely affected political representatives.

Given that observation, it seems justified and desirable to bring all these groups together and initiate a common dialogue or discourse. This is particularly necessary if highly controversial risks are at stake. Organizing a common platform for mutual exchange of ideas, arguments, and concerns does not suffice, however, to ensure fair, efficient, and competent results (Lynn, 1986). Mixing all these knowledge and value sources into one compound implies the danger that each group trespasses its legitimate boundary of expertise. If perceptions replace assessments and the rhetoric of powerful agents replace value input by those who have to bear the risks, the discourse produces distorted and nonlegitimated results. Risk managers need an organizational model that assigns specific roles to each contributor and ensures cooperation and integration. Each contribution must be embedded in a dialogue setting that guarantees mutual exchange of arguments and information and provides all participants with opportunities to insert and challenge claims. The goal is to create active understanding among all participants (Webler, 1995).

In an ideal world, the legitimate decision makers would need a profile of different options spanning the factual consequences and the value-driven concerns. Each profile should include the likely consequences of each option (including doing nothing), the remaining uncertainties, the informed preferences of the people who will be affected by the decision, and the long-term implications for society or even humanity as a whole (if one thinks about global risks such as climate change).

Such a process of option generation, assessment, and evaluation should not be left to one's own intuition – at least if the stakes are high (McDaniels, 1996). In this respect, it is prudent to make sure that the best scientists are part of the decision process and that the affected publics have an opportunity to feed their preferences and concerns into the deliberation process. Feedback from constituents is a major condition on which the whole notion of representative government rests (Rushefsky, 1984; Burns and Überhorst, 1988; Lynn, 1990; Lacob, 1992).

Another issue to consider is the implementation of direct democracy on the local or regional level. Many constitutions include provisions for direct democratic codetermination (Webler and Renn, 1995). While the institutions of representative democracy have become a fact of life in all Western democracies, especially on the higher levels, there are opportunities for direct citizen participation leading to decisions with binding authority in various countries. Among the most publicized are the *Landsgemeinde* found in a few small Swiss Cantons (Wehrli-Schindler, 1987) and popular referendums and initiatives, which are found in many countries. In addition, there is still a great amount of direct participation at local levels of government, such as the town meetings found in the New England states of the United States. Democratic service by nonpoliticians also lives on when people serve on local boards and committees. Democracy has been a tug-of-war between forces that seek more broad-based citizen participation in decision making and those that seek greater power for representatives, interest groups, and their appointees.

In societies where people have major trust in the representative bodies of government, direct elements of democracy may not be essential for gaining public support (Löfstedt, 2001). In societies, however, where distrust and skepticism with respect to the neutrality and openness of the political representatives prevail, opportunities for direct participation may help to stabilize political support and increase satisfaction in the political system. Again, it should be emphasized that hardly anyone who supports the notion of direct democracy believes that stakeholders or randomly selected citizens know better than the experts or the elected officials what is good for society. The rationale for direct democracy is to provide an opportunity for the people directly affected by a decision to reflect the potential outcomes, to review the knowledge available, and to make a prudent choice based on knowledge and values (Langton, 1978; Barber, 1984; Ethridge, 1987; Burns and Überhorst, 1988; Dryzek, 1994). Whether such prudent judgments are made by representatives of stakeholder groups, citizen juries, or elected officials is a matter of political culture, regional traditions, people's expectations, and the issue in question. The process of balancing the pro and cons is essentially identical regardless of who is making these judgments as long as scientific input and broad deliberative processes are both guaranteed. I am convinced, however, that such deliberative processes can meet their promise only if representatives of the four major societal systems (i.e., politics, economics, science, and social structure), are active participants.

3.6. Internal Requirements for Deliberative Processes

The discussion so far has focused on the potential of analytic-deliberative processes, their advantages and disadvantages, and instruments to implement such a process. In this section, we will deal with the internal structure of deliberation within any given model, be it citizen advisory committee, citizen panels, or consensus conferences. There is a need for an internal structure that facilitates common understanding, rational problem solving, and fair and balanced treatment of arguments. The success or failure of a discourse depends on many factors. Among the most influential are (Renn and Webler, 1998, pp. 57ff.):

1. *A clear mandate for the discourse participants.* Models of deliberative democracy require a clear and unambiguous mandate of what the deliberation process should produce or deliver (Armour, 1995). Because discourses are informal instruments, there should be a clear understanding that the results of such a discourse cannot claim any legally

binding validity (unless it is part of a legal process such as arbitration). All the participants, however, should begin the discourse process with a clear statement that specifies their obligations or promises of voluntary compliance after an agreement has been reached.

2. *Openness of result.* A discourse will never accomplish its goal if the decision has been made (officially or secretly) and the purpose of the communication effort is to "sell" this decision to the other parties. Individuals have a good sense of whether a decision maker is really interested in their point of view or if the process is meant to pacify potential protesters (Fiorino, 1989b).

3. *A clear understanding of the options and permissible outcomes of such a process.* The world cannot be reinvented by a discourse, nor can historically made decisions be deliberately reversed. All participants should be clearly informed about the ranges and limits of the decision options that are open for discussion and implementation. If, for example, the technology is already in existence, the discourse can only focus on issues such as emission control, monitoring, emergency management, or compensation. But the range of permissible options should be large enough to provide a real choice situation to the participants (Kasperson, 1986).

4. *A predefined time table.* It is necessary to allocate sufficient time for all the deliberations, but a clear schedule including deadlines is required to make the discourse effective and product-oriented.

5. *A well-developed methodology for eliciting values, preferences, and priorities.* The need for efficiency in risk management demands a logically sound and economical way to summarize individual preferences and integrate them into a group decision (either agreement on dissent, majority and minority positions, tolerated consensus, true consensus or compromise). Formal procedures, such as multiobjective or multiattribute utility analysis could serve as tools for reaching agreements (Edwards, 1954; Chen et al., 1979; von Winterfeldt and Erwards, 1986; Bojorques-Tapia, Ongay-Delhumeau, and Ezcurra, 1994; Maguire and Boiney, 1994; McDaniels, 1996). My colleagues and I have used MAU procedures in most of our experimental deliberative trials (see Renn and Webler, 1998, pp. 88–92).

6. *Equal position of all parties.* A discourse needs the climate of a "powerless" environment (Habermas, 1971, 1991). This does not mean that every party has the same right to intervene or claim a legal obligation to be involved in the political decision-making process. However, the internal rules of the discourse must be strictly egalitarian; every participant must have the same status in the group and the same rights to speak, make proposals, or evaluate options (Kemp, 1985). Two requirements must be met. First, the decision about the procedure and the agenda must rely on consensus; every party needs to

agree. Second, the rules adopted for the discourse are binding for all members, and no party is allowed to claim any privileged status or decision power. The external validity of the discourse results are, however, subject to all legal and political rules that are in effect for the topic in question.

7. *Neutrality of the facilitator of the discourse.* The person who facilitates such a process should be neutral in his or her position on the respective risk management issue and respected and authorized by all participants (Bacow and Wheeler, 1984; Moore, 1986). Any attempt to restrict the maneuverability of the facilitator, moderator, or mediator should be strictly avoided.

8. *A mutual understanding of how the results of the discourse will be integrated in the decision-making process of the regulatory agency.* As a predecisional tool, the recommendations cannot serve as binding decisions in most cases. Rather, they should be regarded as consultative report similar to the scientific consultants who articulate technical recommendations to the legitimate public authorities. Official decision makers need to acknowledge and process the reports by the discourse panelists, but they are not obliged to follow their advice. However, the process will fail its purpose if deviations from the recommendations are neither explained nor justified to the panelists.

A second set of discourse requirements pertaining to the behavior of the participants is necessary for facilitating agreement or at least a productive discussion. Some of these requirements follow:

9. *Willingness to learn.* All parties have to be ready to learn from each other. This does not necessarily imply that they must be willing to change their preferences or attitudes (Webler, Kastenholz, and Renn, 1995b; Daniels and Walker, 1996). Conflicts can be reconciled on the basis that parties accept other parties' positions as a legitimate claim without giving up their own point of view. Learning in this sense entails:

   - recognition of different forms of rationality in decision making (Perrow, 1984; Habermas 1989);
   - recognition of different forms of knowledge, be it systematic, anecdotal, personal, cultural, or folklore wisdom (Habermas, 1971);
   - willingness to subject oneself to the rules of argumentative disputes (i.e., provide factual evidence for claims); obey the rules of logic for drawing inferences; disclose one's own values and preferences vis-à-vis potential outcomes of decision options, and others (Webler, 1995).

10. *Resolution of allegedly irrational responses.* Reflective and participatory discourses frequently demonstrate a conflict between two contrasting modes of evidence: the public refers to anecdotal and personal evidence mixed with emotional reactions, whereas the professionals play out their systematic and generalized evidence based on abstract knowledge (Lynn, 1986; Dietz and Rycroft, 1987). A dialogue between these two modes of collecting evidence is rarely accomplished because experts regard the personal evidence as a typical response of irrationality. The public representatives perceive the experts often as uncompassionate technocrats who know all the statistics but couldn't care less about a single life lost. This conflict can only be resolved if both parties are willing to accept the rationale of the other party's position and to understand and maybe even empathize with the other party's view (Tuler, 1996). If over the duration of the discourse some familiarity with the process and mutual trust among the participants have been established, role-playing can facilitate that understanding. Resolving alleged irrationalities means to discover the hidden rationality in the argument of the other party.

11. *De-moralization of positions and parties.* The individuals involved in a discourse should agree in advance to refrain from moralizing each other or each other's position (Bacow and Wheeler, 1984). Moral judgments on positions or persons impede compromise. Something cannot be 30 percent good and 70 percent bad; either it is good, bad, or indifferent. As soon as parties start to moralize positions, they cannot make trade-offs between their allegedly moral position and the other parties' immoral position without losing face (Scheuch, 1980). A second undesired result of moralizing is the violation of the equality principle stated earlier. Nobody can assign equal status to a party that is allegedly morally inferior. Finally, moralizing masks deficits of knowledge and arguments. Even if somebody knows nothing about a subject or has only weak arguments to support his or her position, assigning blame to other actors and making it a moral issue can help to win points. The absence of moralizing other parties or their position does not mean to refrain from using ethical arguments, such as "this solution does not seem fair to the future generation" or "we should conserve this ecosystem for its own sake." Ethical arguments are essential for resolving environmental disputes.

Given the problems of recruitment, the difficulty in legitimizing deliberative face-to-face interactions to the outside world, and the multitude of internal requirements and rules, it is quite obvious that designing procedures and instruments that promise to meet all the quality criteria that apply to analytic-deliberative processes constitutes a tremendous challenge. The following section introduces several candidates for meeting this challenge.

4. PROCEDURES AND INSTRUMENTS OF DELIBERATIVE DEMOCRACY

4.1. Need for Integrated Procedures

The main line of argument so far has been that deliberative elements in risk management are necessary and vital for gaining legitimacy because the nature of risk decisions is both knowledge- and value-driven. All complex risk decisions demand resources of knowledge and expertise from scientists and local knowledge carriers. This analytical element includes not only technical expertise but also social experience and economic balancing, which translates the contributions of a risk source to expected benefits and risks. The deliberative element refers to the input of stakeholders and social groups, but it is also the overarching principle that guides the integration of expertise, public values, and economic balancing. The combined analytic and deliberative process gains legitimacy by integrating it to the established legal procedures, for example demanding consultations with stakeholders or using citizen panels prior to the official vote or regulatory decision making.

What are the possibilities, procedures, and techniques to implement the idea of an analytic-deliberative process? Because the process should at least have two components, inclusion of analytic expertise and deliberation over interpretations and evaluation, a single type of discourse is usually not sufficient. Analytic-deliberative processes demand a multistage, multilevel procedure consisting of several integrated components. In the following paragraphs, we will list the most popular components that could be combined to a sequence meeting the requirements of an analytic-deliberative process (i.e., effective, efficient, legitimate, and socially acceptable). We will order these components according to the major function they might provide: reducing complexity, handling uncertainty, or dealing with ambiguity. These three challenges necessitate three different types of discourse: cognitive, reflective, and participatory (Renn, 1999c). Table 9.1 provides an overview of the three types of discourses, their function for risk management, and the potential procedures or instruments associated with each discourse type. The following section will briefly introduce the variety of different procedures and instruments used to implement deliberation only briefly.[9]

4.2. Instruments for Reducing Complexity

Resolving complexity requires deliberation among experts. I have given this type of deliberation the title "cognitive or knowledge discourse"

TABLE 9.1. *Discourse Types for Risk Management Challenges*

Challenge	Discourse Type	Objective	Function	Examples of Procedures
Complexity	Cognitive	Effectiveness	Agreement on causal relations and effective measures	Expert panels, expert hearings, meta-analysis, Delphi, etc.
Uncertainty	Reflective	Efficiency	Balancing costs of underprotection versus costs of overprotection facing uncertain outcomes	Negotiated rule making, mediation, Round Tables, stakeholder meetings, etc.
Ambiguity	Participatory	Acceptability	Resolving value conflicts and assuring fair treatment of concerns and visions	Citizen advisory committees, citizen panels, citizen jury, consensus conferences, public meetings, etc.
Integration	Multistage	Legitimacy	Integration in legitimate decision and policy context	Combination of procedures and instruments

(Renn, 1999c). Within a *cognitive discourse*, experts (not necessarily scientists) argue over the factual assessment with respect to the main characteristics of the risk under investigation. The objective of such a discourse is the most adequate description or explanation of a phenomenon (e.g., the question, which physical impacts are to be expected by the emission of specific substances?). The more complex, the more multidisciplinary, and the more uncertain a phenomenon appears to be, the more it is necessary to organize a communicative exchange of arguments among experts. The goal is to achieve a homogeneous and consistent definition and explanation of the phenomenon in question as well as a clarification of dissenting views. The discourse produces a profile of the risk in question on the selected criteria. In addition, a cognitive discourse may reveal that there is more uncertainty and ambiguity hidden in the case than the regulators had initially suspected. If the cognitive discourse includes natural as well as social scientists, future controversies and risk debates may be anticipated. Risk controversies would occur as less of a surprise.

Instruments or procedures that could meet the requirements of a cognitive discourse are expert hearings, expert workshops, science shops, expert panels or advisory committees, and consensus conferences as practiced in the medical field (van Valey and Petersen, 1987; Coppock, 1986; Roqueplo, 1995, Renn, 1995; Boehmer-Christiansen, 1997). If anecdotal knowledge is needed, one can refer to focus groups, panels of volunteers, and simple surveys (Milbrath, 1981; Dürrenberger, Kastenholz, and Behringer, 1999). More sophisticated methods to reduce complexity for difficult risk issues include Delphis, Group Delphi, meta-analytical workshops, and coping exercises (Webler et al., 1991; Wachlin and Renn, 1999).

4.3. Instruments for Dealing with Uncertainty

If risks are associated with high uncertainty, scientific input is only the first step of a more complex evaluation procedure (Charnley and Eliott, 2000). It is still essential to compile the relevant data and the various arguments for the positions of the different science camps. Procedures such as the "Pedigree Scheme" by Funtowicz and Ravetz might be helpful to organize the existing knowledge (Funtowicz and Ravetz, 1990).

In a second step, information about the different types of uncertainties must be collected and brought into a deliberative arena. Representatives of affected stakeholders and public interest groups need to be identified, invited, and informed about the issue in question (Yosie and Herbst, 1998, pp. 644ff.). The objective of the deliberation is to find the right balance between too little and too much precaution. There is no scientific answer to this question, and even economic balancing procedures are of limited value because the stakes are uncertain. I have coined this type of deliberation *reflective discourse* (Renn, 1999c). Beyond the clarification of knowledge (similar to the cognitive) reflective discourse deals with the assessment of trade-offs between the competing extremes of over- and underprotection. Reflective discourses are mainly appropriate as means to decide on risk-averse or risk-prone approaches to risk regulations. This discourse provides answers to the question of how much uncertainty one is willing to accept for some future opportunity (WBGU, 2000, pp. 13ff.). Is taking the risk worth the potential benefit? The classic question of how safe is safe enough comes to play into this type of discourse. Policy makers, representatives of major stakeholder groups, and scientists should take part in this type of discourse. Political or economic advising committees, who propose or evaluate political options, could also be established to advise this core-group (Applegate, 1998). Major instruments for reflective discourses are Round Tables, negotiated rule making, mediation, arbitration, discursive value trees, and others.[10]

4.4. Instruments for Coping with Ambiguity

The last type of deliberation, which I have called *participatory discourse*, is focused on resolving ambiguities and differences about values and preferences (Renn, 1999c). Established procedures of legal decision making, but also novel procedures such as citizen advisory committees and direct citizen participation, belong to this category. Participatory discourses are mainly appropriate as means to search for solutions that are compatible with the interests and values of the people affected and to resolve conflicts among them (Lynn, 1990). This discourse involves weighting of the criteria and an interpretation of the results. Issues of fairness and environmental justice, visions on future technological developments and societal change, and preferences about desirable lifestyles and community life play a major role in these debates. Preferred instruments here are citizen panels or juries (randomly selected), citizen voluntary advisory committees or councils, public consensus conferences, citizen action groups, and other participatory techniques.[11]

4.5. Synthesis of Components

For a process to be called analytic-deliberative, I propose to create a sequential chain consisting of at least one instrument or method from each type of discourse. Any one of the three discourse types may be omitted if the underlying problem (complexity, uncertainty, or ambiguity) is not an issue. If, for example, ambiguities are missing, participatory discourse methods may not be necessary. If the issue appears to be simple and known to science, cognitive discourse methods would be a waste of time. For the more controversial risk issues, however, a chain of at least three modules representing cognitive, reflective, and participatory discourse is recommended. It is difficult, if not impossible, to meet all three discourse functions in one single procedure. Cognitive questions, such as the right extrapolation method for using animal data should not be resolved in a participatory setting with citizens and stakeholders. Similarly, value conflicts should not be resolved in cognitive discourse settings with experts only. Bringing experts, stakeholders, and citizens to one table has been tried many times, but usually without much success because such a setting makes the deliberations overly complex and unmanageable (cf. English et al., 1993; Renn and Webler, 1998, p. 69). It seems rather advisable to separate the treatment of complexity, uncertainty, and ambiguity in different discourse activities because they are based on different forms of argumentation and resolution (Habermas, 1987b; Webler, 1995). Hence, a combination of several instruments or models is needed.

The selection of a specific sequence and the choice of components depend on the risk issues, the context, and the regulatory structure and culture of the country or state in which such a process is planned. Different countries have developed different traditions and different preferences when it comes to deliberative processes. Appendix 9.1 provides a brief review of the deliberative traditions in the United States and Europe. As specific as the concrete hybrid procedure may be to accommodate the respective political culture, it will or should certainly include one component of each discourse type if the problems of complexity, uncertainty, and ambiguity are present. A risk management agency may, for example, organize an expert hearing to understand the knowledge claims that are relevant for the issue. Afterward it might initiate a Round Table for negotiated rule making in order to find the appropriate trade-offs between the costs of over- and underregulation in the face of major uncertainties. The agency could then convene several citizen advisory committees to explore the potential conflicts and dissenting interpretations of the situation and the proposed policy options. Each component builds upon the results of the previous component.

However, the requirements of an analytic-deliberative process are not met by running three or more different components in parallel and feeding the output of each component as input into the next component. Each type of discourse has to be embedded in an integrated procedural concept. In addition, there is a need for continuity and consistency throughout the whole process. Several participants may be asked to take part in all three discourse activities, and an oversight committee may be necessary to provide the mandated links. One example for such an integrated hybrid model is the cooperative discourse approach that several colleagues and I have developed and tested during the last two decades (Renn et al., 1993, Renn and Webler, 1998; Renn, 1999a). We will describe this model in more detail in the next section.

5. AN EXAMPLE FOR AN INTEGRATED HYBRID MODEL: THE COOPERATIVE DISCOURSE

5.1. Model Description

As argued previously, a fair, competent, and cost-effective analytic-deliberative process will not come about through the sole application of individual procedures or instruments, but by the realization of an integrated model that is responsive to its setting. Such a hybrid model must combine elements of different modules presented in the last section in order to arrive at an overall participation process that is effective, efficient, legitimate, and socially acceptable. In addition, such a model needs to produce feasible outcomes that are appropriate to the problem and the cultural setting. By "appropriate" I mean that it finds the proper balance between flexibility and fixed structure and the proper level of focus. Clearly, an integrated model must combine technical expertise and rational decision making with public values and preferences (Stern, 1991; Bohnenblust and Slovic, 1998). Toward these ends, many different compound models have been developed, and some have been tested.

As one example for an integrated model of citizen participation among many potential candidates, I want to highlight the model of cooperative discourse. This model reflects the basic discourse requirements and offers a structure to make them operational in a given policy context. The main features of the model follow:

- First, all parties affected by a decision should have an equal opportunity to participate in the decision-making process.
- Second, each party participating in the discourse has equal rights and duties and is obliged to provide evidence for its claims.
- Third, it is required that the best available knowledge be integrated into the decision process to ensure competence.
- Last, a rational procedure of decision making, which is basically derived from formal decision analysis, is envisioned (Edwards, 1954; Chen et al., 1979) but is oriented toward a multi-actor, multi-value, and multi-interest situation (Vlek, 1996).

To integrate these multidimensional aspects of decision making into a practical procedure, the model assigns specific tasks to different groups in society. These groups represent three forms of knowledge:

- knowledge based on technical expertise (cognitive discourse);
- knowledge derived from social interests and advocacy (reflective discourse);
- knowledge based on common sense and personal experience (participatory discourse).

These three forms of knowledge and corresponding discourse functions are integrated into a sequential procedure in which different actors of society are given specific tasks that correspond to their specific knowledge potentials. The model entails three consecutive steps (Renn et al., 1993). The first step in policy or decision making refers to the identification of objectives or goals that the process should serve once a problem is identified or a political program is established (Merkhofer, 1986, p. 186). The identification of concerns and objectives is best accomplished by asking all relevant stakeholder groups (i.e., socially organized groups that are or perceive themselves to be affected by the decision) to reveal their values and criteria for judging different options. This can be done by a process called value-tree-analysis (Keeney, Renn, and von Winterfeldt, 1987; von Winterfeldt, 1987).The evaluative criteria derived from the value-trees are then operationalized and transformed into indicators by the research team or an external expert group.

After different policy options and criteria are established, the second step follows. Experts representing varying academic disciplines and viewpoints about the issue in question are asked to judge the performance of each option on each indicator (cognitive discourse). For this purpose, my colleagues and I have developed a special method called the *Group* Delphi (Webler et al., 1991; Renn and Webler 1998, pp. 77–80). It is similar to the original Delphi exercise but is based on group interactions instead of individual written responses.

The last step is the evaluation of each option profile by one group or several groups of randomly selected citizens in a participatory discourse. We refer to these panels as "Citizen Panels for Policy Evaluation and Recommendation" (Renn et al., 1993; Renn and Webler, 1998, pp. 81ff.). The objective is to provide citizens with the opportunity to learn about the technical and political facets of policy options and to enable them to discuss and evaluate these options and their likely consequences according to their own set of values and preferences. The idea is to conduct a process loosely analogous to a jury trial with experts and stakeholders as witnesses and advisors on procedure as "professional" judges. For meaningful and productive discourse, the number of participants is limited to about twenty-five. Discourse proceeds in citizen panels with the research team as discussion leaders who guide the group through structured sessions of information, personal self-reflection, and consensus building.

The whole process is supervised by a group of official policy makers, major stakeholders, and moral agents (such as representatives of churches). Their task is to oversee the process, test and examine the material handed out to the panelists, review the decision frames and questions, and write a final interpretation of the results. This oversight committee represents a major component of the reflective discourse. It provides the integrative envelope over the different sequential elements and creates the necessary

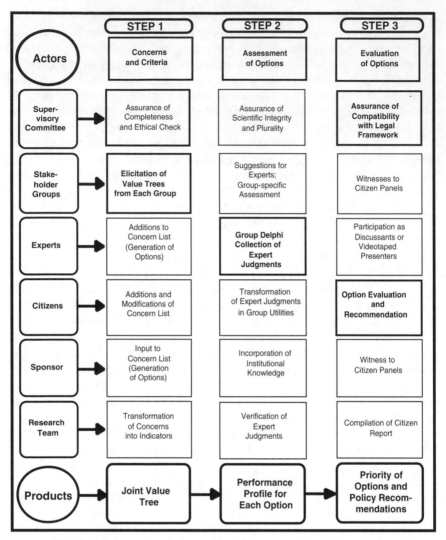

FIGURE 9.3. Basic concept and elements of the cooperative discourse model.

link to the legitimate decision makers. It cannot and should not change the recommendations of the panels but put them into the right frame so that it can be processed by the political institutions that manage or regulate risks.

The functions and procedure of our policy model are illustrated in Figure 9.3. It shows that all actors involved (the experts, the stakeholder groups, the citizens, the sponsor, the oversight committee, and the research team) play a role in each step, but their influence is directed toward the type of knowledge and rationality that they can offer best (these are highlighted in boldface in Figure 9.3). The stakeholders are the principal source for building value-trees, but the other parties may augment the joint tree. Experts are principally responsible for constructing performance profiles for each option, taking into consideration the institutional knowledge of the sponsor and the specific knowledge of the various stakeholder groups. The major task of the citizen panels is to evaluate options and generate or modify policies assisted by expert and stakeholder witnesses. The role of the sponsoring risk management agency is limited to making suggestions about options, providing testimony to the citizen panels, and participating in the oversight committee. The oversight comittee has the task of exercising external control over the whole process, providing the necessary links between the three steps, and ensuring the compatibility of the results with the political institutional frames in which the results are going to be processed. Finally, the research team has the primary task of providing first drafts of the three products (joint value-tree, performance profiles, and citizen report), to gain approval for these products from the respective actors, and to feed them back into the process. This division of labor introduces checks and balances into the process and constitutes a structural order that is logical and transparent.

5.2. Experiences with the Cooperative Discourse Model

Applications of the cooperative discourse model in Germany, Switzerland, and the United States emerged from the early experiences with citizen panels in urban planning in different German cities and communities (Dienel, 1978, 1989). Based on these experiences, the authors and several other researchers experimented with the cooperative discourse method, first in Germany and later in other countries. The following paragraphs describe several large-scale applications in three countries:

• The most comprehensive study dealt with the evaluation of national energy policies in Germany. In August 1982, the German Ministry of Research and Technology initiated a large research project to investigate the preferences of the German population with respect to four energy policy options developed by a parliamentary commission in 1979 (Renn et al., 1985; Renn, 1986). The government was interested in eliciting reliable information on which energy scenario was most appealing to the population and on what basis citizens would evaluate the policy options laid out in each scenario. A research team directed by the author conducted a three-year study to collect data on public preferences and to analyze the motivations and underlying reasons for the judgment process of evaluating the predefined energy scenarios. The study operated with twenty-four citizen panels (each including approximately twenty-five participants) drawn from seven communities in different parts of

West Germany. The panels unanimously rejected a high reliance on energy supply and opted for an energy policy that emphasized energy conservation and efficient use of energy. Nuclear energy was perceived as undesirable but – at least for an intermediate time period – as a necessary energy source. The panelists recommended stricter environmental regulation for fossil fuels even if this meant higher energy prices. They developed a priority list for policies and drafted recommendations for implementing high-priority policies (Dienel and Garbe, 1985).

- A regional study was conducted from 1994 to 1996 in the northern part of the black forest (Southern Germany). The objective was to have stakeholders and citizens take part in planning a waste-management program (Schneider et al., 1998; Renn et al., 1999). A reflective discourse based on the model of Round Tables was organized in 1994 to develop waste reduction policies and to assess the recycling potential of the area (using the method of value trees). The same group also was asked to find the most suitable technical solution for waste processing before final disposal. After these decisions were made, 200 randomly selected citizens from potential host communities were asked to find the most appropriate site for the types of facilities that had been previously chosen by the representatives of the Round Table. The most outstanding result was that panelists were even willing to approve a siting decision that would affect their own community. All ten citizen panels reached a unanimous decision based on tolerated consensus, which involved the recommendation to construct a small state-of-the-art incinerator in the center of the most populated town within the region. The reason for this surprising recommendation was that citizens wanted to present a visual reminder to their fellow citizens. They should not forget the need to reduce waste, to burden those who contribute most to the problem, and to put the incinerator near the political power center as a "clever" means to ensure compliance with the environmental standards. ("Since the mayor resides right next door to the facility he will make sure that nothing harmful will happen.") The decision was given to the regional planning board, which approved the recommendations with some minor modifications. The responsible county parliaments and the city council of the largest city within the region, however, have been reluctant to accept the recommendation as of now.

- In 1992, The Building Department (*Baudepartement*) of the canton Aargau (northern part of Switzerland) asked Tom Wesle and myself (at that time affiliated with the Swiss Federal Institute of Technology) to organize a cooperative discourse for siting one or several landfills in the eastern part of the canton. The mandate of the research team was to organize a cooperative discourse with four citizen panels. These panels had the task first to develop criteria for comparing the different sites; second, to evaluate the geological data that were collected during that period; third, to eliminate the sites that should not be further considered; and fourth, to prioritize the remaining sites with respect to suitability to host a landfill (Webler, 1994). Four citizen panels were formed, each consisting of two representatives from each potential site community. With the exception of one community, every town sent eight people to the panels. Not a single one of these people dropped out during the process. Between January and June 1993, the panels met between seven and nine times before they attended a two-day workshop to come up with the final decision. All participants rated each site on the basis of their self-selected evaluative criteria, their personal impressions, the written and oral information, and the results of a cognitive discourse with experts on the basis of a Group Delphi. All four panels composed a list of prioritized sites for the landfill. The most remarkable outcome was that each panel reached a unanimous decision based on tolerated consensus. In December of 1993, the result of the participation process was made public. The canton government approved the results and entered the next phase of the licensing procedure. As of today, the selected site is still considered, but the erection of a landfill has been postponed because the amount of waste has sharply decreased over the last few years.

- There has been one major attempt to implement the original version of the cooperative discourse in the United States. Using randomly selected citizens for policy making and evaluation is not alien to the United States. The Jefferson Center in Minneapolis has conducted numerous projects with citizen juries (Crosby et al., 1986; Crosby, 1995). Several community planners have experimented with the model of citizen panels, which were composed to reflect a representative sample of the population (cf. Kathlene and Martin, 1991). In our case study, the Department of Environmental Protection of New Jersey asked Ortwin Renn and Thomas Webler in 1988 to apply the cooperative discourse model to sewage sludge management problems (Renn et al., 1989). The objective of the project was to give citizens of Hunterdon County, New Jersey, the opportunity to design the regulatory provisions for an experimental sludge application project on a Rutgers University research farm located in Franklin Township (New Jersey). Although much smaller in scale, the project provided many new insights and experiences that partially confirmed the European observations and partially documented the need for adjustments to the U.S. political culture. The citizen panels were conducted on two consecutive weekends. The desired goal was to elicit recommendations for regulatory provisions that should be included in the permit for the land application of sewage sludge on the site in question. The factual issues were discussed in a Group Delphi with eight sludge experts (Webler et al., 1991). The results of the Delphi were fed into the deliberation process of the panels. The envisioned program for the citizens panel was radically altered after the participants, in particular the land owners abutting the site, made it clear that they rejected the project of land application and that they felt more comfortable conducting their own meetings without assistance of a third party. The citizens met several times without the assistance of a facilitator and formulated recommendations that were forwarded to the sponsor (New Jersey Department of Environmental Protection). In addition to the policy recommendation to reject the proposal of land application, the process provided valuable information about citizen concerns and values. Whereas most of our consulted experts were convinced that citizen concerns focused on issues such as odor, traffic, and contamination of groundwater, the value-tree analysis of the citizens revealed that their major concerns were the expected change of community image from an agricultural community to a "waste dump" and the long-term effects of pollutants on farmland (Webler et al., 1995a).

In summary, the applications of cooperative discourse method provided some evidence and reconfirmation that the theoretical expectations linked to this method can be met on the local, regional, and also national level. It has been proven a valid instrument to elicit preferences and educated responses of citizens in a rather short time period. Evaluation studies by independent scholars confirmed that the objectives of effectiveness, efficiency and social acceptability were largely met in the Swiss as well as German case studies. The U.S. study was not externally reviewed. The evaluators agreed that the main interests and value groups were adequately represented, that the process and the results were cost-effective, and that the outcomes of the process were judged as reasonable and competent by external technical experts (Rymann, 1993; Buser, 1995; Vorwerk and Kämper, 1997; Roch, 1997; Rippe and Schaber, 1999). However, these evaluators were rather skeptical about the legitimization function of these trials. In all cases, the results of the deliberation were not implemented, at least not to its full extent. The evaluations emphasized the need for a stronger role of the supervisory committee (which was installed in the Swiss case) and the inclusion of the actual decision makers in these committees as a means to bridge the gap between the deliberative bodies and the political risk-management institutions.

6. EVALUATION OF DELIBERATIVE MODELS FOR RISK MANAGEMENT

Although the literature is full of good advice on how to conduct deliberative processes and what to do and not to do, the empirical basis for such advice is weak (comparative empirical reviews in Talbot, 1983; Bingham, 1984; Susskind and Ozawa, 1985; Buckle, and Thomas-Buckle, 1986; Frey and Oberholzer-Gee, 1996; Moore, 1996; Creighton et al., 1998). In addition, most empirical analyses are based on one model or one type of application.

Several applications such as hearings, mediation, citizen advisory committees, and expert panels have received substantial empirical attention, while others such as citizen juries or consensus conferences are rarely subjected to systematic evaluation. Basically missing are comparative evaluations of hybrid procedures, the only type of procedures that can satisfy the requirements of an analytic-deliberative process. It is therefore essential for further research and model evolution to organize systematic evaluations of these processes and to compare the results of these evaluations.

There is a lack of clear and unambiguous criteria, however, to evaluate these applications. Most evaluative criteria refer to the subjective satisfaction of the actors involved in the process. Subjective satisfaction is certainly an important, but insufficient, criterion for evaluating success or failure of deliberations (Rohrmann, 1992; Webler, 1995; Linder and Vatter, 1996). People will tend to rate processes positively if they are heavily invested in them – emotionally or just by sacrificing their time. In addition, many processes are able to convey a good feeling but may lead to a very unsatisfactory outcome, and the other way around. There is obviously a need for objective indicators of evaluation that relate to the product and the process independent of the subjective interpretation of the persons involved (Wondollek, 1985).

There are only few examples of nonsubject-centered approaches to evaluating public participation. Ray Kemp used Habermas's definition of the ideal speech situation as a measure against which to compare the discourse that occurred in the public inquiry process to permitting the British uranium reprocessing facility (Kemp, 1985). Daniel Fiorino developed performance criteria for public participation from the theory of participatory democracy and evaluated several generic models of participation (1989a, 1990). Frank Laird has supplemented Fiorino's criteria with another set from the theory of pluralist democracy and evaluated the same models (1993). Although not developed for the purpose of evaluation, the competing values theory by Quinn and Rohrbaugh (1981, 1983) provide another set of interesting criteria for measuring the success or failure of deliberative processes. These criteria are related to four potential organizing principles: flexibility and control on one hand, and internal versus external focus on the other hand. These conflicting principles are associated with corresponding criteria: legitimacy, participatory quality, accountability, and efficiency. Similar criteria were suggested by Renn et al. (1999, pp. 239ff.); these criteria refer to fairness, competence, legitimacy, and efficiency. These four criteria reflect mirror images of the contribution of the four systems: legitimacy and efficiency are two criteria that are identical in both settings, competence refers to effectiveness, and fairness to social cohesion (the subjective satisfaction also belongs to this objective).

Based on several critical remarks on the first publication (Renn et al., 1999), I have revised the evaluation criteria. Vorwerk and Kämper (1999)

TABLE 9.2. *Criteria of Evaluation*

1.	**Fairness**	
1a	Substantive	Fair access to participatory process
1b	Procedural	Equal opportunity to make claims and reject claims during deliberations
2.	**Competence**	
2a	Cognitive	Systematic knowledge on general cause-effect relationships; anecdotal knowledge on unique cause-effect relationships
2b	Reflective	Basic understanding of the meaning and relevance of a knowledge set with respect to a problem or a potential solution
2c	Normative	Basic understanding of the legal, social, and ethical rules that should be applied to the problem or the evaluation of options
3.	**Efficiency**	
3a	Instrumental	Proportion between cost of procedure and the overall stakes associated with the decision options
3b	Procedural	Efficiency of decision-making procedures within the deliberative process

pointed out that the criterion *legitimacy* constitutes a compound requirement that encompasses fairness, competence, and efficiency. If all of these three are met, the process can assume legitimacy within the formal political framework in which it is integrated. In response to this criticism, I reduced the number of criteria from four to three. These three criteria are described in Table 9.2.

Fairness in this framework of evaluation is described in terms of how equally available the various activities in discourse are to the participants (Webler, 1995; Frey and Oberholzer-Gee, 1996; Linnerooth–Bayer and Fitzgerald, 1996). The first subcriterion of fairness is equal access of all affected people to the participation activity, the second subcriterion refers to fair procedures in conducting the discourse among the participants. This aspect of fairness is key to producing a forum where equality and popular sovereignty can emerge and personal competence can develop. When participation is fair, everyone takes part on an equal footing. This means that not only are people provided equal opportunities to determine the agenda, the rules for discourse, and to speak and raise questions, but also equal access to knowledge and interpretations (Webler, 1995, 1999).

Political equality and popular sovereignty also make an argument for competence. Competence is defined neither from the quality of the outcome or the level of satisfaction among the participants, but according to adherence of the participants to rules for knowledge presentation and selection (Webler, 1995). Risk decisions rely on the best expertise about potential consequences of different options available. Without this knowledge, risk decisions cannot be effective. Participants may act on erroneous beliefs or even prejudices. This is why a high level of competence is extremely important in all deliberative procedures. Competence provides a structure that enables participants to use discourse in such a way that all participants reach a sufficient amount of understanding to be aware of the potential consequences of each course of action, including knowledge about uncertainties and ambiguities.

Knowledge includes three dimensions: *cognitive* or factual knowledge about cause-effect relationships, *reflective* knowledge about meaningful interpretation and argumentation leading to balanced judgments, and *normative* knowledge about legal or value orientations that may guide alternative courses of action (corresponding to the three types of discourse mentioned previously). A competent analytic-deliberative process should provide a structure that guarantees the adequate discourse procedure to deal with these three challenges.

Efficiency relates to the proportion between cost and outcome (Renn et al., 1999, pp. 282ff.). The authors proposed to divide this criterion into two subcriteria: the first one measures the overall cost of the procedure in relation to the potential amount of costs and benefits that are at stake. If the deliberation is about siting a multimillion-dollar facility, such as a hazardous waste incinerator, the process can cost considerably more than a process about siting a minor facility such as a recycling station costing only several thousand dollars. The second subcriterion refers to the efficiency of the process itself. Is much time and effort of participants wasted owing to lax moderation or a lack of readily available material? Both indicators can be operationalized in quantitative terms.

These criteria and subcriteria are suggestions for evaluating analytic-deliberative processes. We have used them already as guidelines for internal and external evaluations of the various applications of the cooperative discourse model (Köch, 1997; Vorwerk and Kämper, 1997; Renn et al., 1999, pp. 239ff.). We hope that they may be useful in the future to gain more empirical knowledge about analytic-deliberative processes and to help to improve the present situation.

7. CONCLUSIONS

The objective of this chapter was to address and discuss the need and potential for integrating analysis and deliberation and to introduce and critique different approaches of deliberative decision making. Organizing and structuring discourses on risk go beyond the good intention to have the public involved in risk decision making. The mere desire to initiate a two-way-communication process and the willingness to listen to public concerns are not sufficient (Hadden, 1989; Lynn, 1990). Discursive processes need a structure that ensures the integration of technical expertise, regulatory requirements, and public values. Decisions on risk need to reflect effective regulation, efficient use of resources, legitimate means of action, and social acceptability.

These inputs can be provided by the different systems of society: efficiency by economic markets, knowledge on effectiveness by scientists and experts, legitimacy by the political institutions, and social acceptance by including social actors. The objective is to find an organizational structure that each system contributes to the deliberation process the type of expertise and knowledge that can claim legitimacy within a rational decision-making procedure (von Schomberg, 1995; Renn, 1998). It does not make sense to replace technical expertise with vague public perceptions nor is it justified to have the experts insert their own value judgments into what ought to be a democratic process.

Recently there has been much concern in the professional risk community that opening the risk management arena to public groups would lead to a dismissal of factual knowledge and to an inefficient spending of public money (Rossi, 1997; Cross, 1992; Okrent, 1998). In my opinion, these concerns are not warranted if one looks at actual experiences with discursive models of risk communication. Only a few voices want to restrict scientific input to risk management. The role of scientific analysis in risk management should not be weakened but rather strengthened. Profound scientific knowledge is required in risk management, especially with regard to the main criteria of risk evaluation: probability of occurrence, extent of damage, and remaining uncertainties. This knowledge must be assessed and collected by scientists and risk professionals who are recognized as competent authorities in the respective risk field. The systematic search for the state of the art leads to a knowledge base that provides the data for each of the evaluation criteria. This scientific evaluation procedure entails a collective cognitive basis for the ongoing deliberation process (Yankelovich, 1991).

Placing emphasis on the analytic part of the process does not contradict the deliberative character of the whole decision-making process. Although systematic evaluations of analytic-deliberative processes are largely missing and empirical data about success or failure of such processes are only available for specific components, most reviewers agree that ignorance and misperceptions were not the major problems in participatory settings (Bingham, 1984; Creighton, 1991). On the contrary, even participants from the lay public were not only willing to accept but furthermore demanded that the best technical estimate of the risks under discussion should be employed for the decision-making process (Burns and Überhorst, 1988; Renn, 1998). These participants also insisted, however, that other dimensions apart from expected values should enter the deliberation process.

After the potential contributions of the expert communities, the stakeholder groups, and members of the affected public had been recognized and acknowledged in such settings, a process of mutual understanding and constructive decision making started to unfold. Such a discursive process may not always lead to the desired results, but the experiences so far justify a fairly optimistic outlook. The main lesson from these experiences has been that scientific expertise, rational decision making, and public values can be reconciled if there is a serious attempt to integrate them. The transformation of the risk arena into a cooperative risk discourse seems to be an essential and ultimately inevitable step to improve risk policies and risk management.

This line of argument is particularly true for far-reaching risk management decisions. These decisions are associated with high complexity, uncertainty, and ambiguity. These three challenges require suitable instruments for management and decision making. I have argued that resolving complexity necessitates a cognitive discourse (mostly among experts), dealing with uncertainty requires a reflective discourse (involving major stakeholders), and coping with ambiguity leads to a participatory discourse (integrating representatives of the affected parties and publics).

Deliberative processes are helpful for reducing complexity, essential for handling uncertainty, and mandatory for dealing with ambiguity. Uncertainty and ambiguity cannot be resolved by expertise only, even if this expertise is uncontested. In situations of high uncertainty, economic balancing between over- and underprotection requires subjective assessments. Furthermore, the interpretation of ambiguous consequences necessitates the input of public prefrences and values. Neither agency staff nor scientific advisory groups are able or legitimized to represent the full scope of public preferences and values. This is a compelling reason for broadening the basis of decision making and including those who must "pay" and those who are affected by the likely consequences.

How can and should risk managers collect public preferences, integrate public input into the management process, and assign the appropriate roles to technical experts, stakeholders, and members of the public? This chapter suggested using the distinction between cognitive, reflective, and participatory discourse. The objective is to design an integrated procedure that chains one or more of the available instruments belonging to each of the three discourse domains to the next. Depending on the subject, the political culture, and the institutional mandate, one can choose among a whole set of potential discourse instruments that fall into each of the three categories. One candidate for such a hybrid procedure has been the model of cooperative discourse that has been developed and tested over the last two decades.

Based on our experiences with the model of cooperative discourse in several countries, I have come to the conclusion that the model provides a theoretically ambitious and practically feasible method to put an analytic-deliberative process into action. It provides an opportunity for all four contributors (i.e., representatives of the scientific, social, political, and economic system), to take part in the decision-making process. It ensures that the state of the art in systematic knowledge is included in the deliberation; it ensures a fair representation of public interests; it emphasizes reflective and normative competence; and it builds upon well-proven techniques of efficient decision making such as the application of multiattribute scaling techniques (see review in Rippe and Schaber, 1999).

Beyond the functional argument that decision making in complex, uncertain, and ambiguous matters requires the input of all four societal systems in a well-structured format, many proponents of deliberative democracy also refer to normative arguments to support their case. Involvement of stakeholders and the public at large in deliberative forms of decision making is one of the primary normative goals of democracy. In classical democratic theory, public involvement is morally and functionally essential to the emergence and the sustenance of the two central values of democracy: *popular sovereignty* and *political equality* (N. Rosenbaum, 1978). In this sense, analytic-deliberative processes can also contribute to the enhancement of democratic culture within a given political and regulatory framework (Ruckelshaus, 1985; Zilleßen, 1993).

Needless to say, deliberation is not a panacea for all risk problems. If done improperly, it may actually increase overall risk levels, lead to inefficiencies, stabilize existing power distributions, and make ignorance and incompetence the guiding principles for decision making. Deliberation may also prolong decision making and immobilize institutions. All these risks of deliberation should not be easily dismissed. A competent, accountable agency decision is still better than a superficial consensus among affected parties. At the same time, however, the many arguments in favor of deliberation and its theoretical underpinnings provide ample evidence for its potential contribution to improve risk evaluation and management. It is still an open question whether deliberation can deliver what it promises in theory. The empirical account is still open and incomplete. Being active in developing and testing analytic-deliberative processes, I am confident, however, that over time we can theoretically prove not only the merits and potentials but also the practical feasibility and superiority of analytic-deliberative processes in different political cultures and among a variety of regulatory styles. I am truly convinced that the time for analytic-deliberative processes is here.

APPENDIX 9.1: HISTORY OF DELIBERATION

Ortwin Renn, Tom Webler, and Ragnar Löfstedt. In the United States, public participation of a consolatory nature has a long historic tradition. It is difficult to put a date on when the public and stakeholders became involved in the policy-making process. Lowi (1979) argues that participation has been a recurrent theme in American history. Demands for participation has increased over time as the government has expanded. Referenda, a form of public participation, as well as the arranging of formal interest representation at the national level were put in place in the early part of the twentieth century (Lowi, 1979). Participatory procedures were already in effect in the early 1930s in the area of agricultural policy making (Daneke, Garcia, and Delli Priscoli, 1983, p. 12) and during the 1930s with the Tennessee Valley Authority (Rossi, 1997). By the close of the 1930s, the basic forms of public participation were well established and used (Attorney General's Commission on Administrative Procedure, 1941).

Following World War II, participation continued to grow in popularity after the passing of the Administrative Procedure Act of 1946, which called for due process and the public's right to comment and gave opportunity for hearings. Although the act did not clarify the objectives of participation, it was an important milestone because it officially mandated norms for agency conduct (Daneke et al., 1983). Public participation was first encouraged by the government with the Revised Housing Act of 1954 and later by the Economic Opportunity Act of 1964, which sought "maximum feasible participation" in community development. All of these acts were termed the "old" school of participation, in which participation was seen as a privilege, in which only the organizations that had resources could partake (Fiorino, 1989).

The "new" school of public participation did not come about until the formation of the participation movement of the federal administrative process in the 1960s and 1970s (Fiorino, 1989, p. 504). These laws were passed with the enthusiastic belief that participation could lead to a more just and economically prosperous society. With the passing of the Great Society Era, participation was still valued, but more as a watchdog activity than as a means to liberate citizens from poverty, exploitation, and injustice. During these twenty years, public participation was seen as a necessary element in federal statutes, viewed as an important contribution to democracy and to the quality of the decision-making process itself (e.g., Cramton, 1972; Fischer, 1993). What distinguished the new school from the old school was that now participation was seen as a right.

The first experiment with mass public participation took place in the presidency of Lyndon Johnson and his so-called Great Society effort, where it was used in defining and establishing antipoverty programs (Fiorino, 1989; Rossi, 1997). In the 1970s, participation in the policy-making process grew in popularity. It gained significant impetus following the passage of the National Environmental Policy Act of 1969, which called for public participation in the preparation of Environmental Impact Assessments (Dietz, 1988), which were enforced by the U.S. federal courts (Atherton,

1977). The major piece of legislation that most profoundly influenced the political culture of the United States and continues to drive and shape public participation until today was the Freedom of Information Act (FOIA) of 1974. Today, nearly every federal agency is required to include some form of participation in their regulatory decision-making procedures (DeSario and Langton, 1987). In addition, the so-called reformation of the appeal law in late 1960s and early 1970s led to the use of fairness and equity measures to protect new classes of interests under an expanding government (Breyer, Sunstein, and Stewart, 2001). The greatest impetus to the growth of participation in the 1970s was via the growth of federal statutory innovations. Of all the participation provisions enacted in the 1970s, 60 percent were based on some government statute (Rosenbaum, 1979).

In the 1980s, the popularity of the use of participation in the policy-making process declined in the U.S. government. The Reagan Administration cut funding to participation initiatives, seeing that participation in the policy-making process was actually a way for the opposition to partake in it, usually in the disguise of environmental groups. Rather it advocated the use of cost-benefit analysis as an approach for setting environmental standards (Fiorino, 1989a). This changed in the 1990s, when once again participation grew in popularity on both the national and state levels. On the national level, federal reform proposals, for example, have been used to make agencies more responsible to smaller businesses, while states such as Florida and North Carolina have adopted participation techniques for negotiated rule making (Rossi, 1997).

Public participation has been a major topic of debate and controversy in all European countries from the beginning of the nineteenth century to this date (Webler and Renn, 1995). The early development of democracy in the aftermath of the American and French revolutions was characterized by the gradual integration of citizens into the political system. In the beginning, voting was exclusively reserved for the rich and powerful. The fast change of constitutions during the French Revolution from a class-oriented voting system (according to wealth) to an egalitarian system, which was later transformed into a radical one-party rule, served as a model for most European nations. After the restoration in the early nineteenth century, democratic reforms were gradually introduced to most European countries. These include incorporation of bills of rights into national constitutions, division of power, equal access to voting privileges, permission of parties and labor unions, and other related measures. Progress in broadening the base for political power among all members of society took time. The right of women to vote or to be eligible for political office did not become effective in several Swiss Cantons until the 1970s (Fahrni, 1992).

In addition to the continuous movement of citizens to fight for equal rights in the political arena, social attention turned to the economic system since the middle of the nineteenth century. Labor movements, particularly in northern Europe, demanded more power to influence decisions of corporations (Leminsky, 1977, pp. 284ff). The quest for "codetermination" has been a predominant demand of the German and Swedish unions from the 1920s until today. In contrast to American unions, which have been primarily interested in issues of wages and working conditions, most European unions placed much emphasis on taking part in management decisions. In exchange for sharing power, they were willing to play a moderating role in labor-management conflicts. In Germany, large corporations in mining and steel are required to have an equal number of shareholders and representatives of labor on their board of directors.

Direct involvement of citizens in political decision making beyond the conventional modes of voting, party involvement, and economic codetermination has been less pronounced in most European countries until fairly recently. The more centralized European countries (mostly in the south) reserve most political power to the national governmental, but leave room for individual action under the patronage of the local representative of the central government. The countries with strong federal traditions rely heavily on procedural rules when allowing citizens to take part in political decisions. Direct democracy is embedded in a complex framework of bureaucratic rules and regulations that ensure predictability of outcomes even at the expense of losing time.

Although social movements and citizen initiatives have been advocating more direct influence on political decision making since the early 1920s, their protest was not effective until different ecological movements evolved at the beginning of the 1970s. The road to more direct participation was partially prepared by the student movements of the late 1960s, even though the goals and objectives of the student protest were not widely shared among the larger population. This situation changed dramatically with the emergence of ecologically motivated citizen initiatives. These new organizations became major political forces in West Germany, Sweden, Denmark, the Netherlands, and several other European countries. Many citizens throughout Europe, in particular the young generation, sympathized with or even participated in these movements (Watts, 1987). Central to the self-identity of these new movements was the demand for more direct participation, particularly on the local level. New methods of political articulation such as large demonstrations, boycotts, nonviolent actions, and blocking construction of environmentally controversial facilities (e.g., nuclear power plants) became familiar patterns of the political climate in most countries. The political culture of most European countries were first shaken by the popular protest and then adapted slowly to the new demands (Pinkau and Renn, 1998).

Driven by popular demand, most governments throughout Europe enhanced the possibilities for direct citizen input in governmental decision making or became more lenient in extending the existing opportunities for public involvement. Many countries such as Switzerland introduced the right of organizations to litigate against licensing decisions even if their members were not directly affected. Other countries such as Sweden, Great Britain, the Netherlands, Belgium, and Germany gave ecological organizations more opportunities to influence policies and to be either part of the neocorporatist club or to become catalysts for political change (O'Riordan and Wynne, 1987). The European Community promulgated directives to all member states that extended the right of the public to be better informed about hazards and to become more involved in local hazard management (Renn, 1995). As a result of these activities, Europe has become a major testing ground for new methods of citizen participation, primarily because the new social movements were much more of a shock for the established parties and power players than in the United States, where political protest and social movements have been a common characteristic for as long as the country has existed. Some of the most innovative proposals to give citizens more power in influencing political decisions originated in Europe and were then imported to the United States. At the same time, alternate methods of conflict resolution such as mediation and bargaining that have been used in the United States for at least two decades were introduced in Europe only recently and have changed the former closed-door policy-making arena.

European as well as American countries share a tradition of democratic values that are focused on the right of individuals to codetermine their own social and political environment. The degree of direct participation may vary from one political culture to another, but the basic values of deliberative democracy – first, that collective decisions require inputs from all those who are affected and, second, that a balanced exchange of arguments between different stakeholders occurs – is echoed in almost all democratic systems in the world.

Notes

1. See similar definitions in Kates and Kasperson (1983); Hohenemser, Kates and Slovic (1983); and Rosa (1997). In economic theory, risk refers to both gains and losses. Since I am dealing here with risks to the environment and human health, I believe that the confinement to negatively evaluated consequences is more in line with the average understanding of risk in this context. One should note, however, that the labeling of consequences as positive or negative refers to genuine social judgments and cannot be derived from the nature of the hazard (Wynne, 1992b).
2. See reviews in Nelkin and Pollak, 1979, 1980; Brooks, 1984; Brickman, Jasanoff, and Ilgen, 1985; O'Riordan and Wynne, 1987; Thomas, 1990; Pinkau and Renn, 1998; Perry 6, 2000. There is also ample literature on specific approaches to risk regulation such as: muddling through (Lindbloom 1959, 1965); scientific advisory boards (Dietz and Rycroft, 1987; Wynne, 1992a; Renn, 1995; Harter,

Orenstein, and Dalton, 1998; Heyvaert, 1999); mediation and negotiated rule making (Susskind, Bacow, and Wheeler, 1983; Folberg and Tayor, 1984; Bingham, 1984); and participatory approches (Renn, Webler, and Wiedemann, 1995; Creighton, Dunning, and Delli Priscoli, 1998; Yosie and Herbst, 1998; Rowe and Frewer, 2000).

3. This analysis of societal functions and systems is based on the functional school of sociology and political sciences. Early representatives of this school that are relevant for my arguments have been Parsons, Shils, Easton, and Lowi (Parsons and Shils, 1951; Parson, 1951, 1963, 1967, 1971; Easton, 1965; Lowi, 1964). This school of though has been heavily criticized for its stationary and equilibrium-based assumptions (see Coser, 1956). Modern functionalists have responded to this criticism by adding a dynamic and change-inducive component to the theoretical framework (Alexander, 1985; Alexander and Colomy, 1990). My approach in this chapter has been inspired by the (neo)functional school of Bielefeld in Germany (Luhmann, 1982, 1984, 1990, 1993; Münch, 1982; Willke, 1995).

4. For our purpose, we can work with the definition of civil society put forth by J. Alexander: Civil society is defined as "the realm of interaction, institutions, and solidarity that sustains public life of societies outside the worlds of economy and state" (Alexander, 1993, p. 797).

5. See overviews in Burke, 1968; Nelkin, 1977; Checkoway and van Til, 1978; Guild, 1979; Nelkin and Pollak, 1979; Langton, 1981; Daneke, Garcia, and Delli Priscoli, 1983; Bingham, 1984; Brickman et al.; 1985; Jasanoff, 1986; Vogel, 1986; Ethridge, 1987; Maguire and Boiney, 1994; Renn et al., 1995; Majone, 1998; Creighton et al., 1998; Applegate, 1998; Harter et al., 1998; Rowe and Frewer, 2000).

6. Reviews of the implications of a constructivist versus realist concept of risk can be found in Bradbury, 1989 and Shrader-Frechette, 1991. A pronounced constructivist approach to risk management can be found in Rayner, 1987, 1990; Hillgartner, 1992; Luhmann, 1993; Adams, 1995; or a recent German book Japp, 1996. Realist perspectives in the social sciences on risk and environment can be found in Catton, 1980; Dunlap, 1980; Rosa, 1997. Several analysts pose themselves between the two poles, for example, Beck, 1992, or Shrader-Frechette, 1991.

7. Similar assumptions are made in the cultural theory of risk; compare Douglas and Wildavsky, 1982; Rayner, 1990; Thompson, Ellis, and Wildavsky, 1990; Dake, 1991; Adam, 1995.

8. Authors who favor more direct democracy accept the overall framework of political representation, but would like to limit the influence of the representational bodies or to enrich their decision-making routines by adding elements of direct democracy (see in this line of reasoning in Bachrach, 1967; Barber, 1984; Dryzek, 1994; Sclove, 1995).

9. Reviews of the pros and cons of each suggested procedure or instrument can be found in the literature (Langton, 1981; Kweit and Kweit, 1981, 1987; Amy, 1983; Bacow and Wheeler, 1984; Crosby, Kelly, and Schaefer, 1986; Kraft, 1988; Burns and Überhorst, 1988; Laird, 1993; Sclove, 1995; Susskind and Fields, 1996; see reviews in Pollak, 1985; Fiorino, 1990; Renn et al., 1995; Creighton et al., 1998; Rowe and Frewer, 2000).

10. There is a considerable collection of empirical and theoretical studies on stakeholder participation methods – most notably on mediation, alternate dispute resolution, and negotiated rule making (see O'Connor, 1978; Cormick, 1980; Mernitz, 1980; Godschalk and Stiftle, 1981; Talbot, 1983; Susskind et al., 1983; Bingham, 1984; Folberg and Taylor, 1984; Buckle and Thomas-Buckle, 1986; Edwards, 1986; Peritt, 1986; Pruitt, 1986; Moore, 1986; Bingham et al., 1987; Doniger, 1987; Delli Priscoli, 1989; Baughman, 1995; Fiorino, 1995; Hadden, 1995; Moore, 1996; Amy, 1987; Bingham et al., 1987; Doniger, 1987; Delli Priscoli, 1989; Baughman, 1995; Fiorino, 1995; Hadden, 1995; Moore, 1996). Additional literature refers to value tree analysis (Keeney, Renn, and von Winterfeldt, 1987; von Winterfeldt, 1987; Renn and Webler, 1998, p. 77) or other forms of stakeholder involvment (Brooks, 1984; Creighton, 1991; English et al., 1993; Renn et al., 1995; Wondelleck, Manring, and Crowfoot, 1996).

11. The literature includes studies on citizen advisory groups (Laksmanan, 1990; Lynn and Kartez, 1995; Vari, 1995; Applegate, 1998), consensus conferences (Andersen, 1995; Durant and Joss, 1995; Joss, 1997; Marris and Joly, 1999), public hearings (Checkoway, 1981), citizen panels or citizen juries (Crosby et al., 1986; Stewart et al., 1994; Crosby, 1995; Armour, 1995; Dienel and Renn, 1995; Renn and Webler, 1998, pp. 81–96), and citizen initiatives (Forester and Stitzel, 1989; Claus, 1995; Linnerooth-Bayer, 1995). Overviews can be found in Mazmanian, 1976; Langton, 1978; Rosener, 1982; Ethridge, 1987; Susskind and Cruishank, 1987; Fiorino, 1990; Lynn, 1990; Stolwijk and Canny, 1991; Renn et al., 1995; Rowe and Frewer, 2000).

References

Adams, J. 1995. *Risk*. UCL Press, London.

Alexander, J., ed. 1985. *Neofunctionalism*. Sage, Beverley Hills, CA.

Alexander, J. 1993. Return of civil society, *Contemporary Sociology*, 22, 797–803.

Alexander, J., and Colomy, P. 1990. Neofunctionalism: Reconstructing a theoretical tradition. In G. Ritzer, ed., *Frontiers of Social Theory: The New Syntheses*, pp. 33–67. Columbia University Press, New York.

Amy, D. J. 1983. Environmental mediation: An alternative approach to policy stalemates. *Policy Sciences*, 15: 345–65.

Amy, D. J. 1987. *The Politics of Environmental Mediation*. Cambridge University Press, Cambridge and New York.

Andersen, I.-E. 1995. *Feasibility Study on New Awareness Initiatives*. Studying the Possibilities to Implement Consensus Conferences and Scenario Workshops. Interfaces, Science, Technology, Society. European Commission, DG Research, Copenhagen and Brussels.

Apel, K.-O. 1992. Diskursethik vor der Problematik von Recht und Politik: Können die Rationalitätsdifferenzen zwischen Moralität, Recht und Politik selbst noch durch Diskursethik normativ-rational gerechtfertigt werden? In K.-O. Apel and M. Kettner, eds., *Zur Anwendung der Diskursethik in Politik, Recht und Wissenschaft*, pp. 29–61. Suhrkamp, Frankfurt/Main.

Applegate, J. 1998. Beyond the usual suspects: The use of citizens advisory boards in environmental decisionmaking. *Indiana Law Journal*, 73: 903.

Armour, A. 1995. The citizen's jury model of public participation. In O. Renn, Th. Webler, and P. Wiedemann, eds., *Fairness and Competence in Citizen Participation. Evaluating New Models for Environmental Discourse, pp.* 175–88. Kluwer, Dordrecht and Boston.

Aron, J. B. 1979. Citizen participation at government expense. *Public Administration Review,* 39: 477–85.

Ash, A., ed. 1997. *Beyond Market and Hierarchy: Interactive Governance and Social Complexity.* Elgar, Cheltenham.

Atherton, C. C. 1977. Legal requirements for environmental impact reporting. In J. McEvoy III and T. Dietz, eds., *Handbook for Environmental Planning: The Social Consequences of Environmental Change.* Wiley Interscience, New York.

Attorney General's Commission on Administrative Procedure. 1941. *Administrative Procedure in Government Agencies.* ACAP, Washington, DC.

Bachrach, P. 1967. *The Theory of Democratic Elitism: A Critique.* Little Brown, Boston.

Bacow, L. S., and Wheeler, M. 1984. *Environmental Dispute Resolution.* Plenum: New York.

Bailey, K. D. 1994. *Sociology and the New Systems Theory. Toward a Theoretical Synthesis.* State University of New York Press, Albany.

Barber, B. R. 1984. *Strong Democracy: Participatory Politics for a New Age.* University of California Press, Berkeley.

Baughman, M., Mediation. In O. Renn, Th. Webler, and P. Wiedemann eds., *Fairness and Competence in Citizen Participation. Evaluating New Models for Environmental Discourse, pp.* 253–66. Kluwer, Dordrecht and Boston.

Beck, U. 1992. *Risk Society: Toward a New Modernity,* translated by Mark A. Ritter. Sage, London.

Benhabib, S. 1992. Autonomy, modernity, and community: Communitarianism and critical theory in dialogue. In A. Honneth, T. McCarthy, C. Offe, and A. Wellmer, eds., *Cultural-Political Interventions in the Unfinished Project of Enlightenment, pp.* 39–61. MIT Press, Cambridge, MA.

Bennet, P. G. 2000. Applying the precautionary principle: A conceptual framework. In M. P. Cottam, D. W. Harvey, R. P. Paper, and J. Tait, eds., *Foresight and Precaution,* Vol. 1, pp. 223–7, A. A. Balkema, Rotterdam and Brookfield.

Bingham, G. 1984. *Resolving Environmental Disputes: A Decade of Experience.* The Conservation Foundation, Washington, DC.

Bingham, G., Anderson, F. R., Silberman, R. G., Habicht, F. H., Zoll, D. F., and Mays, R. H. 1987. Applying alternative dispute resolution to government litigation and enforcement cases. *Administrative Law Review,* 1 (Fall): 527–51.

Boehmer-Christiansen, S. 1997. Reflections on scientific advice and EC transboundary pollution policy. *Science and Public Policy,* 22 (3): 195–203.

Bohman, J. 1997. Deliberative democracy and effective social freedom: Capabilities, resources, and opportunities. In J. Bohman and W. Rehg, eds., *Deliberative Democracy. Essays on Reason and Politics, pp.* 321–48. MIT Press, Cambridge, MA.

Bohman, J. 1998. Survey article: The coming of age of deliberative democracy. *The Journal of Political Philosophy,* 6 (4): 400–25.

Bohnenblust, H., and Slovic, P. 1998. Integrating technical analysis and public values in risk based decision making. *Reliability Engineering and System Safety,* 59, 151–9.

Boholm, A. 1998. Comparative studies of risk perception: A review of twenty years of research. *Journal of Risk Research*, 1 (2): 135–63.

Bojorquez-Tapia, L. A., Ongay-Delhumeau, E., and Ezcurra E. 1994. Multivariate approach for suitability assessment and environmental conflict resolution. *Journal of Environmental Management*, 41: 187–98.

Borcherding, K., Rohrmann, B., and Eppel, T. 1986. A psychological study on the cognitive structure of risk evaluations. In B. Brehmer, H. Jungermann, P. Lourens, and G. Sevon, eds., *New Directions in Research on Decision Making*, pp. 254–62. North-Holland, Amsterdam.

Bradbury, J. A. 1989. The policy implications of differing concepts of risk. *Science, Technology, and Human Values*, 14 (4): 380–99.

Breyer, S. G. 1993. *Breaking the Vicious Circle. Toward Effective Risk Regulation*. Harvard University Press, Cambridge, Mass.

Breyer, S. G., Sunstein, C. R., and Stewart, R. B. 2001. *Administrative Law and Regulatory Policy: Problems, Text, and Cases*. Aspen, New York.

Brickman, R., Jasanoff, S., and Ilgen, T. 1985. *Controlling Chemicals: The Politics of Regulation in Europe and the United States*. Cornell University Press, Ithaca, NY.

Brion, D. 1988. An essay on LULU, NIMBY, and the problem of distributive justice. *Environmental Affairs*, 15: 437–503.

Brooks, H. 1984. The resolution of technically intensive public policy disputes. *Science, Technology, & Human Values*, 9 (Winter) 39–50.

Brulle, R. 1992. Jurgen Habermas: An exegesis for human ecologists. *Human Ecology Bulletin*, 8 (Spring/Summer): 29–40.

Buckle, L. G., and Thomas-Buckle, S. R. 1986. Placing environmental mediation in context: Lessons from "failed" mediations. *Environmental Impact Assessment Review*, 6: 55–70.

Burke, E. 1968. Citizen participation strategies. *Journal of the American Institute of Planners*, 35: 287–94.

Burns, T. R., and Überhorst, R. 1988. *Creative Democracy: Systematic Conflict Resultion and Policymaking in a World of High Science and Technology*. Praeger, New York.

Buser, M. 1995. *Die Wertbaumanalyse im Projekt Aargau: eine kritische Bestandsaufnahme*. Manuscript. Swiss Institute of Technology, Zürich.

Catton, W. R. 1980. *Overshoot: The Ecological Basis of Revolutionary Change*. University of Illinois Press, Urbana.

Charnley, G. 2000. *Democratic Science: Enhancing the Role of Science in Stakeholder-Based Risk Management Decision-Making*. Report of Health Risk Strategies. Washington, DC, July.

Charnley, G., and Elliott, E. D. 2000. Risk versus precaution: A false dichotomy. In M. P. Cottam, D. W. Harvey, R. P. Paper, and J. Tait, eds., *Foresight and Precaution*, Vol. 1, pp. 209–12. Balkema, Rotterdam and Brookfield.

Checkoway, B. 1981. The politics of public hearings. *Journal of Applied Behavioral Science*, 17 (4): 566–82.

Checkoway, B., and Van Til, J. 1978. What do we know about citizen participation? A selective review of research. In S. Langton, ed., *Citizen Participation in America*. Lexington Books, Lexington, MA.

Chen, K., Mathes, J. C., Jarboe, K., and Wolfe, J. 1979. Value oriented social decision analysis: Enhancing mutual understanding to resolve public policy issues. *IEEE Transactions on Systems, Man, and Cybernetics*, SMC-9: 567–80.

Chess, C., Dietz, Th., and Shannon, M. 1998. Who should deliberate when? *Human Ecology Review*, 5 (1), 60–8.

Clarke, L., and Short, J. F. 1993. Social organization and risk: Some current controversies. *Annual Review of Sociology*, 19: 375–99.

Claus, F. 1995. The Varresbecker Bach participatory process: The model of citizen initiatives. In O. Renn, Th. Webler, and P. Wiedemann, eds., *Fairness and Competence in Citizen Participation. Evaluating New Models for Environmental Discourse*, pp. 189–201. Kluwer, Dordrecht and Boston.

Coglianese, C. 1997. Assessing consensus: The promise and performance of negotiated rule making. *Duke Law Journal*, 46: 1255–1333.

Coglianese, C. 1999. Limits of consensus. *Environment*, 28: 28.

Coglianese, C. N. D. *Is Consensus an Appropriate Basis for Regulatory Policy?* Manuscript. Harvard University, Kennedy School of Government, Boston.

Cohen, J. 1997. Procedure and substance in deliberative democracy. In J. Bohman and W. Rehg, eds., *Deliberative Democracy. Essays on Reason and Politics*, pp. 407–37. MIT Press, Cambridge, MA.

Collingridge, D. 1996. Resilience, flexibility, and diversity in managing the risks of technologies. In C. Hood and D. K. C. Jones, eds., *Accident and Design. Contemporary Debates in Risk Management*, pp. 40–5. UCL-Press, London.

Connor, D. 1993. A generic design for public involvement programs. *Constructive Citizen Participation*, 21 (June & September): 1–2.

Coppock, R. 1985. Interactions between scientists and public officials: A comparison of the use of science in regulatory programs in the United States and West Germany, *Policy Sciences*, 18: 371–90.

Cormick, G. W. 1980. The "theory" and practice of environmental mediation. *Environmental Mediation and Conflict Management*, Special Issue of *The Environmental Professional*, 2 (1): 24–33.

Coser, L. A. 1956. *The Function of Social Conflict*. Free Press, New York.

Covello, V. T. 1983. The perception of technological risks: A literature review. *Technological Forecasting and Social Change*, 23: 285–97.

Cramton, R. C. 1972. The why, where and how of broadened public participation in the administrative process, *Georgetown Law Journal*, 60: 525–32.

Creighton, J. L. 1991. A comparison of successful and unsuccessful public involvement: A practioner's viewpoint. In C. Zervos, ed., *Risk Analysis. Prospects and Opportunities*, pp. 135–141. Plenum Press, New York.

Creighton, J. L., Dunning, C. M., and Delli Priscoli, J., eds. 1998. *Public Involvement and Dispute Resolution: A Reader on the Second Decade of Experience at the Institute of Water Resources*. U.S. Army Corps of Engineers. Institute of Water Resources, Fort Belvoir.

Crosby, N. 1995. Citizen juries: One solution for difficult environmental questions. In O. Renn, Th. Webler, and P. Wiedemann, eds., *Fairness and Competence in Citizen Participation. Evaluating New Models for Environmental Discourse*, pp. 157–74. Kluwer, Dordrecht and Boston.

Crosby, N., Kelly, J. M., and Schaefer, P. 1986. Citizen panels: A new approach to citizen participation. *Public Administration Review*, 46: 170–8.

Cross, F. B. 1992. The risk of reliance on perceived risk. *Risk – Issues in Health and Safety*, 3: 59–70.

Cross, F. B. 1998. Facts and values in risk assessment. *Reliability Engineering and Systems Safety*, 59: 27–45.

Cupps, D. S. 1977. Emerging problems of citizen participation. *Public Administraion Review*, 37: 478–87.

Dahl, R. A. 1989. *Democracy and Its Critics*. Yale University Press, New Haven, CT.

Dahl, R. A. 1994. A democratic dilemma: System effectiveness versus citizen participation. *Political Science Quarterly*, 109 (1): 23–34.

Dake, K. 1991. Orienting dispositions in the perceptions of risk: An analysis of contemporary worldviews and cultural biases. *Journal of Cross-Cultural Psychology*, 22: 61–82.

Daneke, G. A., Garcia, M. W., and Delli Priscoli, J., eds. 1983. *Public Involvement and Social Impact Assessment*. Westview Press, Boulder, CO.

Daniels, G. A., and Walker, G. B. 1996. Collaborative learning: Improving public deliberation in ecosystem-based management. *Environmental Impact Assessment Review*, 16: 71–102.

Davis, C. 1986. Public involvement in hazardous waste siting decisions. *Polity*, 19 (2): 296–304.

Delli Priscoli, J. 1989. Public involvement, conflict management: Means to EQ and social objectives. *Journal of Water Resources Planning and Management*, 115 (1): 31–42.

De-Marchi, B., and Ravetz, J. R. 1999. Risk management and governance: A post-normal science approach. *Futures*, 31: 743–57.

DeSario, J., and Langton, S. 1987. Toward a metapolicy for social planning. In J. DeSario and S. Langton, eds., *Citizen Participation in Public Decision Making*, pp. 205–21. Greenwood Press, Westport, CT.

Dienel, P. C. 1978. *Die Planungszelle*. Westdeutscher Verlag, Opladen.

Dienel, P. C. 1989. Contributing to social decision methodology: Citizen reports on technological projects. In C. Vlek and G. Cvetkovich eds., *Social Decision Methodology for Technological Projects*, pp. 133–51. Kluwer Academic, Dordrecht.

Dienel, P. C., and Garbe, D. 1975. *Zukünftige Energiepolitik. Ein Bürgergutachten*, HTV Edition. Technik und Sozialer Wandel, Munich.

Dienel, P. C., and Renn, O. 1995. Planning cells: A gate to "fractal" mediation. In O. Renn, Th. Webler, and P. Wiedemann, eds., *Fairness and Competence in Citizen Participation. Evaluating New Models for Environmental Discourse*, pp. 117–40. Kluwer, Dordrecht and Boston.

Dietz, T. 1987. Theory and method in social impact assessment. *Social Inquiry*, 57: 54–67.

Dietz, T. 1988. Social impact assessment as applied to human ecology: Integrating theory and method. In R. J. Bordent and J. Jacobs, eds., *Human Ecology: Research and Application*, Society for Human Ecology, University of Maryland Press, College Park.

Dietz, T., and Rycroft, R. W. 1987. *The Risk Professionals*. Russell Sage Foundation, New York.

Dietz, T., Stern, P. C., and Rycroft, R. W. 1989. Definitions of conflict and the legitimation of resources: The case of environmental risk. *Sociological Forum*, 4: 47–69.

Doniger, D. D. 1987. Negotiated rulemaking at the EPA: Examples of wood stove emissions and truck engine emissions. *Environmental Law Reporter,* 17 (July): 10251–4.

Douglas, M., and Wildavsky, A. 1982. *Risk and Culture.* University of California Press, Berkeley.

Downs, A. 1957. *An Economic Theory of Democracy.* Harper Collins, London.

Drottz-Sjöberg, B.-M. 1991. *Perception of Risk. Studies of Risk Attitudes, Perceptions, and Definitions.* Center for Risk Research, Stockholm.

Dryzek, J. S. 1994. *Discursive Democracy. Politics, Policy, and Political Science.* 2nd ed. Cambridge University Press, Cambridge.

Dunlap, E. 1980. Paradigmatic change in social science: From human exemptionalism to an ecological paradigm. *American Behavioral Scientist,* 24: 5–14.

Durant, J., and Joss, S. 1995. *Public Participation in Science.* Science Museum. London.

Dürrenberger, G., Kastenholz, H., and Behringer, J. 1999. Integrated assessment focus groups: Bridging the gap between science and policy? *Science and Public Policy,* 26 (5): 341–9.

Earle, T. C., and Cvetkovich, G. T. 1995. *Social Trust. Toward a Cosmopolitan Society.* Greenwood, Westport, CT.

Earle, T. C., and Cvetkovich, G. T. 1999. Social trust and culture in risk management. In G. T. Cvetkovich and R. Löfstedt, eds., *Social Trust and the Management of Risk,* pp. 9–21. Earthscan, London.

Easton, D. 1965. *A Framework for Political Analysis.* Prentice-Hall, Englewood Cliffs, NJ.

Eder, K. 1992. Politics and culture: On the sociocultural analysis of political participation. In A. Honneth, T. McCarthy, C. Offe, and A. Wellmer, eds., *Cultural-Political Interventions in the Unfinished Project of Enlightenment,* pp. 95–120. MIT Press, Cambridge, MA.

Edwards, H. T. 1986. Alternative dispute resolution: Panacea or anathema? *Harvard Law Review,* 99: 668–81.

Edwards, W. 1954. The theory of decision making. *Psychological Bulletin,* 51: 380–414.

Elliot, M. 1984. Improving community acceptance of hazardous waste facilities through alternative systems for mitigating and managing risk. *Hazardous Waste,* 1: 397–410.

Elster, J. 1989. *Solomonic Judgments. Studies in the Limitation of Rationality.* Cambridge University Press, Cambridge.

English, M. R., Gibson, A. K., Feldman, D. L., and Tonn, B. E. 1993. *Stakeholder Involvement: Open Processes for Reaching Decisions about the Future Uses of Contaminated Sites.* University of Tennessee, Knoxville.

Ethridge, M. 1987. Procedures for citizen involvement in environmental policy: An assessment of policy effects. In J. DeSario and S. Langton, eds., *Citizen Participation in Public Decision Making,* pp. 115–32. Greenwood Press, Westport, CT.

Etzioni, A. 1968. *The Active Society: A Theory of Societal and Political Processes.* The Free Press, New York.

Evers, A., and Nowotny, H. 1987. *Über den Umgang mit Unsicherheit. Die Entdeckung der Gestaltbarkeit von Gesellschaft.* Suhrkamp, Frankfurt/Main.

Fahrni, D. 1992. *An Outline History of Switzerland: From the Origins to the Present Day.* Pro Helvetia Arts Council, Zürich.

Fiorino, D. J. 1989a. Environmental risk and democratic process: A critical review. *Columbia Journal of Environmental Law,* 14 (2): 501–47.

Fiorino, D. J. 1989b. Technical and democratic values in risk analysis. *Risk Analysis,* 9 (3): 293–9.

Fiorino, D. J. 1990. Citizen participation and environmental risk: A survey of institutional mechanisms. *Science, Technology, & Human Values,* 15 (2): 226–43.

Fiorino, D. J. 1995. Regulatory negotiation as a form of public participation. In O. Renn, Th. Webler, and P. Wiedemann, eds., *Fairness and Competence in Citizen Participation. Evaluating New Models for Environmental Discourse,* pp. 223–38. Kluwer, Dordrecht and Boston.

Fischer, F. 1981. Citizen participation and the democratization of policy expertise: From theoretical inquiry to practical cases. *Policy Sciences,* 26: 165–82.

Fischhoff, B. 1985. Managing risk perceptions. *Issues in Science and Technology,* 2 (1): 83–96.

Fischhoff, B. 1996. Public values in risk research. In Annals of the American Academy of Political and Social Science, Special Issue, H. Kunreuther and P. Slovic, eds., *Challenges in Risk Assessment and Risk Management,* pp. 75–84. Sage, Thousand Oaks, CA.

Fisher, R., and Ury, W. 1981. Getting to yes: Negotiating agreement without giving in. Penguin Books, New York.

Folberg, J., and Taylor, A. 1984. *Mediation. A Comprehensive Guide to Resolving Conflicts without litigation.* Jossey-Bass Publishers, San Francisco.

Forester, J., and Stitzel, D. 1989. Beyond neutrality: The possibilities of activist mediation in public sector conflicts. *Negotiation Journal,* 5 (July): 251–64.

Freudenburg, W. R. 1983. The promise and the peril of public participation in social impact assessment. In G. A. Daneke, M. W. Garcia, and J. Delli Priscoli, eds., *Public Involvement and Social Impact Assessment,* pp. 227–34. Westview, Boulder, CO.

Freudenburg, W. R., and Pastor, S. K. 1992. Public responses to technological risk: Toward a sociological perspective. *Sociological Quarterly,* 33 (3): 389–412.

Frey, B., and Oberholzer-Gee, F. 1996. Fair siting procedures: An empirical analysis of their importance and characteristics. *Policy Analysis and Management,* 15: 353–76.

Friedman, J., ed. 1995. *The Rational Choice Controversy.* Yale University Press, New Haven, CT.

Funtowicz, S. O., and Ravetz, J. R. 1990. *Uncertainty and Quality in Science for Policy.* Kluwer, Dordrecht.

Godschalk, D. R., and Stiftle, B. 1981. Making waves: Public participation in state water planning. *Journal of Applied Behavioral Analysis,* 17: 597–614.

Graham, J. D., and Wiener, J. B. 1995. *Risk vs. Risk: Tradeoffs in Protecting Health and the Environment.* Harvard University Press, Cambridge, MA.

Granovetter, M. 1992. Economic action and social structure: The problem of embeddedness. In M. Granovetter and R. Swedberg, eds., *The Sociology of Economic Life,* pp. 53–81. Westview Press, Boulder, CO, and Oxford.

Gregory, R., Lichtenstein, S., and Slovic, P. 1993. Valuing environmental resources: A constructive approach. *Journal of Risk and Uncertainty,* 7: 177–97.

Grima, A. P. 1983. Analyzing public inputs to environmental planning. In G. A. Daneke, M. W. Garcia, and J. Delli Priscoli, eds., *Public Involvement and Social Impact Assessment*, pp. 111–19. Westview Press, Boulder, CO.

Guild, N. K. 1979. *Technology on Trial: Public Participation in Decision-Making Related to Science and Technology*. OECD, Paris.

Gusman, S. 1983. Selecting participants for a regulatory negotiation. *Environmental Impact Assessment Review*, 4: 195–202.

Habermas, J. 1968. *Technik und Wissenschaft als 'Ideologie.'* Suhrkamp, Frankfurt.

Habermas, J. 1970. Towards a theory of communicative competence. *Inquiry*, 13: 363–72.

Habermas, J. 1971. *Knowledge and Human Interests*. Beacon Press, Boston.

Habermas, J. 1984. *Theory of Communicative Action. Vol. I. Reason and the Rationalization of Society*. Beacon Press, Boston.

Habermas, J. 1987a. *Theory of Communicative Action. Vol. II. Reason and the Rationalization of Society*. Beacon Press, Boston.

Habermas, J. 1987b. *The Philosophical Discourse of Modernity*. Polity Press, Cambridge.

Habermas, J. 1989. The public sphere: An encyclopedia article, translated by S. Lennox and F. Lennox. In S. E. Bronner and D. M. Kellner, eds., *Critical Theory and Society: A Reader*, pp. 136–42. Routledge, London.

Habermas, J. 1991. *Moral Consciousness and Communicative Action*, translated by C. Lenhardt and S. Weber Nicholson, 2nd edition. MIT Press, Cambridge, MA.

Hadden, S. 1989. *A Citizen's Right-to-Know: Risk Communication and Public Policy*. Westview Press, Boulder, CO.

Hadden, S. 1995. Regulatory negotiation as citizen participation: A critique. In O. Renn, Th. Webler, and P. Wiedemann, eds., *Fairness and Competence in Citizen Participation. Evaluating New Models for Environmental Discourse*, pp. 239–52. Kluwer, Dordrecht and Boston.

Hajer, M. 1995. *The Politics of Environmental Discourse*. Oxford University Press, Oxford.

Hampel, J., and Renn, O., eds. 2000. *Gentechnik in der Öffentlichkeit. Wahrnehmung und Bewertung einer umstrittenen Technologie*, 2nd edition. Campus, Frankfurt/ Main.

Harrison, K., and Hoberg, G. 1994. *Risk, Science and Politics*. McGill-Queen's University Press, Montreal.

Harter, P., Orenstein, S., and Dalton, D. 1998. *Better Decisions through Consultation and Collaboration*. U.S. Environmental Protection Agency, Washington, DC.

Hartmann, C. 1983. The voter initiative as a form of citizen participation in Swiss transportation policy. In L. Susskind, M. Elliot, and Associates, *Paternalism, Conflict, and Coproduction. Learning from Citizen Action and Citizen Participation in Western Europe*, pp. S157–76. Plenum, New York and London.

Held, D. 1995. *Democracy and the Global Order: From the Modern State to Cosmopolitan Governance*. Polity Press, Cambridge.

Helten, E. 1998. Umwelt, Verkehr, Technik – Welchen Preis hat der Fortschritt? In Gesamtverband der Deutschen Versicherungswirtschaft, ed., *Risiko. Wieviel Risiko braucht die Gesellschaft?*, pp. 192–207. Verlag Versicherungswirtschaft, Karlsruhe.

Heyvaert, V. 1999. The changing role of science in environmental regulatory decision-making in the European Union. *Law and European Affairs*, 9 (3–4): 426–43.

Hirst, P. 1994. *Associative Democracy: New Forms of Economic and Social Governance*. Polity Press, Cambridge.

Hillgartner, S. 1992. The social construction of risk objects: Or, how to pry open networks of risk. In J. F. Short and L. Clarke, eds., *Organizations, Uncertainties, and Risk*, pp. 39–53. Westview, Boulder, CO, 1992.

Hillgartner, S., and Bosk, C. L. 1988. The rise and fall of social problems: A public arenas model. *American Journal of Sociology*, 94: 53–78.

Hohenemser, C., Kates, R. W., and Slovic, P. 1983. The nature of technological hazard. *Science*, 220: 378–84.

Horlick-Jones, T. 1998. Meaning and contextualization in risk assessment. *Reliability Engineering and Systems Safety*, 59: 79–89.

Huntington, S. 1970. The democratic distemper. *The Public Interest*, 41: 9–38.

Hyman, E. L., and Stiftel, B. 1988. *Combining Facts and Values in Environmental Impact Assessment*. Westview Press, Boulder, CO.

Jaeger, C., Renn, O., Rosa, E., and Webler, Th. 2001. *Risk, Uncertainty and Rational Action*. Earthscan, London.

Japp, K. 1996. *Soziologische Risikotheorie*. Juventa, Weinheim and Munich.

Jasanoff, S. 1982. Science and the limits of administrative rule-making: Lessons from the OSHA cancer policy. *Osgoode Hall Law Journal*, 20: 536–61.

Jasanoff, S. 1986. *Risk Management and Political Culture*. Russell Sage Foundation, New York.

Jasanoff, S. 1991. Acceptable evidence in a pluralistic society. In D. G. Mayo and R. D. Hollander, eds., *Acceptable Evidence*, pp. 29–47. Oxford University Press, Oxford.

Jasanoff, S. 1993. Bridging the two cultures of risk analysis. *Risk Analysis*, 13 (2): 123–9.

Jasanoff, S. 1998. The political science of risk perception. *Reliability Engineering and Systems Safety*, 59: 91–9.

Johnson, B. B. 1999. Exploring dimensionality in the origins of hazard-related trust. *Risk Research*, 2 (4): 325–54.

Joss, S. 1997. *Experiences with Consensus Conferences*. Paper at the International Conference on Technology and Democracy. Center for Technology and Culture. University of Oslo, Norway. January 17–19, 1997. Science Museum, London.

Kasperson, R. E. 1974. Participating in public affairs: Theories and issues. In R. Kasperson and M. Breitbart, eds., *Participation, Decentralization, and Advocacy Planning*. Resource Paper 25, pp. 1–16. Association of American Geographers, Washington, DC.

Kasperson, R. E. 1986. Six propositions for public participation and their relevance for risk communication. *Risk Analysis*, 6 (3): 275–81.

Kasperson, R. E., and Kasperson, J. X. 1983. Determining the acceptability of risk: Ethical and policy issues. In J. T. Rogers and D. V. Bates, eds., *Assessment and Perception of Risk to Human Health*, Conference Proceedings, pp. 135–55. Royal Society of Canada, Ottawa.

Kates, R. W., and Kasperson, J. X. 1983. Comparative risk analysis of technological hazards. A review. *Proceedings of the National Academy of Sciences*, 80: 7027.

Kathlene, L., and Martin, J. 1991. Enhancing citizen participation: Panel designs, perspectives, and policy formation. *Policy Analysis and Management*, 10 (1): 46–63.

Keeney, R. L. 1992. *Value Focused Thinking. A Path to Creative Decision Making.* Harvard University Press, Cambridge, MA.

Keeney, R. L. 1996. The role of values in risk management. In Annals of the American Academy of Political and Social Science, Special Issue, H. Kunreuther and P. Slovic, eds., *Challenges in Risk Assessment and Risk Management*, pp. 126–24. Sage, Thousand Oaks, CA.

Keeney, R. L., Renn, O., and von Winterfeldt, D. 1987. Structuring West Germany's energy objectives. *Energy Policy*, 15 (4): 352–62.

Kemp, R. 1985. Planning, political hearings, and the politics of discourse. In J. Forester, ed., *Critical Theory and Public Life*, pp. 177–201. MIT Press, Cambridge, MA.

Kitschelt, H. 1980. *Kernenergiepolitik. Arena eines gesellschaftlichen Konflikts.* Campus, Frankfurt and New York.

Kitschelt, H. 1986. New social movements in West Germany and the United States. *Political Power and Social Theory*, 5: 286–324.

Klinke, A., and Renn, O. 1999. *Prometheus Unbound. Challenges of Risk Evaluation, Risk Classification, and Risk Management.* Working Paper No. 153 of the Center of Technology Assessment. Center of Technology Assessment, Stuttgart.

Klinke, A., and Renn, O. 2001. Precautionary principle and discursive strategies: classifying and managing risks. *Journal of Risk Research*, 4 (2): 159–74.

Klinke, A., and Renn, O. 2002. A new approach to risk evaluation and management: Risk-based, precaution-based and discourse-based management. *Risk Analysis*, 22 (6): 71–94.

Knight, J. 1998. The bases of cooperation. Social norms and the rule of law. *Journal of Institutional and Theoretical Economics*, 154 (4): 754–63.

Knorr-Cetina, K. D. 1981. *The Manufacture of Knowledge: An Essay on the Constructivist and Contextual Nature of Science.* Pergamon Press, Oxford.

Kraft, M. 1988. Evaluating technology through public participation: The nuclear waste disposal controversy. In M. E. Kraft and N. J. Vig, eds., *Technology and Politics*, pp. 253–77. Duke University Press, Durham, NC.

Krimsky, S. 1979. Citizen participation in scientific and technological decision making. In S. Langton, ed., *Citizen Participation Perspectives. Proceedings of the National Conference on Citizen Participation*, Washington, DC, Sept. 28–Oct. 1, 1978. Tufts University Lincoln Filene Center for Citizenship and Public Affairs, Medford.

Kunreuther, H. 1995. Voluntary siting of noxious facilities: The role of compensation. In O. Renn, Th. Webler, and P. Wiedemann, eds., *Fairness and Competence in Citizen Participation. Evaluating New Models for Environmental Discourse*, pp. 283–95. Kluwer, Dordrecht and Boston.

Kunreuther, H., and Slovic, P. 1996. Science, values, and risk. In Annals of the American Academy of Political and Social Science, Special Issue, H. Kunreuther and P. Slovic, eds., *Challenges in Risk Assessment and Risk Management*, pp. 116–25. Sage, Thousand Oaks, CA.

Kweit, M. G., and Kweit, R. W. 1981. *Implementing Citizen Participation in a Bureaucratic Society.* Praeger, New York.

Kweit, M. G., and Kweit, R. W. 1987. The politics of policy analysis: The role of citizen participation in analytic decision making. In J. DeSario and S. Langton, eds., *Citizen Participation in Public Decision Making*, pp. 19–38. Greenwood Press, Westport.

Lacob, M. 1992. The Enlightenment redefined: The formation of modern civil society. *Social Research*, 59: 475–95.

Laird, F. 1993. Participatory analysis: Democracy and technological decision making. *Science, Technology, and Human Values*, 18 (3): 341–61.

Laksmanan, J. 1990. An empirical argument for nontechnical public members on advisory committees: FDA as a model. *Risk – Issues in Health and Safety*, 1: 61–74.

Langton, S. 1978. Citizen participation in America: Current reflections on the state of the art. In S. Langton, ed., *Citizen Participation in America*, pp. 1–12. Lexington Books, Lexington.

Langton, S. 1981. Evolution of a federal citizen involvement policy. *Policy Studies Review*, 1: 369–78.

Latour, B., and Woolgar, S. 1979. *Laboratory Life: The Social Construction of Scientific Facts*. Sage, London.

Leminsky, G. 1977. Bürgerbeteiligung, Mitbestimmung und Gewerkschaften – Eine vorläufige Übersicht. In H. Matthöfer, ed., *Bürgerbeteiligung und Bürgerinitiativen*, Argumente in der Energiediskussion, Vol. 3, pp. 282–93. Neckar Verlag, Villingen.

Lijphart, A. 1997. Unequal participation: Democracy's unresolved dilemma. *American Political Science Review*, 91 (1): 1–14.

Lindbloom, C. 1959. The science of muddling through. *Public Administration Review*, 19: 79–99.

Lindbloom, C. 1965. *The Intelligence of Democracy. Decision Making Through Mutual Adjustment*. Basic Books, New York.

Linder, W., and Vatter, A. 1996. Kriterien zur Evaluation von Partizipationsverfahren. In K. Selle, ed., *Planung und Kommunikation*, pp. 181–8. Bauverlag, Wiesbaden and Berlin.

Lindner, C. 1990. *Kritik der partizipatorischen Demokratie*. Westdeutscher Verlag, Opladen.

Linnerooth-Bayer, J. 1995. The Varresbecker Bach participatory process: An evaluation. In O. Renn, Th. Webler and P. Wiedemann, eds., *Fairness and Competence in Citizen Participation. Evaluating New Models for Environmental Discourse*, pp. 203–21. Kluwer, Dordrecht and Boston.

Linnerooth-Bayer, J., and Fitzgerald, K. B. 1996. Conflicting views on fair siting processes: Evidence from Austria and the U.S. *Risk – Health, Safety & Environment*, 7 (2): 119–34.

Löfstedt, R. 1997. Evaluation of two siting strategies: The case of two UK waste tire incincerators. *Risk – Health, Safety and Environment*, 8 (1): 63–77.

Löfstedt, R. 2001. *Trust and Risk Management*. Manuscript. Kings College, London.

Lowi, T. J. 1964. Four systems of policy, politics, and choice. *Public Administration Review*, 32: 298–310.

Lowi, T. 1979. *The End of Liberalism: The Second Republic of the United States*. Norton, New York.

Luhmann, N. 1982. The world society as a social system. *International Journal of General Systems*, 8: 131–8.

Luhmann, N. 1983. *Legitimation durch Verfahren.* Suhrkamp, Frankfurt.

Luhmann, N. 1984. *Soziale Systeme: Grundriß einer allgemeinen Theorie.* Suhrkamp, Frankfurt.

Luhmann, N. 1986. The autopoiesis of social systems. In R. F. Geyer and J. van der Zouven, eds., *Sociocybernetic Paradoxes: Observation, Control and Evolution of Self-Steering Systems,* pp. 172–92. Sage, London.

Luhmann, N. 1989. *Ecological Communication.* Polity Press, Cambridge.

Luhmann, N. 1990. Technology, environment, and social risk: A systems perspective. *Industrial Crisis Quarterly,* 4: 223–31.

Luhmann, N. 1993. *Risk: A Sociological Theory.* Aldine de Gruyter, New York.

Lynn, F. M. 1986. The interplay of science and values in assessing and regulating environmental risks. *Science, Technology and Human Values,* 11 (2): 40–50.

Lynn, F. M. 1987. Citizen involvement in hazardous waste sites: Two North Carolina success stories. *Environmental Impact Assessment Review,* 7: 347–61.

Lynn, F. M. 1990. Public participation in risk management decisions: The right to define, the right to know, and the right to act. *Risk – Issues in Health and Safety,* 1: 95–101.

Lynn, F. M., and Kartez, J. D. 1995. The redemption of citizen advisory committees: A perspective from critical theory. In O. Renn, Th. Webler, and P. Wiedemann, eds., *Fairness and Competence in Citizen Participation. Evaluating New Models for Environmental Discourse,* pp. 87–102. Kluwer, Dordrecht and Boston.

MacLean, D. 1986. Social values and the distribution of risk. In D. MacLean, ed., *Values at Risk,* pp. 75–93. Rowman and Allanheld, Totowa.

Maguire, L. A., and Boiney. L. G. 1994. Resolving environmental disputes: A framework for incorporating decision analysis and dispute resolution techniques. *Journal of Environmental Management,* 42: 31–48.

Majone, G. 1979. Process and outcome in regulatory decision-making. *American Behavioral Scientist,* 22 (5): 561–83.

Majone, G. 1989. *Evidence, Argument and Persuasion in the Policy Process.* Yale University Press: New Haven, CT, and London.

March, J. 1978. Bounded rationality, ambiguity, and the rationality of choice. *Bell Journal of Economics,* 9: 587–608.

Margolis, H. 1996. *Dealing with Risk. Why the Public and the Experts Disagree on Environmental Issues.* University of Chicago Press, Chicago.

Markowitz, J. 1990. Kommunikation über Risiken – Eine Theorie-Skizze. *Schweizerische Zeitschrift für Soziologie,* 3: 385–420.

Marris, C., and Joly, P.-B. 1999. Between consensus and citizens: Public participation in technology assessment in France. *Science Studies,* 12 (2): 3–32.

Mayntz, R. 1987. Politische Steuerung und gesellschaftliche Steuerungsprobleme. Anmerkungen zu einem theoretischen Paradigma. In T. Ellwein, J. J. Hesse, R. Mayntz, and F. W. Scharpf, eds., *Jahrbuch zur Staats- und Verwaltungswissenschaft,* Vol. 1, pp. 89–100. Nomos, Baden-Baden.

Mazmanian, D. 1976. Participatory democracy in a federal agency. In J. Pierce and H. Doerksen, eds., *Water Politics and Public Involvement,* pp. 127–36. Science Publishers, Ann Arbor, MI.

McCarthy, T. 1975. Translator's introduction. In J. Habermas, *Legitimation Crisis.* Beacon Press, Boston.

McDaniels, T. 1996. The structured value referendum: Eliciting preferences for environmental policy alternatives. *Journal of Policy Analysis and Management*, 15 (2): 227–51.

McDaniels, T. 1998. Ten propositions for untangling descriptive and prescriptive lessons in risk perception findings. *Reliability Engineering and System Safety*, 59: 129–34.

Merkhofer, L. W. 1986. Comparative analysis of formal decision making approaches. In: V. T. Covello, J. Menkes, and J. Mumpower, eds., *Risk Evaluation and Management*, pp. 183–220. Plenum, New York.

Mernitz, S. 1980. *Mediation of Environmental Disputes: A Source-Book*. Praeger, New York.

Metha, M. D. 1998. Risk and decision-making: A theoretical approach to public participation in techno-scientific conflict situations. *Technology in Society*, 20: 87–98.

Milbrath, L. W. 1981. Citizen surveys as citizen participation. *Journal of Applied Behavioral Science*, 17 (4): 478–96.

Mill, J. S. 1873. *Considerations on Representative Government*. Henry Holt and Company, New York.

Moore, C. 1986. *The Mediation Process. Practical Strategies for Resolving Conflict*. Jossey-Bass, San Francisco.

Moore, S. A. 1996. Defining "successful" environmental dispute resolution: Case studies from public land planning in the United States and Australia. *Environmental Impact Assessment Review*, 16: 151–69.

Morell, D., and Magorian, C. 1982. *Siting Hazardous Waste Facilities: Local Opposition and the Myth of Preemption*. Ballinger, Cambrige.

Münch, R. 1982. *Basale Soziologie: Soziologie der Politik*. Westdeutscher Verlag, Opladen.

National Research Council. 1988. *Complex Mixtures*. National Academy Press, Washington, DC.

Nelkin, D. 1977. *Technological Decisions and Democracy*. Sage, Beverly Hills, CA.

Nelkin, D., and Pollak, M. 1979. Public participation in technological decisions: Reality or grand illusion? *Technology Review*, 6 (August/September): 55–64.

Nelkin, D., and Pollak, M. 1980. Problems and procedures in the regulation of technological risk. In C. H. Weiss and A. F. Burton, eds., *Making Bureaucracies Work*, pp. 233–53. Sage, Beverly Hills, CA.

Nicholson, M. 1991. Negotiation, agreement and conflict resolution: The role of rational approaches and their criticism. In R. Väyrynen, ed., *New Directions in Conflict Theory. Conflict Resolution and Conflict Transformation*. Sage, Newbury Park.

O'Connor, D. 1978. Environmental mediation: The state-of-the-art? *Environmental Impact Assessment Review*, 2 (October): 9–17.

O'Hare, M. 1990. The importance of compensation and joint gains in environmental disputes. In W. Hoffmann-Riem and E. Schmidt-ßmann, eds., *Konfliktbewältigung durch Verhandlungen*, pp. 191–204. Nomos, Baden-Baden.

O'Hare, M., Bacow, L., and Sanderson, D. 1983. *Facility Siting and Public Opposition*. Van Nostrand Reinholt, New York.

Olson, M. 1965. *The Logic of Collective Action*. Harvard University Press, Cambridge, MA.

Olson, M. 1984. *Participatory Pluralism: Political Participation and Influence in the United States and Sweden*. Nelson-Hall, Chicago.

Okrent, D. 1996. Risk perception research program and applications: Have they received enough peer review? In C. Cacciabue and I. A. Papazoglou eds., *Probabilistic Safety Assessment and Management*, PSAM3, Vol. II, pp. 1255–60. Springer, Berlin.

Okrent, D. 1998. Risk perception and risk management: On knowledge, resource allocation and equity. *Reliability Engineering and Systems Safety*, 59: 17–25.

O'Riordan, T., and Wynne, B. 1987. Regulating environmental risks: A comparative perspective. In P. R. Kleindorfer and H. C. Kunreuther, eds., *Insuring and Managing Hazardous Risks: From Seveso to Bhopal and Beyond*, pp. 389–410. Springer, Berlin.

Parson, T. E. 1951. *The Social System*. Free Press, Glencoe, IL.

Parsons, T. E. 1963. On the concept of political power. *Proceedings of the American Philosophical Society*, 17: 352–403.

Parsons, T. E. 1967. *Sociological Theory and Modern Society*. Free Press, New York.

Parsons, T. E. 1971. *The System of Modern Societies*. Prentice-Hall, Englewood Cliffs, NJ.

Parsons, T. E., and Shils, E. A., eds. 1951. *Toward a General Theory of Action*. Cambridge University Press, Cambridge.

Perritt, H. H. 1986. Negotiated rulemaking in practice. *Journal of Policy Analysis and Management*, 5 (Spring): 482–95.

Perrow, C. 1984. *Normal Accidents: Living with High Risk Technologies*, Basic Books, New York.

Perry 6. 2000. The morality of managing risk: Paternalism, prevention and precaution, and the limits of proceduralism. *Risk Research*, 3 (2), 135–65.

Peters, R. G. 1995. A study of factors determining perceptions of trust and credibility in environmental risk communication. Ph.D. Thesis. Columbia University School of Public Health, New York.

Petts, J. 1979. The public-expert interface in local waste management decisions: Expertise, credibility, and process. *Public Understanding of Science*, 6 (4): 769–81.

Phillips, L. D. 1979. *Introduction to Decision Analysis*, Tutorial Paper 79-1. London School of Economics and Political Science, London.

Pidgeon, N. F. 1997. The limits to safety? Culture, politics, learning and manmade disasters. *Journal of Contingencies and Crisis Management*, 5 (1), 1–14.

Pinkau, K., and Renn, O., eds., 1998. *Environmental Standard Setting*. Kluwer, Dordrecht and Boston.

Pollak, M. 1985. Public participation. In H. Otway and M. Peltu, eds., *Regulating Industrial Risk*, pp. 76–94. Butterworths, London.

President's Council on Sustainable Development. 1997. *Lessons Learned from Collaborative Approaches*. President's Council on Sustainable Development, Washington, DC.

Priddat, B. P. 1996. Risiko, Ungewißheit und Neues: Epistemiologische Probleme ökonomischer Entscheidungsbildung. In G. Banse, ed., *Risikoforschung zwischen Disziplinarität und Interdisziplinarität. Von der Illusion der Sicherheit zum Umgang mit Unsicherheit*, pp. 105–24. Edition Sigma, Berlin.

Primack, J., and von Hippel, F. 1974. *Advice and Dissent: Scientists in the Political Arena*. Basic Books, New York.

Pruitt, D. G. 1986. Trends in the scientific study of negotiation and mediation. *Negotiation Journal*, 2 (July): 237–44.

Quinn, R. E., and Rohrbaugh, J. W. 1981. A competing values approach to organizational analysis. *Public Productivity Review*, 5: 141–59.

Quinn, R. E., and Rohrbaugh, J. W. 1983. A spatial model of effectiveness criteria: Towards a competing values approach to organizational analysis. *Management Science*, 29: 369.

Raiffa, H. 1994. *The Art and Science of Negotiation*. 12th edition. Cambridge University Press, Cambridge.

Rayner, S. 1987. Risk and relativism in science for policy. In V. T. Covello and B. B. Johnson, eds., *The Social and Cultural Construction of Risk*, pp. 5–23. Reidel, Dordrecht.

Rayner, S. 1990. *Risk in Cultural Perspective: Acting under Uncertainty*. Kluwer, Dordrecht and Norwell.

Rayner, S., and Cantor, R. 1987. How fair is safe enough? The cultural approach to societal technology choice. *Risk Analysis*, 7: 3–10.

Reagan, M., and Fedor-Thurman, V. 1987. Public participation: Reflections on the California energy policy experience. In J. DeSario and S. Langton, eds., *Citizen Participation in Public Decision Making*, pp. 89–113. Greenwood, Westport, CT.

Renn, O. 1986. Decision analytic tools for resolving uncertainty in the energy debate. *Nuclear Engineering and Design*, 93 (2 and 3): 167–80.

Renn, O. 1992. The social arena concept of risk debates. In S. Krimsky and D. Golding, eds., *Social Theories of Risk*, pp. 169–97. Praeger, Westport, CT.

Renn, O. 1995. Style of using scientific expertise: A comparative framework. *Science and Public Policy*, 22 (June): 147–56.

Renn, O. 1998. The role of risk communication and public dialogue for improving risk management. *Risk Decision and Policy* 3 (1): 5–30.

Renn, O. 1999a. A model for an analytic-deliberative process in risk management. *Environmental Science and Technology*, 33 (18): 3049–55.

Renn, O. 1999b. Participative technology assessment: Meeting the challenges of uncertainty and ambivalence. *Futures Research Quarterly*, 15 (3): 81–97.

Renn, O. 1999c. Diskursive Verfahren der Technikfolgenabschätzung. In Th. Petermann and R. Coenen, eds., *Technikfolgenabschätzung in Deutschland. Bilanz und Perspektiven*, pp. 115–30. Campus, Frankfurt am Main.

Renn, O., and Levine, D. 1990. Credibility and trust in risk communication. In R. Kasperson and P. J. Stallen, eds., *Communicating Risk to the Public*, pp. 175–218. Kluwer, Dordrecht.

Renn, O., and Webler, Th. 1992. Anticipating conflicts: Public participation in managing the solid waste crisis. *GAIA Ecological Perspectives in Science, Humanities, and Economics*, 1 (2): 84–94.

Renn, O., and Webler, Th. 1998. Der kooperative Diskurs – Theoretische Grundlagen, Anforderungen, Möglichkeiten. In O. Renn, H. Kastenholz, P. Schild, and U. Wilhelm, eds., *Abfallpolitik im kooperativen Diskurs. Bürgerbeteiligung bei der Standortsuche für eine Deponie im Kanton Aargau*, pp. 3–103. Hochschulverlag AG, Zürich.

Renn, O., Albrecht, G., Kotte, U., Peters, H. P., and Stegelmann, H. U. 1985. *Sozialverträgliche Energiepolitik. Ein Gutachten für die Bundesregierung*, HTV Edition. Technik und Sozialer Wandel, Munich.

Renn, O., Goble, R., Levine, D., Rakel, H., and Webler, Th. 1989. *Citizen Participation for Sludge Management*, Final Report to the New Jersey Department of Environmental Protection. CENTED, Clark University, Worcester.

Renn, O., Schrimpf, M., Büttner, Th., Carius, R., Köberle, S., Oppermann, B., Schneider, E., and Zöller, K. 1999. *Abfallwirtschaft 2005. Bürger planen ein regionales Abfallkonzept*. Nomos, Baden-Baden.

Renn, O., Webler, Th., and Wiedemann, P. 1995. The pursuit of fair and competent citizen participation. In O. Renn, Th. Webler, and P. Wiedemann, eds., *Fairness and Competence in Citizen Participation. Evaluating New Models for Environmental Discourse*, pp. 339–68. Kluwer, Dordrecht and Boston.

Renn, O., Webler, Th., Rakel. H., Dienel, P. C., and Johnson, B. 1985. Public participation in decision making: A three-step-procedure. *Policy Sciences*, 26: 189–214.

Rip, A. 1985. Experts in public arenas. In H. Otway and M. Peltu, eds., *Regulating Industrial Risks*, pp. 94–110. Butterworth, London.

Rip, A. 1992. The development of restrictedness in the sciences. In N. Elias, H. Martins, and R. Whitley, eds., *Scientific Establishments and Hierarchies*, pp. 219–38. Kluwer, Dordrecht and Boston.

Rippe, K. P., and Schaber, P. 1999. Democracy and environmental decision-making. *Environmental Values*, 8: 75–98.

Roch, I. 1997. *Evaluation der 3. Phase des Bürgerbeteiligungsverfahrens in der Region Nordschwarzwald*, Research Report Nr. 71. Center of Technology Assessment, Stuttgart.

Rohrmann, B. 1992. The evaluation of risk communication effectiveness. *Acta Psychologica*, 81: 169–92.

Rohrmann, B. 1999. *Risk perception research: Review and documentation*. Studies in Risk Communication, No. 68. Research Center Jülich, Jülich.

Rohrmann, B., and Renn, O. 2000. Introduction. In O. Renn and B. Rohrmann, eds., *Cross-Cultural Risk Perception*, pp. 5–32. Kluwer, Dordrecht and Boston.

Roqueplo, P. 1995. Scientific expertise among political powers, administrators and public opinion. *Science and Public Policy*, 22 (3): 175–82.

Rosa, E. A. 1988. NAMBY PAMBY and NIMBY PIMBY: Public issues in the siting of hazardous waste facilities. *Forum for Applied Research and Public Policy*, 3: 114–23.

Rosa, E. A. 1997. Metatheoretical foundations for post-normal risk. *Journal of Risk Research*, 1 (1): 15–44.

Rosenbaum, N. 1976. *Citizen Involvement in Land Use Governance*. The Urban Institute, Washington, DC.

Rosenbaum, N. 1978. Citizen participation and democratic theory. In S. Langton, ed., *Citizen Participation in America*, pp. 43–54. Lexington Books, Lexington.

Rosenbaum, W. 1979. Elitism and citizen participation. In S. Langton, ed., *Citizen Participation Perspectives. Proceedings of the National Conference on Citizen Participation* (Sept. 28–Oct. 1, 1978), Washington, DC. Tufts University Lincoln Filene Center for Citizenship and Public Affairs, Medford.

Rosener, J. 1982. Making bureaucracy responsive: A study of the impacts of citizen participation and staff recommendations on regulatory decision making. *Public Administration Review*, 42: 339–45.

Rossi, J. 1997. Participation run amok: The costs of mass participation for deliberative agency decisionmaking. *Northwestern University Law Review*, 92: 173–249.

Rowe, G., and Frewer, L. J. 2000. Public participation methods: A framework for evaluation. *Science, Technology & Human Values*, 225 (1): 3–29.

Ruckelshaus, W. 1985. Risk, science and democracy. *Issues in Science and Technology*, 1 (3): 19–38.

Rushefsky, M. 1984. Institutional mechanisms for resolving risk controversies, In: S. G. Hadden, ed., *Risk Analysis, Institutions, and Public Policy*, pp. 133–148, Associated Faculty Press, Port Washington, NY.

Rymann, C. 1993. *Demokratische Evaluation eines Deponiestandortes im östlichen Kantonsteil des Kantons Aaregau.*Kantonsschule Wohlen, Aarau.

Sager, T. 1994. *Communicative Planning Theory*. Aldershot, Avebury.

Scharpf, F. W. 1991. Die Handlungsfähigkeit des Staates am Ende des zwanzigsten Jahrhunderts. *Politische Vierteljahresschrift*, 32 (4): 621–34.

Schellnhuber, H.-J. 1999. "Earth system" analysis and the second copernican revolution. *Nature*, 402 (December): C19–C23.

Scheuch, E. K. 1980. Kontroverse um Energie – ein echter oder ein Stellvertreterstreit. In H. Michaelis, ed., *Existenzfrage Energie*, 279–293. Econ, Düsseldorf.

Schmidt, M. 2000. *Demokratietheorien*, 3rd edition. Leske + Budrich, Opladen.

Schneider, E., Oppermann, B., and Renn, O. 1998. Implementing structured participation for regional level waste management planning. *Risk – Health, Safety & Environment*, 9 (Fall): 379–95.

Schoenbrod, D. 1983. Limits and dangers of environmental mediation: A review essay. *New York University Law Review*, 58 (December): 1453–76.

Sclove, R. 1995. *Democracy and Technology*. Guilford Press, New York.

Scott, W., and Meyer, W. 1991. The organization of societal sectors: Proposition and early evidence. In W. Powell and P. J. DiMaggio, eds., *The New Institutionalism in Organizational Analysis*, pp. 108–40. University of Chicago Press, Chicago.

Seligman, A. B. 1993. *The Idea of Civil Society*. Free Press, New York.

Shils, E. 1991. The virtues of civil society. *Government and Opposition*, 26 (2): 3–20.

Shrader-Frechette, K. 1985. *Risk Analysis and Scientific Method: Methodological and Ethical Problems with Evaluating Societal Risks*. Reidel, Dordrecht.

Shrader-Frechette, K. 1990. Scientific method, anti-foundationalism, and public policy. *Risk – Issues in Health and Safety*, 1: 23–41.

Shrader-Frechette, K. 1991. *Risk and Rationality. Philosophical Foundations for Populist Reforms*. University of California Press, Berkeley.

Sjöberg, L. 1999a. Risk perception in western Europe. *Ambio*, 28 (6): 543–9.

Sjöberg, L. 1999b. Consequences of perceived risk. *Journal of Risk Research*, 2: 129–49.

Skillington, T. 1997. Politics and the struggle to define: A discourse analysis of the framing strategies of competing actors in a "new" participatory forum. *British Journal of Sociology*, 48 (3): 493–513.

Slovic, P. 1987. Perception of risk. *Science*, 236 (4799): 280–5.

Slovic, P. 1992. Perception of risk: Reflections on the psychometric paradigm. In D. Golding and S. Krimsky, eds., *Theories of Risk*, pp. 117–52. Praeger, London.

Slovic, P. 1993. Perceived risk, trust and democracy. *Risk Analysis*, 13: 675–82.

Slovic, P., Fischhoff, B., and Lichtenstein, S. 1981. Perceived risk: Psychological factors and social implications. *Proceedings of the Royal Society*, A376: 17–34.

Starr, Ch., and Whipple, C. 1980. Risks of risk decisions. *Science*, 208 (4403): 1116.

Stern, P. C. 1991. Learning through conflict: A realistic strategy for risk communication. *Policy Sciences*, 24: 99–119.

Stern, P. C., and Fineberg, V. 1996. *Understanding Risk: Informing Decisions in a Democratic Society*. National Research Council, Committee on Risk Characterization, National Academy Press, Washington, DC.

Stewart, J., Kendall, E., and Coote, A. 1994. *Citizen Juries*. Institute for Public Research, London.

Stirling, A. 1998. Risk at a turning point? *Journal of Risk Research*, 1 (2), 97–109.

Stirling, A. 1999. *On Science and Precaution in the Management of Technological Risks*. Final report of a project for the EC Forward Studies Unit under auspices of the ESTO Network. Report EUR 19056 EN. European Commission, Brussels.

Stolwijk, J. A. J., and Canny, P. F. 1991. Determinants of public participation in the management of technological risk. In M. Shubik, ed., *Risk, Organizations, and Society*, pp. 33–48. Kluwer, Dordrecht and Boston.

Susskind, L. E., and Cruishank, J. 1987. *Breaking the Impasse: Consensual Approaches to Resolving Public Disputes*. Basic Books, New York.

Susskind, L. E., and Fields, P. 1996. *Dealing with an Angry Public: The Mutual Gains Approach to Resolving Disputes*. The Free Press, New York.

Susskind, L. E., and Ozawa, C. 1985. Mediating public disputes: Obstacles and possibilities. *Journal of Social Issues*, 41: 145–59.

Susskind, L. E., Bacow, L., and Wheeler, M. 1083. *Resolving Environmental Regulatory Disputes*. Schenkman, Cambridge.

Talbot, A. 1983. *Settling Things: Six Case Studies in Environmental Mediation*. Conservation Foundation/Ford Foundation: Washington, DC.

Thomas, J. C. 1990. Public involvement in public management: Adapting and testing a borrowed theory. *Public Administration Review*, 50: 435–45.

Thompson, M., Ellis, W., and Wildavsky, A. 1990. *Cultural Theory*. Westview Press, Boulder, CO.

Thompson, P. D. 1999. Agriculture, biotechnology, and the political evaluation of risk. *Policy Studies Journal*, 17 (1): 97–108.

Tuler, S. 1996. *Meanings, Understandings, and Interpersonal Relationships in Environmental Policy Discourse*. Doctoral Dissertation. Clark University, Worcester.

Tuler, S., and Webler, Th. 1995. Process evaluation for discoursive decision making in environmental and risk policy. *Human Ecological Review*, 2: 62–74.

Tuler, S. and Webler, Th. 1999. Designing an analytic deliberative process for environmental health policy making in the U. S. nuclear weapons complex. *Risk – Health, Safety & Environment*, 10 (1): 65–87.

U.S. Environmental Protection Agency (EPA). 1983. *Community Relations in Superfund: A Handbook*. EPA: Washington, DC.

U.S. Environmental Protection Agency (EPA). 1987. *Unfinished Business: A Comparative Assessment of Environmental Problems*. EPA: Washington, DC.

U.S. Environmental Protection Agency (EPA). 2000. Policy on alternative dispute resolution. *Federal Register*, 65 (249), December 27: 81858–60.

van Asselt, M. B. A. 2000. *Perspectives on Uncertainty and Risk*. Kluwer, Dordrecht and Boston.

van den Daele, W. 1992. Scientific evidence and the regulation of technical risks: Twenty years of demythologizing the experts. In N. Stehr and R. V. Ericson, eds., *The Culture and Power of Knowledge. Inquiries into Contemporary Societies*, pp. 323–40. de Gruyter, Berlin.

van den Daele, W. 2000. Interpreting the precautionary principle – Political versus legal perspectives. In M. P. Cottam, D. W. Harvey, R. P. Paper, and J. Tait eds., *Foresight and Precaution*, Vol. 1, pp. 213–21. A. A. Balkema, Rotterdam and Brookfield.

van Valey, T. L., and Petersen, J. C. 1987. Public service science centers: The Michigan experience. In J. DeSario and S. Langton, eds., *Citizen Participation in Public Decision Making*, pp. 39–63. Greenwood Press, Westport.

Vari, A. 1989. Approaches towards conflict resolution in decision processes. In C. Vlek and G. Cvetkowich, eds., *Social Decision Methodology for Technological Projects*, pp. 74–94. Kluwer, Dordrecht.

Vari, A. 1995. Citizens' advisory committee as a model for public participation: A multiple-criteria evaluation. In O. Renn, Th. Webler, and P. Wiedemann, eds., *Fairness and Competence in Citizen Participation. Evaluating New Models for Environmental Discourse*, pp. 103–16. Kluwer, Dordrecht and Boston.

Viscusi, W. K. 1998. *Rational Risk Policy. The 1996 Arne Ryde Memorial Lectures.* Oxford University Press, Oxford and New York.

Vlek, C. A. 1996. A multi-level, multi-stage and multi-attribute perspective on risk assessment, decision-making, and risk control. *Risk, Decision, and Policy*, 1 (1): 9–31.

Vlek, C. A., and Cvetkovich, G. 1989. Social decision making on technological projects: Review of key issues and a recommended procedure. In C. Vlek and G. Cvetkovich, eds., *Social Decision Methodology for Technological Projects*, pp. 297–322. Kluwer, Dordrecht.

Vogel, D. 1986. *National Styles of Regulation: Environmental Policy in Great Britain and the United States.* Cornell University Press, Ithaca, NY.

von Schomberg, R. 1992. Argumentation im Kontext wissenschaftlicher Kontroversen. In K.-O. Apel and M. Kettener, eds., *Zur Anwendung der Diskursethik in Politik, Recht, Wissenschaft*, pp. 260–77. Suhrkamp, Frankfurt/Main.

von Schomberg, R. 1995. The erosion of the valuespheres. The ways in which society copes with scientific, moral and ethical uncertainty. In R. von Schomberg, ed., *Contested Technology. Ethics, Risk and Public Debate*, pp. 13–28. International Centre for Human and Public Affairs, Tilburg.

von Winterfeldt, D. 1987. Value tree analysis: An introduction and an application to offshore oil drilling. In P. R. Kleindorfer and H. C. Kunreuther, eds., *Insuring and Managing Hazardous Risks: From Seveso to Bhopal and Beyond*, pp. 349–76. Springer, Berlin.

von Winterfeldt, D., and Edwards, W. 1986. *Decision Analysis and Behavioral Research.* Cambridge University Press, Cambridge.

Vorwerk, V., and Kämper, E. 1997. *Evaluation der 3. Phase des Bürgerbeteiligungsverfahrens in der Region Nordschwarzwald.* Working Report No. 70. Center of Technology Assessment, Stuttgart.

Wachlin, K. D., and Renn, O. 1999. Diskurse an der Akademie für TA in Baden-Württemberg: Verständigung, Abwägung, Gestaltung, Vermittlung. In S. Bröchler, G. Simonis, and K. Sundermann, eds., *Handbuch Technikfolgenabschätzung*, Vol. 2, pp. 713–22. Sigma, Berlin.

Waldo, J. 1987. Win/win does work. *Timber-Fish-Wildlife. A Report from the Northwest Renewable Resouirces Center*, 1 (1): 7.

Waller, T. 1995. Knowledge, power and environmental policy: Expertise, the lay public and water management in the western United States. *The Environmental Professional*, 7: 153–66.

Warren, M. E. 1993. Can participatory democracy produce better selves? Psychological dimensions of Habermas discursive model of democracy. *Political Psychology*, 14: 209–34.

Watts, N. S. J. 1987. Mobilisierungspotential und gesellschaftliche Bedeutung der neuen sozialen Bewegungen. In R. Roth and D. Rucht, eds., *Neue Soziale Bewegungen in der Bundesrepublik Deutschland*, pp. 47–67. Campus, Franfurt/Main.

WBGU (German Advisory Council on Global Change). 2000. *World in Transition: Strategies for Managing Global Environmental Risks*. Springer, Berlin.

Webler, Th. 1994. *Experimenting with a New Democratic Instrument in Switzerland: Siting a Landfill in the Eastern Part of Canton Aargau*. Working Paper. Polyproject: Safety of Technological Systems. Swiss Institute of Technology, Zürich.

Webler, Th. 1995. "Right" discourse in citizen participation. An evaluative yardstick. In O. Renn, Th. Webler, and P. Wiedemann, eds., *Fairness and Competence in Citizen Participation. Evaluating New Models for Environmental Discourse*, pp. 35–86. Kluwer, Dordrecht and Boston.

Webler, Th. 1999. The craft and theory of public participation: A dialectical process. *Risk Research*, 2 (1): 55–71.

Webler, Th., and Renn, O. 1995. A brief primer on participation: Philosophy and practice. In O. Renn, Th. Webler, and P. Wiedemann, eds., *Fairness and Competence in Citizen Participation. Evaluating New Models for Environmental Discourse*, pp. 17–34. Kluwer, Dordrecht and Boston.

Webler, Th. and Tuler, S. 1999. Integrating technical analysis with deliberation in regional watershed management planning: Applying the National Research Council approach. *Policy Studies Journal*, 27 (3): 530–43.

Webler, Th., Kastenholz, H., and Renn, O. 1995b. Public participation in impact assessment: A social learning perspective. *Environmental Impact Assessment Review*, 15: 443–63.

Webler, Th., Levine, D., Rakel, H., and Renn, O. 1991. The Group Delphi: A novel attempt at reducing uncertainty. *Technological Forecasting and Social Change*, 39 (3): 253–63.

Webler, Th., Rakel, H., Renn, O., and Johnson, B. 1995a. Eliciting and classifying concerns: A methodological critique. *Risk Analysis*, 15 (3): 421–36.

Wehrli-Schindler, B. 1987. *Demokratische Mitwirkung in der Raumplanung*. SVPW, Bern.

Weidner, H. 1993. *Mediation as a Policy for Resolving Environmental Disputes with Special References to Germany*. Manuscript of the Series "Mediationsverfahren im Umweltschutz." Science Center, Berlin.

Weingart, P. 1979. Das 'Harrisburg-Syndrom' oder die De-Professionalisierung der Experten. Preface to H. Nowotny, *Kernenergie: Gefahr oder Notwendigkeit,* pp. 9–17. Suhrkamp, Frankfurt am Main.

Weingart, P. 1983. Verwissenschaftlichung der Gesellschaft – Politisierung der Wissenschaft. *Zeitschrift für Soziologie,* 12: 225–41.

Wiesendahl, E. 1987. Neue soziale Bewegungen und moderne Demokratietheorie. In R. Roth and D. Rucht, eds., *Neue Soziale Bewegungen in der Bundesrepublik Deutschland,* pp. 364–84. Campus, Franfurt/Main.

Willke, H. 1985. *Systemtheorie III. Steuerungstheorie.* UTB Fischer, Stuttgart and Jena.

Wondolleck, J. M. 1985. The importance of process in resolving environmental disputes. *Environmental Impact Assessment Review,* 5 (December): 341–56.

Wondelleck, J. M., Manring, N. J., and Crowfoot, J. E. 1996. Teetering at the Top of the ladder: The experience of citizen group participants in alternative dispute resolution processes. *Sociological Perspectives,* 39 (2): 249–62.

Wynne, B. 1984. Public perceptions of risk. In J. Aurrey, ed., *The Urban Transportation of Irradiated Fuel,* pp. 246–59. Macmillan, London.

Wynne, B. 1980. Sheepfarming after Chernobyl. *Environment,* 31: 11–15, 33–9.

Wynne, B. 1992a. Uncertainty and environmental learning. Reconceiving science and policy in the preventive paradigm. *Global Environmental Change,* 2 (June): 111–27.

Wynne, B. 1992b. Risk and social learning: Reification to engagement. In S. Krimsky and D. Golding, eds., *Social Theories of Risk,* pp. 275–97. Praeger, Westport, CT.

Yankelovich, D. 1991. *Coming to Public Judgment. Making Democracy Work in a Complex World.* Syracuse University Press, Syracuse, NY.

Yosie, T. F., and Herbst, T. D. 1998. Managing and communicating stakeholder-based decision making. *Human and Ecological Risk Assessment,* 4: 643–6.

Zeckhauser, R., and Viscusi, K. W. 1996. The risk management dilemma. Annals of the American Academy of Political and Social Science, Special Issue, H. Kunreuther and P. Slovic, eds., *Challenges in Risk Assessment and Risk Management,* pp. 144–55. Sage, Thousand Oaks, CA.

Zillßen, H. 1993. Die Modernisierung der Demokratie im Zeichen der Umweltpolitik. In H. Zilleßen, P. C. Dienel, and W. Strubelt, eds., *Die Modernisierung der Demokratie,* pp. 17–39. Westdeutscher Verlag, Opladen.

10

Global Change and Transboundary Risks

Joyce Tait and Ann Bruce

1. INTRODUCTION

Risks that are generated under one regulatory jurisdiction and have significant actual or anticipated impacts in another, regionally or globally, are a source of concern for regulators, politicians, and the public. Attempts to control and regulate such risks and to bring in systems of liability for damage are often regarded by industry as restricting competitiveness and inhibiting innovation (House of Lords Select Committee on Science and Technology, 1993; Leonard, 1988). On the other hand, to take a more optimistic view, transboundary risk issues and their mitigation have been at the forefront in stimulating international collaboration among governments and public and private institutions. Success in this area, although limited, has been greater than in many others.

A diverse spectrum of issues can be included under the heading "transboundary risks," prominent recent examples being the Chernobyl nuclear disaster; acid rain in Europe and North America; pollution of seas and of the River Rhine, the Colorado River, and the Great Lakes; bovine spongiform encephalopathy (BSE); genetically modified (GM) crops, hormone treated beef; and global climate change (Linnerooth-Bayer, 2000). The risks that attract international attention are usually distinguished by the scale of their production, their scope and pervasive nature, their origins in industrialized societies, the role of science and technology in their production and control, the greater complexity of interactions and hence the difficulty of modeling their impacts, the sometimes irreversible nature of these impacts and issues of accountability and liability for damage. Attempting to deal with such hazards at a national level would either be

The authors would like to thank Joanne Linnerooth-Bayer, International Institute for Applied Systems Analysis (IIASA), Laxenburg, Austria, and also anonymous referees for reviewing this article and for many helpful suggestions.

ineffective or violate the principles of international equity, and the "polluter pays."

The amount and quality of scientific evidence to support claims of damage or potential hazard is also variable, and public responses to risk, in the transboundary field as in other areas, are usually not proportional to the impacts themselves or to the evidence for them. Regulators may therefore be attempting to develop international controls in advance of strong public interest or concern (as for global climate change) or as a delayed reaction to public or expert concern (as for BSE), or they may be faced with levels of public concern that are not justified by the available evidence (as for GM crops in Europe).

In a parallel and related series of developments, the increasingly rapid pace of technological innovation and the increasing size and power of multinational companies are leading to globalization of production and trading systems accompanied by pressures for further trade liberalization in the "Millennium Round" of negotiations under the World Trade Organisation (WTO). These changes diminish the sense of power and influence of individual citizens and appear to negate local and national democratic processes, raising fundamental questions of sovereignty and governance.

For example, Giddens (1999) in his BBC Reith Lecture series on globalization referred to risk as "the mobilising dynamic of a society bent on change, that wants to determine its own future." On the other hand, he noted that we now live in a world where hazards created by ourselves are as, or more, threatening than so-called natural hazards. Several months later the WTO talks in Seattle were seriously disrupted by public demonstrations against several transboundary risk issues and against the globalization process itself (Kyriakou, 2000). A year later, Chris Patten (2000) gave the first of another series of BBC Reith Lectures on sustainable development. Continuing the globalization theme, he referred to the self-contradictory banner of a lobbyist at Seattle which read "The World Wide Movement Against Globalization" and he raised some questions about the legitimacy of nongovernmental organizations (NGOs), as representatives of public concerns at such negotiations. Disruption by groups and individuals opposed to globalization is a continuing accompaniment to major intergovernmental meetings.

Both lecturers recognize that international issues of risk and sustainable development are intimately bound up with one another and with a broader set of societal concerns about globalization and national and international governance. Along with the ability to generate hazards that cross national boundaries, society is thus facing the need to develop a range of international mechanisms to regulate them.

This chapter gives an overview of the field of transboundary risk regulation and summarizes the state of development of international controls for a representative set of risks. Sections 2 and 3 set the overall context and introduce a system of risk categories each of which implies a particular set of challenges for risk regulation and management. Sections 4 and 5 give examples of transboundary risks in each of our proposed risk categories and describe the development and application of the corresponding regulatory measures.

Section 6 develops our overall approach to analysis and interpretation of transboundary risk issues and demonstrates how the four basic categories can help our understanding of the issues and of related societal interactions, particularly between transboundary risk regulation and global trade issues.

One of the most contentious aspects of international risk debates is the relevance and appropriateness of the precautionary principle. This chapter does not cover the precautionary principle in detail, but it does not ignore it. Table 10.1 summarizes the potential roles of the precautionary principle and its counterpart, the risk-based approach for the four categories of transboundary risk, and this is discussed further in Section 6.3.

It is not possible to provide an exhaustive set of references to the relevant literature on such a broad range of topics. Instead we have provided a route into each topic through books, refereed publications, trade journals, and websites.

2. PROCEDURES AND INSTITUTIONS FOR TRANSBOUNDARY RISK MANAGEMENT

The environmental concerns that gave rise to modern systems of transboundary risk regulation began in the 1960s with the problems of acidification in Scandinavian and North American lakes (Shaw, 1993). At this time, the acid rain issue was interpreted by many Swedes as a form of intervention in their affairs by West Germany and Britain (Linnerooth-Bayer, Loefstedt, and Sjoestedt, 2000). In 1972, Sweden invited the UN Conference on the Human Environment to Stockholm (von Weizsäcker, 1994), and this is usually considered the starting point for concerted development of international environmental regulation. It drew attention to the need for sustainable development and called for a more holistic approach to environmental protection (Tolba and Rummel-Bulska, 1998).

2.1. Regulatory Instruments

Since the 1972 conference, there has been a rapid proliferation of regulatory initiatives of varying international coverage and effectiveness. As described in more detail later, the stimulus for transboundary regulation can be seen as arising either from risks that are attached (literally or conceptually) to products traded across national boundaries or to risks that cross these boundaries accidentally or from nationally permitted emissions of pollutants.

A common approach to transboundary regulation is to begin by negotiating a framework convention setting out general principles, followed by protocols to the convention that define actual commitments. Both convention and protocols only come into force after they have been ratified by a previously agreed number of nations that then become "parties" to the convention or protocol. Both the initial negotiations and achieving ratification can be lengthy processes as negotiators face opposition based on economic, political, or personal issues.

NGOs are often important actors in such processes, encouraging governments to act. On the other hand, they may seek maximum impact at an early stage of negotiation, rather than beginning with achievable objectives that could be followed by more stringent measures, and this can delay and complicate final outcomes.

A number of approaches have been used to speed up the adoption of specific regulatory measures (Tolba and Rummel-Bulska, 1998).

> Provisional implementation of an initiative may be achieved rapidly but only on an interim basis (e.g., the Geneva Convention on Long Range Transboundary Air Pollution – see Section 5.2).
>
> Informal, nonbinding agreements that do not need ratification can be put into place immediately (the so-called soft law). Although not legally enforceable, these may eventually develop into legally binding treaties (e.g., the Cairo Guidelines developed into the Basel Convention on Hazardous Wastes – see Section 4.1).
>
> Lawmaking may be delegated if the situation faces rapid changes in technology or scientific knowledge. The intergovernmental body charged with developing the treaty may be allowed to revise it without the need for ratification (e.g., the Montreal Protocol on Substances that Deplete the Ozone Layer – see Section 5.2).

Sand (1999) notes that the implementation of instruments and agreements dealing with transboundary risks has mainly been influenced by such factors as financial resources, technical and scientific assistance, public information and national reporting duties. Despite concerns about the effectiveness of international environmental agreements, Victor and Skolnikoff (1999) argue that most countries do comply most of the time with their treaty obligations. Compliance appears, however, to be an inevitable outcome of the observation that states generally sign agreements that are in their immediate interest and that can be easily carried out. Reservations and exemptions on agreements can make it easier for countries to sign by allowing them time to adjust to full compliance; on the other hand, allowing too many exemptions may undermine the regulation. Sand shows that international supervisory bodies, noncompliance procedures, and dispute settlement procedures have not played a major role in treaty implementation; on the other, hand Renn and Klinke (2000) argue that improved dispute resolution procedures are needed for resolving cross-border disputes on risk issues.

Most cases of adjudication on transboundary risks depend on common agreement by the parties. One state cannot apply to the International Court of Justice; usually other parties must agree to this. For example, this is the case with the Basel Convention (hazardous waste) and the Vienna Convention (protection of the ozone layer).

2.2. Economic Instruments

Economic instruments are increasingly favored in addition, or as an alternative, to regulatory instruments. Since the mid-1980s the European Community (EC) and Organization for Economic Co-operation and Development (OECD) have declared that they would give more prominence to economic instruments to support environmental policy (von Weizsäcker, 1994) and a variety of different approaches is being considered and adopted (e.g., taxation, tradable permits, environmental charges).

Shortly before the 1972 Stockholm conference, the OECD agreed that the "polluter pays" should be a fundamental guiding principle in most cases involving transboundary pollution-related hazards (Andersen, 1994). The report *Our Common Future* (World Commission on Environment and Development, 1987) marked a further development, supporting the concept of "cleaner technology" as the key to sustainability and recommending the use of economic instruments to encourage the development of such technologies.

Taxation is often favored by governments as a means to deal with environmental pollution problems within national boundaries. However, ecologically motivated tax reforms have not been easy to implement (O'Riordan, 1994, p. 218). Both the U.S. Clinton Administration proposal for an energy tax and the EU carbon/energy tax proposals have met with political opposition, and questions have also been asked about the validity of such national initiatives in the context of WTO rules.

The effectiveness of taxation as an instrument to mitigate environmental or other hazards has also been questioned, particularly by industry. Tax rises may merely encourage taxpayers to pass the additional cost on to their customers or to find other ways to reduce their tax payments without delivering the expected benefits. Distortion of competition, nationally and internationally, and unfair distribution of the tax burden are also potential problems. For example, Pearson and Smith (1991) found that the proposed EC carbon/energy tax would have a larger impact on poorer households.

The development of systems of tradable permits appears to be a more promising approach to international risk regulation (see for example the U.S. sulfur dioxide trading component of the Acid Rain Program (http://www.eap.gov/acidrain/allsys.htm), which formed the basis of emissions trading in the Kyoto Protocol). Tradable permits are discussed in more detail in Sections 5.2 and 6.2.

3. CHARACTERIZING TRANS-BOUNDARY RISKS

Transboundary risks encompass a very wide range of environmental and health-related issues, and there are diverse ways of categorizing these risk issues to capture their source, their nature, the relationships among the countries involved, their potential consequences, and their management and control. Drawing upon empirical evidence of transboundary-risk conflicts, Kasperson and Kasperson (2000) argue that transboundary risks come in different complexes and forms and cannot be viewed as a single problem or challenge. They define four types of transboundary risk:

- *border-impact risks* involving activities, industrial facilities, or developments that affect populations or ecosystems in a border region on both sides of the boundary;
- *point-source transboundary risks* involving point sources of pollution or accident threatening one or more adjoining countries or regions;
- *structural/policy transboundary risks* involving less identifiable, more subtle and diffuse effects associated with state policy, transportation or energy systems, or the structure of the economy; and
- *global environmental risks* involving human activities in a given country that affect many or all other countries through alterations of the global "systemic" environment.

Linnerooth-Bayer (2000) suggests that the concept of "consent" or "voluntariness" is also important for comparing transboundary risk issues. The categorization of transboundary risks developed for this paper is based on her approach, which distinguishes between "involuntary" risks, which cannot be detained at borders and thus cannot be controlled by a system of consents from importing countries (e.g., nuclear pollution of the atmosphere), and "voluntary" risks, which can be controlled by a system of consents by importing authorities (for example GM organisms).

Based on similar arguments, we have identified two basic transboundary risk categories, "traded" and "public" risks, and two subcategories within each of these (Table 10.1). Sections 3.1 and 3.2 summarize the basic characteristics of the four types, and they are developed further using a series of examples in Sections 4 and 5.

All classifications reflect the purposes and perspectives of those who develop them. Ours derives from a systemic, interdisciplinary perspective on policy analysis with the purpose of contributing to the understanding and improvement of regulatory processes. The four categories do not cover all possible transboundary risks, and they are not necessarily mutually exclusive. Risks can also shift from one category to another as circumstances change (see Section 6.2). Indeed, a useful feature of this framework is the opportunity it gives to track changes in stakeholder perceptions of risks and to give a better understanding of their implications.

3.1. Traded (Voluntary) Risks

Traded risks cross national boundaries as a result of commercial transactions, usually with the knowledge and consent of the official national bodies charged with risk regulation. In theory, therefore, it is possible for national authorities to prevent products bearing these risks from crossing borders if they choose to do so.

National actions are thus the basic regulatory mechanisms, permitting or blocking trade. However, this national action needs to be backed up by internationally coordinated standards. Transboundary regulation in these cases is about controlling the mechanisms of trade, avoiding the use of national regulations to discriminate unfairly among suppliers. Alternatively, regulation can operate by assigning property rights in what would otherwise be regarded as a "free good," enabling trade to take place (see Section 6.2).

Within this category there are two major classes that raise different issues and require different approaches to risk management.

- *Product-based.* In this case, the product itself is considered potentially hazardous – food contaminated by pesticide residues or radioactivity, defective cars, international trade in waste products, GM food or seeds (insofar as they may affect the health of consumers or cause environmental damage). National authorities can, at least theoretically, detect products giving rise to such risks at the frontier and can restrict entry.
- *Production system based.* In this case, the risk is symbolically attached to the traded product (e.g., the fishing method used for tuna; unsustainable logging methods for timber products; intensive farming methods for food products). Generally, there is no regulatory basis for national authorities to refuse to import such products. However, if the origin of the products is identified by labeling, individual consumers can subsequently exercise their right "not to buy" and can put pressure

TABLE 10.1. *Categorizing Transboundary Risks*

	Traded Risks		Public Risks	
	Product-Based	Production System-Based	Contained	Pervasive
Risk Characteristics and Examples	The product per se may be risky (e.g., pesticide residues or fungal carcinogens in food imports, defective imported cars, international trade in waste products, GM seeds and crop commodities)	Risk is "symbolically" attached to products which are traded across boundaries (e.g., fishing method for tuna, intensive farming methods, unsustainable timber exploitation)	Risk is, in principle, contained and only materializes in the case of an accident; therefore, it includes mainly point sources (e.g., nuclear power accidents as at Chernobyl, catastrophic river pollution as at Tisza, marine oil pollution as from tankers or extraction rigs)	Chronic, often insidious risks, from diffuse sources (e.g., acid rain, ozone depletion, greenhouse gases, and global climate change)
Regulation and Control	Sampling and testing of products for conformance with internationally agreed standards; prior informed consent	Labeling and consumer choice or voluntary accreditation schemes; needs industry collaboration	Potentially affected countries rely on another jurisdiction's regulatory system and vigilance, international liability mainly lacking, only legal rights through courts in country of origin	Usually through international agreements requiring quotas or bans or by allocating the equivalent of internationally effective "property rights"

Source of Pressure for Regulation or Management of Risks	Mainly industry and regulators, but public and environmental groups become involved if perceive lack of control	Mainly public interest groups, NGOs	National governments may perceive a threat to their economy or environment; also involved in clean-up in event of an accident; strong interest of environmental groups and often the public	Mainly governments and NGOs; scientific community often very active in identifying risks
Relevance of Risk-Based Approach	Accepted risk assessment procedures available; varying degrees of uncertainty; benefits straightforward to identify	Difficult if not impossible to assess intangibles; often strongly value-based assessment of risks	Complications in assessing low-probability events; access to information in foreign countries is often limited	Sources of pollution and damage often difficult to identify; complexity makes it difficult to determine cause-effect relationships
Relevance of Precautionary Principle (PP) (see Section 6.3)	PP not usually relevant	PP highly relevant in many cases; even where there is no evidence of environmental damage, PP may be used to justify intervention	PP may be relevant but there is no mechanism by which it could be applied	For potentially major impacts with high levels of uncertainty and irreversibility, the PP is relevant

on retail outlets to change to a new supplier of a more environmentally acceptable product. In this case, the risk itself does not cross national boundaries but peoples' concerns do – they want to have an influence on what happens in other countries, often from an altruistic basis of concern for global environmental sustainability.

3.2. Public (Involuntary) Risks

The term "public risks" as used here is the converse of the economist's "public good" – just as it is difficult to exclude individuals from using public goods, it is difficult to exclude them from the effects of public risks. After a hazardous entity has crossed a national boundary, measures can be taken *within the national boundary* to protect a population or a natural environment from its effects (e.g., adding lime to lakes to mitigate the impacts of acid rain), but the costs of these measures will fall on the *affected country rather than the polluter*. In this sense, these risks are involuntary and the public and their governments, in the absence of an effective international regulatory system, have no choice whether to accept the risk and the associated costs of mitigating it.

The most effective means of risk reduction in such cases is to restrict the production or release of offending materials at source, requiring national commitments to apply internationally agreed upon regulatory standards. It is generally more difficult to get agreement to regulatory approaches of this nature and also more difficult to monitor implementation.

Two distinct types of public risk are particularly relevant in a transboundary context.

Contained risks. Contained risks usually arise from point sources, and, in principle, they do not cross national boundaries to an unacceptable extent unless there is an accidental release of polluting material. Examples include the Chernobyl nuclear accident and the transboundary pollution incidents on the River Tisza, a tributary of the Danube.

Pervasive risks. Pervasive risks are usually chronic and insidious in their effects. The impacts result from complex system interactions, and cause-effect relationships are often difficult to prove. Unlike contained risks, individual sources of pollution are also often difficult to identify. Examples include acid rain, air pollutants causing ozone depletion, and global climate change.

4. TRADED RISKS

As outlined previously, under this heading we have included examples of risks that are linked to products in international trade. Where the risk is potentially attached to the product itself (see Section 4.1) it can be controlled by setting standards (e.g., for freedom from pesticide residues or fungal contamination in food products) or by devising procedures (e.g., to alert governments to trade in toxic wastes or dangerous products). Although the products themselves may present no demonstrable risk, the production system from which they are derived may be regarded, at least by some, as unsustainable or unsafe (see Section 4.2). A defining characteristic of many risk debates in the 1990s has been the attempts by NGOs and other public interest groups to influence production systems in other countries by influencing the purchasing decisions of individual consumers or commercial organisations.

4.1. Traded, Product-Based Risks

Three examples are given here of the regulation of international trade in potentially hazardous products – hazardous waste, pesticide residues in food, and GM organisms. The first two present scientifically demonstrable risks that have been recognized for some time and for which there are established procedures in international trade. Attempts to regulate international trade in GM organisms, on the other hand, are very recent, although some countries have for some time been experiencing a lively debate about the risks of these products (Carr, 2000).

Hazardous Waste. International trade in hazardous waste became an issue because of the increasing trend for individuals or companies in industrialised countries, where disposal of such waste is subject to restrictions and is expensive, to export it without consent to poorer countries where it is often a significant danger to the population or environment (Krueger, 1999).

The Basel Convention (Table 10.2) was the first global environmental treaty to address this issue, and NGOs were heavily involved in stimulating international action. On the other hand, although the United States had taken a lead in promoting global environmental initiatives in the 1970s, in the 1980s it acted more frequently as the leader of a veto coalition that was resistant to the Basel Convention.

During the negotiation process, Greenpeace called for a ban on the production of hazardous substances and African and some other countries were persuaded not to sign the convention because of concerns that it would be difficult for them to use it to prevent illegal dumping, given their limited technical capabilities (Tolba and Rummel-Bulska, 1998). Kempel (1993) suggests that this illustrates one form of power asymmetry, where developing countries acquired a dominant role, making effective use of the media and preventing developed countries from taking a strong position in areas such as notification procedures. In this case, NGOs were blamed for exacerbating the problems of hazardous waste disposal by raising the issue's profile and increasing public resistance to disposal in

TABLE 10.2. *Hazardous Wastes*

Relevant Issues	A large number of different kinds of hazardous waste can adversely affect human health and the environment (e.g., materials contaminated with dioxins, heavy metals, and organic wastes). A key issue is the transport of hazardous wastes from rich to poor countries which are less likely to have the capability to handle the waste safely.[a]
Regulation	Basel Convention on the Control of Transboundary Movement of Hazardous Wastes and their Disposal (1989); in force 1992. The Protocol on Liability and Compensation for Damage Resulting from the Transboundary Movement of Hazardous Wastes and their Disposal, adopted December 1999.[b] Agreed hazardous categories (Annex I) include clinical waste, wastes resulting from the surface treatment of metals and plastics, waste including zinc compounds. Wastes must also possess characteristics listed in Annex III (e.g., explosive, flammable liquid). Wastes not included in the preceding but defined as hazardous wastes by the domestic legislation of the Party of Export, Import or Transit are also included (Article 1). Prior informed consent procedures are used.
Aim	To ensure that hazardous wastes are managed and disposed of in an environmentally sound manner. Principles: movements of wastes are reduced to the minimum; disposal is close to the source of production; and production of waste is minimized. Since 1994, movement of waste from OECD to non-OECD countries has been banned. Provides for a liability regime; also compensation for damage from transboundary movement of hazardous waste, including accidents as a result of illegal traffic.[c]
Organizing Body	Based on United Nations Environment Programme (UNEP)'s Cairo Guidelines, 1987.
Compliance	Relatively few reported cases of illegal traffic but the number of actual cases is probably underreported.[a] Basel Convention cooperates with Interpol.[c] Disputes can be referred to the International Court of Justice or to Arbitration (Article 20).
Area Covered	Global. As of July 1999 it has 130 parties (preamble to the Basel Convention).
Liability	The protocol addresses who is financially responsible in the event of an accident (covering waste generation, export, international transit, import and final disposal). Interim arrangements cover emergencies until the protocol enters into force.[d]

Effectiveness	There was no globally accepted definition of hazardous waste in 1989,[e] so it is difficult to evaluate the effect of the convention. A more important role in environmental security may be to reduce the production of hazardous substances.
Key Drivers	Increasing amount of waste generated; closure of old waste disposal facilities; opposition to new facilities and higher costs of waste disposal in industrialized countries.[a]
Relationship with Trade Issues	Some wastes covered by the convention can be recycled; hazardous wastes may then be categorized as "products" or "goods" and would come under the WTO, and the convention could then be considered a barrier to trade.[f]

[a] Krueger (1999).
[b] http://www.unep.ch/basel/.
[c] http://www.unep.ch/basel/pub/illegaltrafic.html.
[d] http://www.unep.ch/basel/press/liability1299.html.
[e] Tolba and Rummel-Bulska (1998).
[f] Kummer (1994).

the richer countries, encouraging disposal overseas (Tolba and Rummel-Bulska, 1998).

The convention's loopholes were also criticized in the early stages; for example, developed nations were able to export large amounts of hazardous wastes to developing nations by classifying them as suitable for recycling (Ovink, 1995). This loophole was closed in 1994 when a resolution was passed to ban the export of all hazardous wastes from OECD countries to non-OECD countries (see Table 10.2). This ban is now being extended by the "Basel Ban," to cover exports from EU, OECD, and Liechtenstein (known as the Annex VII countries) to all other parties (www.unep.ch/basel/pub/BaselBan.html).

In the future, innovative approaches are needed to reduce hazardous waste generation, and here the Basel Convention sees a key role for the private sector and also civil society in reducing consumption of waste-generating goods (www.unep.ch/basel/).

Protecting Food Quality in International Trade: Pesticide Residues. In some countries there are concerns that imported food may contain excessive residues of approved pesticides or residues of pesticides that are not internationally approved as safe. Many countries have systems for sampling and testing imported produce, and the Codex Alimentarius provides the baseline standard for such systems. It has become the global reference point for consumers, food producers and processors, national food control agencies, and the international food trade (Table 10.3) (*Understanding*

TABLE 10.3. *Pesticide Residues in Food*

Relevant Issues	Concern that pesticide residues in food may have adverse impacts on human health. Environmental and consumer groups accuse Codex Committees of not being impartial because they include corporate representatives.[a] The commission's view is that national governments should involve consumers at a national level, rather than in the international work of the Codex.[b]
Regulation	Codex Alimentarius Standards; FAO provides information based on field trials worldwide; WHO conducts toxicological evaluations.
Aim	To provide international standards; the Joint FAO/WHO Meeting on Pesticide Residues proposes MRLs that are considered by a Codex Committee for recommendation to the Codex Commission for adoption as Codex MRLs.
Organizing Body	FAO/WHO; Codex Alimentarius Commission established in 1961; membership in 1998, 163 countries, representing 97 percent of the world's population.[b]
Compliance	Voluntary, but used as standards by, for example, WTO. The Agreement on the Application of Sanitary and Phytosanitary Measures (SPS) uses Codex standards as a baseline (see Section 6.1).
Area Covered	Global.
Liability	Voluntary so no liability regime. Differing legal formats, administrative and political systems, and sometimes national attitudes and concepts of sovereign rights impede the process of developing legally binding Codex standards.[b]
Effectiveness	Monitoring and testing schemes have only occasionally found MRLs exceeded, but there is concern about the quality of testing procedures. Only a small percentage of foods are tested and a laboratory in the United States was found to produce fraudulent data.[a]
Inducements	Acceptance of Codex MRLs is on the basis of (i) "full acceptance" (governments undertake that any imported or domestic foods will conform to specifications and they will not legislate to hinder distribution of food conforming to this limit); or (ii) "free distribution" (governments only commit to not enacting legislation to hinder distribution of food conforming to the limit). Governments that do not accept MRLs are requested to state why they do not accept them.[a]

Key Drivers	Increasing concern, particularly in northern countries about the health effects of food contaminants, such as pesticides. The use of the Codex standards in WTO have resulted in increasing interest in the Codex meetings, particularly by developing countries.[b]
Relationship with Trade Issues	The WTO encourages use of Codex standards as a baseline in trade disputes. However, for the European Union and the United States, this would result in a lowering of standards.[a]

[a] Hough (1998).
[b] *Understanding the Codex Alimentarius*, http://www.fao.org/docrep/w9114e/.

the Codex Alimentarius, www.fao.org/docrep/w9114e/). In a world where food products of all kinds are increasingly traded on a global basis, it plays an important role in international regulation.

The Codex Alimentarius Commission encourages food traders to adopt a voluntary Code of Ethics for International Trade in Food. It aims to prevent the dumping of poor quality or unsafe food in international markets, to protect the health of consumers, and to ensure fair practices in the food trade. Pesticide residues in food are therefore only a part of its remit.

Food and Agriculture Organization (FAO) and World Health Organization (WHO) provide assistance to developing countries to enable them to take full advantage of the commission's work. They convene expert meetings – for example, the Joint FAO/WHO Expert Committee on Food Additives (JECFA), which considers chemical, toxicological, and other food contaminants. The Codex Committee on Pesticide Residues (CCPR) has an ad-hoc working group, the Joint FAO/WHO Meeting on Pesticide Residues (JMPR), the FAO component of which considers, for example, patterns of pesticide use, the fate of residues, animal and plant metabolism. Based on these data, maximum residue levels (MRLs) are proposed for specific pesticides in individual food and feed items or well-defined groups of commodities. These proposals are considered by the CCPR with the aim of reaching agreement between governments on MRLs for pesticide residues in food and feed commodities moving in international trade.

The Codex contains more than 200 standards for individual foods or groups of foods, but it is difficult for many countries to accept Codex standards in the statutory sense. Differing legal formats and administrative systems, varying political systems and sometimes the influence of national attitudes and concepts of sovereign rights impede the process of harmonisation. The EU and the United States, on the other hand, generally adopt more stringent standards than the Codex, and any attempt to impose standards uniformly worldwide as a means to remove trade barriers would result in a lowering of their standards (Hough, 1998).

The Codex sees itself as primarily a science-based activity, involving experts and specialists from a wide range of disciplines to ensure that its standards withstand rigorous scientific scrutiny. Those selected for membership of expert consultations are appointed in their personal right, not as government representatives or spokespeople for organizations, and they must be pre-eminent in their specialty.

Despite such assurances, Braithwaite and Drahos (2000, p. 401) have pointed out that the biggest funder of the establishment of the Codex Alimentarius Commission was not the U.S. government, but the U.S. food industry. They also note that more corporations have members of delegations to Codex committees than nations and quote a study by the UK National Food Alliance in 1993 which found that there were 445 industry representatives on Codex committees, compared with eight representatives of public interest groups. Industry dominance of Codex continues to be a point of criticism, particularly by public interest pressure groups, but Braithwaite and Drahos caution that we should not give too much significance to raw numbers of participants in the Codex process as strategic bargaining on Codex expert committees tends to be done by government representatives of the key states.

Genetically Modified Organisms. Trade in GM seeds and crops has become one of the most contentious international risk issues. Numerous NGOs and some national governments (in response to public pressure) are calling for much more rigorous controls, a moratorium on genetically modified organisms (GMO) developments, or an outright ban. On the other hand, the companies involved in their development, supported by the United States and some other governments, maintain that they are "substantially equivalent" to the crops from which they were derived and that they have been tested rigorously in the United States and the European Union to ensure that they do not present a risk to the environment (Tomlinson, 2000).

The first binding international agreement addressing situations where GMOs cross national borders was provided by the Cartagena Protocol on Biosafety to the Biodiversity Convention (Table 10.4), agreed in January 2000 when environment ministers and trade negotiators from 138 governments concluded five years of talks (*Bridges Weekly Trade News Digest*,[1] vol. 4, no. 4, 1 Feb. 2000; www.ictsd.org.html/weekly/story1.01-02-00.htm). The main issues for debate in the protracted negotiations were trade issues, the scope of the protocol, the precautionary principle, the relationship to other agreements, and liability (Cosbey and Burgiel, 2000).

TABLE 10.4. *Genetically Modified Organisms*

Relevant Issues	Questions of the food and environmental safety of GMOs has led to calls for controls on their export. It is not yet clear how long it will take to ratify the protocol and how it will work in practice.
Regulation	Cartagena Protocol on Biosafety to the Biodiversity Convention (2000).
Aim	Governments signal willingness to accept LMOs; shipments of commodities containing LMOs to be clearly labeled; stricter Advanced Informed Agreement procedures for LMOs to be introduced intentionally into the environment.
Organizing Body	UNEP. The treaty was opened for signature in May 2000 and will enter into force once it has been ratified by fifty countries.
Compliance	Due to come into force September 2003
Area Covered	Global.
Liability	Liability procedures will be addressed at the first Conference of the Parties (Article 27). Target for completion, 4 years.
Effectiveness	Not yet clear.
Inducements	Article 22 introduces capacity building measures including scientific and technical training in safe management of biotechnology and the use of risk assessment and risk management for biosafety. The exact mechanism for this and the financial responsibility has not been fully expounded.
Key Drivers	The Miami Group[a] of major exporters supported free trade in GM products, avoiding measures which were promoted on environmental grounds but were in reality protectionist trade barriers. The Like-Minded Group (a developing country negotiating coalition) looked for strong protection from imports of LMOs, particularly for countries without relevant well-developed regulatory structures and institutions. The European Union was under pressure from public and the press for a strong protocol to regulate LMOs. The Central and Eastern European countries focused primarily on the practical applicability of the protocol. The Compromise Group[b] of countries developed compromise positions.[c]
Relationship with Trade Issues	Not clear how the protocol will function in relation to trade issues. The preamble emphasizes that "this Protocol shall not be interpreted as implying a change in the rights and obligations of a Party under any existing international agreements" but also understands that "the above recital is not intended to subordinate this Protocol to other international agreements."

[a] Argentina, Australia, Canada, Chile, the United States, and Uruguay.
[b] Japan, Mexico, Norway, Singapore, South Korea, Switzerland, and New Zealand.
[c] Cosbey and Burgiel (2000).

The inclusion of the precautionary principle in the protocol was seen by many groups as a key point. The European Union's approach to implementation of the precautionary principle (CEC, 2000) includes the requirement that measures should be proportionate to the level of protection, nondiscriminatory in application, and based on consideration of the costs and benefits of action or lack of action. Environmental groups were disappointed by the strict requirements for risk assessment in the protocol and the lack of consultation with themselves when putting the guidelines together (*Bridges*, vol. 4, no. 5, 8 Feb. 2000).

Under the Protocol, governments will signal whether or not they are willing to accept imports of agricultural commodities that include living modified organisms (LMOs)[2] by communicating their decision to the world community via an internet-based Biosafety Clearing House. In addition, shipments of commodities that may contain LMOs are to be clearly labeled. Stricter Advanced Informed Agreement procedures will apply to seeds, live fish, and other LMOs that are to be intentionally introduced into the environment. In these cases, the exporter must provide detailed information to each importing country in advance of the first shipment, and the importer must then authorize the shipment. The aim is to ensure that recipient countries have both the opportunity and the capacity to assess risks involving such GMOs.

The following points from the convention relate specifically to risk issues.

- "Risk assessments undertaken pursuant to this Protocol shall be carried out in a scientifically sound manner . . . taking into account recognised risk assessment techniques." (Article 15.1)
- "Where there is uncertainty regarding the level of risk, it may be addressed by requesting further information on the specific issues of concern or by implementing appropriate risk management strategies and/or monitoring the living modified organism in the receiving environment." (Annex III, f)
- "The Parties . . . may take into account, . . . socio-economic considerations arising from the impact of living modified organisms on the conservation and sustainable use of biological diversity, especially with regard to the value of biological diversity to indigenous and local communities." (Article 26.1)

Concern has been expressed that concessions made in the Biosafety Protocol could compromise its objectives. The Miami Group (Argentina, Australia, Canada, Chile, Uruguay, and the United States) was very forceful in seeking to water down labeling requirements. They succeeded insofar as the protocol applies only to LMOs so that no segregation is required for nonliving GMOs, to the disappointment of the like-minded group of developing countries and environmental NGOs. Ambiguity over the relationship of the protocol with WTO regulations was also of concern (*Bridges* vol. 4, no. 5, 8 Feb. 2000).

4.2. Traded, Production System-Based Risks

For issues in this category, the products themselves are not hazardous and so the product-based approach cannot legitimately be used to restrict trade. Concerns arise among individual citizens or public interest groups over what is perceived to be unsustainable exploitation of a natural resource (i.e., the concern is about the system of production, not the product).

Timber. Concerns about the sustainability of timber production methods can arise within the country concerned, in which case the issue is dealt with, if at all, by national policy. However, where the timber is exported, this opens up opportunities for other nations and organizations to exert an influence on production methods across national boundaries.

Among traded production system-based risks for timber, the International Tropical Timber Agreement (ITTA) attempts to control exploitation of tropical timber (see Table 10.5). European consumer boycotts against tropical timber led the International Tropical Timber Organisation (ITTO) to develop standards and performance measures, although the wide discrepancy between the aims of developed and developing countries undermines the basis of this agreement. Considerations of risk and sustainability are usually seen as peripheral to national and international political considerations.

In 1990, there was general agreement that the ITTA had not decreased the rate of deforestation, and at the Group of Seven economic summit, the United States proposed a new convention on the world's forests. This was not accepted by Europeans who saw it as a tactic to delay negotiations on a climate change agreement (see Section 5.2), and most states (producers and consumers) argued for a global forest agreement to be part of a protocol to a framework convention on climate change.

By 1991, the following positions had been adopted (Porter and Welsh Brown, 1991):

- Japan advocated a World Charter of Forests with nonbinding principles for forest management;
- Canada proposed a legal instrument that would depend on voluntary action through national plans;
- the United States suggested a framework convention on the model of the Vienna Convention on ozone.

In the preparatory stages to the 1992 UN Conference on Economics and Development (UNCED) summit, Malaysia (with support from the Group of Seventy-Seven) demanded that tropical forest countries be compensated

TABLE 10.5. *Unsustainable Timber Production*

Relevant Issues	Loss of (particularly tropical) forest has resulted in loss of biodiversity, increased global warming, increased local flooding, droughts, and silting of rivers.[a]
Regulation	International Tropical Timber Agreement (1983), negotiated for a limited time, came into force in 1985. A successor agreement came into force in 1997.[b]
Aim	To provide an effective framework for co-operation and consultation between countries producing and consuming tropical timber; to contribute to sustainable development; to promote expansion and diversification of international trade in tropical timber by improving structural conditions in the international market; to support research and development for improved forest management and wood utilization. All tropical timber entering international trade should come from sustainably managed sources by 2000.[c]
Organizing Body	United Nations Conference on Trade and Development (UNCTAD) set up the ITTO as part of the ITTA in 1987.
Compliance	Agreement requires reporting of resource and trade data rather than compliance information. In 1990, of 46 parties only 15 (12 importing and 3 producing nations) submitted the required data on harvesting and trading.[b]
Area Covered	Countries producing or consuming tropical timber.
Liability	Not legally binding.
Effectiveness	Widely regarded as ineffective.

[a] Porter and Welsh Brown (1991).
[b] Sand (1999).
[c] http://wwwitto.or.jp/inside/about.html.

by developed countries for all costs of compliance to any convention that commited developing countries to halting or slowing-down deforestation. Malaysia (again with the support of the Group of Seventy-Seven) also opposed negotiation of a forest agreement until developed countries had committed themselves to reduce energy consumption and to provide support for developing countries to control their emissions. In the case of the Forest Principles proposed at the Rio Conference, negotiators could not even reach agreement on a framework convention, and there is now a proliferation of initiatives to deal with issues related to sustainable timber production.

Partly because of the lack of impact of the ITTA and partly because of the pervasive effect of forest management practices on climate, soil erosion, and other ecosystems, forest management has been included in a wide range of other proposals and declarations. Most developing countries at the Rio Conference strongly resisted a binding treaty on forests because they saw this as an attempt by richer countries to control their economies and resources, but this has led to a renewed interest in various certification schemes. A wide range of initiatives (at least eleven) are developing a common understanding of what is meant by sustainable forest management but none has begun to address the technical feasibility or practicability of measuring the large number of indicators generated (Kimmins, 1997).

Intensive/Conventional Farming Systems. The transboundary risks posed by *products* used in intensive farming systems, for example GM crops, were discussed in Section 4.1, and the farming systems themselves cannot be considered as transboundary risks. Some individuals and groups regard them as risks to human health and the environment at national levels, but they are not the subject of any international conventions or protocols.

They are included in this paper because of their focal position in debates about globalization of food production and trading systems and about the increasing dominance of multinational companies in controlling these systems. Since 1998, the European public and NGOs have led the way in what may become a worldwide debate about future food production systems, which may even influence the pace and trajectory of the globalization process itself.

The link is made with transboundary risk issues by the fact that European NGOs have focused on the risk regulation of GM seeds and crops as their entry point into this debate. Although the risks cited are most often those to the health of consumers of GM produce or to the environment in which the crops are grown, it is clear from much that is written by these groups that the real targets are intensive farming systems themselves (Tait, 2001; National Consumer Council, 1998; www.gecko.ac.uk; www.truefood.org/news/index.html; www.greenpeace.org/~geneng/). The fact that GM crops could make intensive farming systems more sustainable than purely pesticide-based systems (Tait and Morris, 2000) is regarded by such groups as part of the problem, not part of the solution.

The importing of GM crops into Europe in the late 1990s, without widespread notification and without labeling, gave the pressure groups the lever with which to mobilize European public opinion, regardless of the justification for their concerns in risk regulation terms. Public pressure through demonstrations at supermarkets and threats to boycott produce mobilized the food processors and retailers. Most major newspapers including the quality press, sensing a boost to circulation, mounted "anti-GM" campaigns reinforcing the already negative public opinion.

An important strand of the NGO strategy has been to attach negative attributes to food containing imported GM products and to demand labeling of all such products so as to allow consumers to express their attitudes through their purchasing decisions. The campaign initially focused on imported GM soya and is being gradually extended to other products including meat produced by animals reared on GM feed and clothing made from GM cotton. An important parallel to this strategy is the boosting of the positive attributes of organic produce.

The issues that these events raise for transboundary risk regulation in the context of world trade negotiations are explored in more detail in Section 6.1. Another important aspect of this debate is the role played by the precautionary principle (see Section 6.3). Because of the lack of scientific data on which to base decisions to restrict the import of GM crops into Europe, the precautionary principle is invoked, and a battery of scientific tests extending over a period of years is demanded before they can be deemed safe. Some such requests are justified, but the form of precaution being advocated goes far beyond that envisaged by most risk regulators (CEC, 2000) and is in danger of bringing the principle itself into disrepute, destroying its usefulness in other contexts.

5. PUBLIC RISKS

5.1. Public, Contained Risks

Public risks in the contained category, as noted in Section 3.2, do not present problems from an international perspective unless there is an accident. Indeed, the setting up of international conventions and agreements usually follows on from such events (i.e., regulation is reactive rather than anticipatory).

Nuclear Installations. Although there had been a latent awareness of the transboundary risks associated with nuclear power, for much of the period since the 1960s public concerns focused on military uses, particularly weapons testing, as a source of transboundary risk on a global scale. The Chernobyl nuclear accident in 1986 focused the concerns of both public and regulators on the potential dangers of civilian uses of nuclear power in some parts of the world.

The first international indications of the accident were increased levels of radioactivity in Sweden arising from a source somewhere in the Soviet Union, but the Soviet government did not inform the International Atomic Energy Authority (IAEA) until 72 hours after the accident (Sands, 1988). The accident was blamed on operators testing a safety device, but Western experts also highlighted design flaws in the reactor. Subsequently, blame has also been placed higher up in the Soviet administration (Medvedev, 1990). The radioactive cloud traveled over 2,000 km across more than twenty countries (Consortium of Opposing Local Authorities Special

Briefing Number 21, March 1996). Medvedev (1993) states that before the disaster the Soviet environmental movement was very weak and divided, but after it environmental and antinuclear movements began to emerge, combining antinuclear concerns with other national grievances in Belarus and Ukraine.

The Chernobyl accident generated concern in Western countries over the possibility of further accidents in similar reactors. Many of those that caused concern were located near national borders; for example, Austria was concerned about Mohovce and Bohunice in Slovakia, Temelin and Dukovany in the Czech Republic, Krsko in Slovenia, and Paks in Hungary. However, attention was not restricted to borders between Eastern and Western Europe. Denmark became anxious about the Barse-baeck nuclear power station in Sweden, and Germany about the French nuclear power station Catonom (Linnerooth-Bayer and Loefstedt, 1996).

After the disintegration of the Soviet Union, Western countries, notably Sweden, Finland, and Austria, along with the European Bank for Recon-struction and Development, invested in measures to improve safety of the most hazardous nuclear power stations in Eastern Europe (Linnerooth-Bayer et al., 2000). Sweden, for example, financed improvements in Ignalina (Lithuania) and Sosnovyi Bor (near St. Petersburg), an effort that was con-troversial from the point of view of policy makers and the public in both donor and recipient countries (Loefsted and Jankauskas, 2000). Lithuania was concerned that conditional aid impinges on their national sovereignty and that the money will be used to pay the fees of Swedish consultants; policy makers and individuals have expressed frustration that aid was not given to what they view as more serious local environmental problems (Linnerooth-Bayer, 2000). The Swedish consultants were critical of efforts to continue the operation of an inherently unsafe plant and concerned about the possibility of Swedish liability should an accident occur.

The Chornobyl accident thus drew attention to the need for clearer inter-national rules on responsibility and liability in such cases and the OECD, European Union, and IAEA began to push for stricter international stan-dards to prevent such accidents in the future. As a result, executives in the nuclear industry shifted from an attitude of "not my brother's keeper" to one of "everything my brother does is going to affect me" (Rees, 1994, p. 4; quoted in Braithwaite and Drahos, 2000).

In 1986, there was no multilateral treaty requiring the provision of information on radioactive materials (Sands, 1988), and the two interna-tional agreements established by the IAEA after the accident concerned early notification and emergency assistance. The Convention on Early Notification of a Nuclear Accident (1986) stipulates how and when a nation's authorities should notify other nations about a nuclear accident (Sjostedt, 1993). It places specific obligations on the parties to the conven-tion and establishes a notification system for specified types of nuclear accident that could be of radiological safety significance for another state. The Convention on Assistance in the Case of a Nuclear Accident or Radiological Emergency (1986) sets out an international framework for co-operation among parties and, with the IAEA, to facilitate prompt assistance and support in the event of nuclear accidents or radiological emergencies, including when and how nations should provide assistance, who the competent authorities are, the role of the IAEA, confidentiality and public statements, and procedures for settling disputes (Sjostedt, 1993; http: // www.iaea.org/ns/rasanet/ programme/ radiationsafety/ convimp/ emeressys.htm).

The two conventions were adopted surprising rapidly, four weeks after the start of the negotiations and four months after the Chernobyl accident (Sands, 1988). The aim in devising them was to obtain signatures quickly rather than to achieve the optimum result, and the text contained serious loopholes such as failure to define "early notification." Although most participants added reservations to the convention text, it did generate political goodwill and commitment, given that the issue was not characterized by scientific or technical uncertainty and the outcome did not result in immediate disadvantages accompanied by a long wait for associated advantages as is true of some conventions (Sjostedt, 1993). Only three countries were required to ratify this convention to speed up the adoption process (Sand, 1999), but by 1999, eighty-two states were party and three organizations had agreed to be bound by the Early Notification Convention, with significant additional numbers having agreed to be bound by the Assistance Convention.

The Convention on Nuclear Safety (1996) was developed under the auspices of the IAEA as an international co-operation mechanism to maintain safety in nuclear installations and ratified by fifty states by April 1999. It covers technical co-operation to maintain effective defenses against radiological hazards, the purpose being to protect individuals, society, and the environment from harmful effects of ionizing radiation, to prevent accidents with radiological consequences, and to mitigate such consequences in the event of an accident. Trends likely to affect nuclear safety were identified (*www.acus.org/Publications/policypapers/energy/NuclearPower.pdf*) as

- deregulating electricity markets with resulting ownership changes and increasing competition;
- maintaining competence (e.g., in countries with small nuclear programmes or where nuclear power is being phased out); and
- exhibiting a lack of economic resources.

The OECD Nuclear Energy Agency assists member countries to maintain and develop the scientific, technical, and regulatory knowledge base needed to assess the safety of nuclear reactors and other civilian installations at all stages in their life cycles (www.nea.fr/html/nsd/).

Transboundary Pollution of Major Rivers. In the same year as the Chernobyl accident, a fire in the Sandoz chemical factory in Switzerland caused very serious pollution of the river Rhine, with transboundary impacts on France, Germany, and the Netherlands. However, this case did not result in the development of international regulatory procedures. Indeed, the clean-up was financed mainly by downstream countries, particularly the Netherlands (Linnerooth-Bayer and Loefstedt, 1996). This is an example of violation of the "polluter pays" principle as was also the case for bilateral arrangements to ensure the safety of nuclear facilities.

In January and February 2000, another series of major pollution incidents occurred on a European river system, the River Tisza, which flows into the Danube. These cases may yet stimulate the development of international instruments and procedures to govern such pollution incidents, but their long-term environmental or policy implications are not yet clear.[3]

The River Tisza incident originated in northwest Romania where a dam at the Aurul gold mine in the Baia Mare area overflowed, releasing about 100,000 cubic meters of water contaminated with cyanide, copper, and zinc into the River Tisza. Large numbers of fish were killed and in Hungary and Northern Yugoslavia, the incident has been a major ecological disaster, killing much of the wildlife in the rivers and also affecting birds and animals that feed on them. (Hungary was in the process of applying to have the Tisza designated under the Ramsar Convention, a treaty which protects wetlands of international nature conservation importance.) In addition to the environmental catastrophe, the water supplies of 2.5 million people were threatened.

Reported levels of contamination vary. Close to the mine, cyanide reached 800 times the acceptable level. However, it is probable that, by the time it reached the Danube, the cyanide was diluted to safe levels. On the one hand, Bulgaria's stretch of the Danube was claimed to be 40 percent above internationally accepted levels. On the other hand Serbia's Deputy Agriculture Minister claimed that the maximum concentrations were 0.06 mg/L, compared to the maximum allowable concentration of 0.1 mg/L.

Six weeks later, a second spill from the same source resulted in contamination with a further 20,000 tons of waste containing zinc and lead and a third incident followed, spilling waste contaminated with heavy metals from the Baia Borsa mine into the same river system.

The Aurul mine was a US$30 million project developed as a joint venture between the Australian company Esmeralda Exploration Limited (50 percent) and REMIN SA, a Romanian state-owned precious metal mining company with finance provided by N. M. Rothschild and Dresdner Bank. The European Commission was expected to contribute to clean-up costs. Hungary and Serbia, the worst-affected countries, sought compensation from the Romanian Government or an Australian source (public or private), but the Romanian government passed responsibility to the mine's owners. After the second incident, the Hungarian government asked Romania to take immediate action to identify and, if necessary, close industrial plants that pose a serious danger to the environment.

Activists attempted to publicize the scale of the disaster to the Australian public, and there were moves by the Democrat and Green parties to bring in tougher laws on the conduct of Australian mining companies operating abroad. Esmeralda Exploration denied responsibility for the pollution incident and undertook its own evaluation, claiming that "chemistry can be fudged but not physics." The company was suspended from trading on the Australian share market and went into receivership in anticipation of multi-million dollar compensation claims.

5.2. Public, Pervasive Risks

Public, pervasive risks are less dramatic, more subtle, and chronic in their impact, but because of the wider scale of these impacts and their longer-term nature, they have been the cause of more concern and action in terms of internationally based instruments than public, contained risks.

Transboundary Air Pollution. The Geneva Convention on Long Range Transboundary Air Pollution (CLRTAP) does not stipulate any detailed environmental regulations. Instead, it establishes a framework for technical, scientific, and political co-operation within which specific pollution protocols can be created (Table 10.6). Initially developed to deal with the acid rain issue in Europe, it has now been extended by a series of protocols to cover a range of other air pollutants.

Acid rain was a classic case of the solution to one environmental problem creating another, more pervasive one. In the 1960s, industrialized countries reduced local air pollution from power stations and factories by increasing the height of the chimneys to disperse pollutants more widely. The emissions (particularly sulfur dioxide) were transported into higher reaches of the atmosphere, and from there, to other countries as "acid rain" particularly in Northern European countries and Canada. Proposals for international regulation initially faced strong resistance from polluting countries, including the United States, the United Kingdom, and Germany, and a heavy burden of proof was demanded before these countries would accept that they were the source of the problem. A key factor in persuading West Germany to sign the relatively toothless convention in 1979, along with thirty-four other nations, was the scientifically proven impact of acid rain on its own Black Forest (Braithwaite and Drahos, 2000).

The First Sulfur Protocol developed under the Geneva Convention involved twenty-one nations that agreed to reduce emissions of sulfur dioxide by 30 percent, relative to 1980 levels, by 1993. The United States,

TABLE 10.6. *Transboundary Air Pollution*

Relevant Issues	Sulfur dioxide and nitrogen oxides are produced by the burning of fossil fuels (e.g., in power stations, central heating boilers, and vehicles).[a] Pollution impacts include defoliation of trees, declining fish stocks in lakes, and damage to buildings. Acidification of lakes, identified in Sweden in 1967, was raised in the 1972 UN Conference on the Human Environment, Stockholm. OECD began to monitor and model the international exchange of pollutants. Eastern Europe and the USSR were major contributors to pollution and had to be included in negotiations.
Regulation	Geneva Convention on Long-Range Transboundary Air Pollution (1979).[b]
Aim	To protect humans and the environment against air pollution. Protocols exist for sulfur dioxide (1985, 1994), nitrogen oxides (1988), volatile organic compounds (1991), Persistent Organic Pollutants (1998), and heavy metals (1998).[c]
Organizing Body	UN Economic Commission for Europe (UN ECE).
Compliance	The 1979 framework convention does not stipulate detailed commitments; establishes a framework for technical, scientific, and political co-operation within which the protocols have been created; allows signatories to select from a range of regulatory options and base years.[d]
Area Covered	Member states of UN ECE (North America and Europe, including the European region of the former Soviet Union).
Effectiveness	Established a mechanism for monitoring and evaluating pollutant levels (Co-operative Programme for the Monitoring and Evaluation of the Long-Range Transmission of Air Pollution in Europe EMEP).[b] Sulfur dioxide emissions halved in Europe from 1980–95; emissions of nitrogen oxides and ammonia fell by 15 percent.[a]
Inducements	Trust fund established to finance emission reduction programs.[e]
Key Drivers	The damage caused by acidification.

[a] *Euroabstacts*, vol. 37, 5/99.
[b] Shaw (1993).
[c] http://sedac.ciesin.org/pidb/guides/ (check UNEP guide).
[d] Sand (1999).
[e] Tolba and Rummel-Bulska (1998).

the United Kingdom, and Poland did not sign the protocol (Shaw, 1993). The United Kingdom objected to the baseline, and Poland did not have the technology to deliver the reduction (both countries were large emitters of sulfur dioxide).

The Second Sulfur Protocol committed signatories to decrease sulfur dioxide levels by 40 percent by 1993, 60 percent by 1998, and 70 percent by 2003. However, recent research suggests that even if sulfur dioxide emissions are cut by 90 percent, the accumulated acidity will take decades to dissipate (*New Scientist*, 4/3/00, p. 19).

The European negotiations leading up to this protocol were unusually successful examples of international cooperation to reduce transboundary risks. The use of a computer model based on the "critical load" concept[4] facilitated the adoption of a cost-effective solution to meeting European targets. Patt (2000) has noted that differential reductions of pollutants based on value judgments are harder for negotiators to justify than those based on a physical measure of damage to the soils and that negotiators tend to focus on "bright lines," easily quantifiable and justifiable emission goals.

The Nitrogen Protocol to the CLRTAP resulted in an agreement to freeze emissions of nitrogen oxides at 1987 levels by 1994 and to implement reductions from 1996 (Sand, 1999). However, growth in car use is offsetting the technical gains made, for example from catalytic converters. Transport is now the dominant producer of nitrogen oxides in Europe, and the potential for large increases in car ownership in Central and Eastern Europe is a cause for concern (*Euroabstracts*, vol. 37, 5/99).

The protocol on Persistent Organic Pollutants (POPs) (1998) aims to control, reduce, or eliminate discharges, emissions, and losses of organic compounds that are toxic, persistent, bioaccumulative, and prone to long-range atmospheric transport and deposition within the CLRTAP region. Unlike other major pollutants such as sulfur and nitrogen, POPs cannot be easily narrowed down to a few substances, and one of the key issues here is to set up criteria to define POPs to be included in the protocol (Selin and Hjelm, 1999).

Compensation across borders has been a major issue in dealing with transboundary air pollution, particularly the fact that Western European countries can often benefit more by investing in pollution reduction in Eastern Europe than investing these same funds in reducing domestic sources. However, although economically efficient, this approach may not be considered fair; in a Swedish survey, many people felt that it was the responsibility of Eastern European countries to deal with their own pollution even if it meant more risks to the Swedish public (Loefstedt, 1994). We expand on this point in Section 6.2.

Ozone Depletion. Ozone depletion in the upper atmosphere occurs in winter over the polar regions and is greatly exacerbated by persistent halogen-containing compounds of human origin, which are used in refrigeration, fire-fighting, production of aerosols and plastic foams, anaesthesia, and soil sterilization. Initially confined to Antarctica, it now occurs over Arctic regions, affecting heavily populated regions of Europe and North America.

When the protective effects of ozone in the upper atmosphere are removed, living organisms are exposed to increased levels of ultraviolet-B radiation, with impacts on human health (skin cancer, cataracts), plant life (yield reductions in crops, susceptibility to disease, changes in plant structure and pigmentation, alterations in interspecies competition), and aquatic ecosystems (reduction in primary productivity of phytoplankton and changes in the biogeochemical cycling of nitrogen, both of which are likely to affect many organisms further along the marine food chain). Chlorofluorocarbons (CFCs) are among the major causes of ozone depletion and are also implicated in global climate change because of their ability to absorb longwave infrared radiation (www.ciesin.org/TG/OZ/o3depl.htm).

Early discovery and regular monitoring of ozone depletion since 1978 has been one of the benefits of satellite remote sensing technology. Success in developing the Vienna Convention and the subsequent Montreal Protocol (Table 10.7) owes much to the availability of these and other scientific data on long-term trends.

A range of political and commercial factors were important in negotiations on the Montreal Protocol where the United States and the European Community emerged as the main protagonists. The Toronto Group (United States, Canada, Australia, and the Nordic countries) wanted a worldwide ban on the use of CFCs as aerosol propellants but were opposed to other restrictions. The European Community wanted eventual limits on total production of ozone-depleting substances. Developing countries were concerned about any measures that would limit their own development. The United Kingdom, France, and Japan were reluctant to agree to any regulatory measures. European chemical manufacturers saw U.S. industries as being ahead of them in developing substitutes for CFCs and feared losing out. Given that the United States had already banned the use of CFCs in aerosol containers (Szell, 1993), European manufacturers were concerned that further controls could affect products for refrigeration and foam manufacture.

With this protocol, the United States returned to being a leader rather than a blocker of international environmental conventions. With support from significant industry pressure groups, the Montreal Protocol was eventually hailed as a major success. As scientific evidence accumulated, protocol targets were exceeded, and production and consumption of CFCs has virtually ceased in the developed world (Braithwaite and Drahos, 2000, p. 264). However, despite its success, recovery of the ozone layer is expected to be slow. Even without further additions of ozone-depleting chemicals, the effects are expected to peak in 2000 and 2001, and it could take an

TABLE 10.7. *Ozone Depletion*

Relevant Issues	Reduced ozone layer will result in increased exposure to UV-B radiation (implicated in human skin cancer and blindness) adversely affecting marine ecosystems and affecting the early developmental stages of fish and crustaceans.[a] First recognized in 1974.
Regulation	Vienna Convention for the Protection of the Ozone Layer (1985) commited signatories to measures to protect human health and the environment from activities with the potential to have adverse effects on the ozone layer.[b] Nonbinding agreement, established a framework for negotiating the Montreal Protocol on Substances that Deplete the Ozone Layer (1987), amended London (1990), Copenhagen (1992), and Vienna (1995).[c]
Aim	To protect the ozone layer by taking measures to control emissions that deplete it. Specifies phase-out dates for particular substances. 1987 protocol aimed to reduce production of all the five CFC types by 50 percent by 1999 from base levels in 1986. Based on scientific evidence of need, stronger measures were adopted in 1990; governments agreed to phase out CFCs by 2000; additional chemicals included in list of controlled substances.[b]
Organizing Body	UNEP; negotiations began in 1984.
Compliance	Responsibilities differentiated for industrialized and developing countries.
Area Covered	Global (first truly global environmental treaty).[b]
Effectiveness	The protocol (widely considered to have been effective) stipulates that production of CFCs, halons, and carbon tetrachloride to be phased out by 2000, methyl chloroform by 2005, and methyl bromide by 2010.[c] Reduction in rate of growth of atmospheric CFCs and decline in levels of methyl chloroform are leading to greater credibility.[c]
Inducements	From 1990, ten-year grace period for implementation by developing countries with less than 0.3 kg per capita consumption of the substances and trust fund to finance programs. China and India, nonsignatories to the protocol, had the potential to produce large quantities of ozone-depleting substances. These inducements encouraged China to join the protocol in 1991, and India joined soon after.[c]
Key Drivers	Informal consultation, availability of scientific information to convince reluctant partners of the necessity of regulation (the discovery of the Antarctic ozone "hole"), and the force of public opinion alerted by NGOs and the media.[b]

[a] *Euroabstracts*, vol. 37, p. 14–15, 5/99.
[b] Tolba and Rummel-Bulska (1998).
[c] www.sedac.ciesin.org/pidb/guides/.

additional seventy years for the ozone layer to return to its previous state (*Euroabstracts*, vol. 37, p. 14–15, 5/99).

Global Climate Change. Concerns about climate change and the possibility of a human-induced "greenhouse effect" or "global warming" have been expressed gradually more loudly and more coherently by the scientific community since the late 1950s. Other scientific voices have been raised in opposition, claiming that observed climatic fluctuations were the result of natural causes. However, partly owing to the development of ever more sophisticated modeling techniques, supported by data collection on an unprecedented scale, the balance of probability has gradually shifted toward the views of those who claim a major role for greenhouse gases of human origin (e.g. carbon dioxide, methane, nitrous oxide, and halocarbons) (Houghton et al., 1996).

INTERGOVERNMENTAL PANEL ON CLIMATE CHANGE. The Intergovernmental Panel on Climate Change (IPCC) was set up in 1988 under the auspices of UNEP and the World Meteorological Association (WMO) to provide authoritative assessments to governments on the state of knowledge concerning climate change, including the scientific, technical, and socioeconomic information relevant to the understanding of the risks of human-induced climate change (e.g., IPCC, 1995) based mainly on published and peer-reviewed literature.

UN FRAMEWORK CONVENTION ON CLIMATE CHANGE. The IPCC played an important role in establishing the UN Framework Convention on Climate Change (UNFCCC) and has continued to provide advice to the parties to this convention and to the world community. Its second report provided key inputs to the negotiations, which led to the adoption of the Kyoto Protocol (http://www.ipcc.ch/about/). However, its task has not always been easy. Ben Santer, one of the lead authors of the First IPCC Report (1995) was accused of "doctoring" the final report and has been much criticized by the U.S. body representing the oil and automobile industries, the Global Climate Coalition (the "Carbon Club") (*New Scientist*, 4/3/00, p. 43).

Negotiations to develop the UNFCCC began in 1991 under the Intergovernmental Negotiating Committee and in June 1992 at the UN Conference on Environment and Development (UNCED), it was opened for signature, entering into force in March 1994, after ratification by fifty countries (http://sedac.ciesin.org/pidb/guides/) (Table 10.8). It has now been widely adopted with over 180 countries party to the convention.

As a framework treaty, there were no target emission level reductions. The convention sets out principles and general commitments to be adopted in national programs for mitigating climate change, to develop adaptation strategies, and to promote the sustainable management and conservation of greenhouse gas "sinks," such as forests (www.sedac.ciesin.org/pidb/guides/). Article 2 aims to stabilize greenhouse gas concentrations

TABLE 10.8. *Global Climate Change*

Relevant Issue	Greenhouses gases arising from human activity are affecting the global climate in potentially harmful ways.
Regulation	UN Framework Convention on Climate Change (UNFCCC, 1922) and its Kyoto Protocol (1997).[a]
Aim	*UNFCCC*: To stabilize greenhouse gas concentrations in the atmosphere. Sets out principles to be adopted in national programs. No specific commitments.
	Kyoto Protocol: To reduce emission of greenhouse gases to an average of 5 percent below 1990 levels, during 2008–12. No commitments from developing countries. Flexibility provided by (i) International Emissions Trading allowing one country to emit more if another emits less, (ii) Clean Development Mechanism whereby industrialised countries can receive credits against their own commitments if they invest in technologies to reduce emissions in developing countries.
Organizing Body	UNEP.
Area Covered	Global.
Effectiveness	Measures affect the driving force of industrialized economies so this is a very difficult area to change. The trend of increasing greenhouse gas emissions has not changed so far.
Inducements	Greenhouse gas producers gain emission credits if they invest in energy conservation measures in developing countries.
Key Drivers	In the 1990s, driven by the United States, the major greenhouse gas producer. The United States also favored having developing countries, which it saw as competition for U.S. industrial activity, included in the protocol.

[a] Grubb et al. (1999).

in the atmosphere at a level that would prevent dangerous anthropogenic impacts on the climate system, setting a time frame sufficient to allow ecosystems to adapt naturally to climate change, to ensure that food production is not threatened, and to enable economic development to proceed in a sustainable manner (Watson et al., 1998). Articles 4 and 12 require parties to prepare national communications on their implementation of the convention, giving guidelines for their production and review.

The convention adopts the following principles:

- Precautionary principle – regulators should act even in the face of uncertainty.
- Principle of common but differentiated responsibility – industrial countries that have caused the problem should have the greatest responsibility for mitigating effects.

It also sets up institutions to monitor progress and take decisions on rules (conference of the parties) and procedures to monitor developments in science and act accordingly.

KYOTO PROTOCOL. The development of the Kyoto Protocol to the UNFCCC (1997) (see Table 10.8) has been described as an extraordinary process and a remarkable achievement, getting more than 150 countries with divergent interests and perceptions to agree on an issue with major national implications for all. In an era of economic globalization, it was an agreement struck by governments. NGOs wanted stronger commitments but opposed almost all the transfer mechanisms. However, widespread public concern and activism on this issue in developed countries has yet to emerge. Most of U.S. industry opposed the process and spent up to $100 million on fighting it. However, governments wanted the protocol to succeed and did what they considered was possible to protect the atmosphere while also protecting their own interests (Grubb, Vrolijk, and Brack, 1999).

The eventual impact of the protocol depended on its ratification by a minimum of fifty-five parties. This was not a high hurdle, but there were many countries for which ratification was not a foregone conclusion (Grubb et al., 1999, pp. 253–7). There was, for example, the question whether the U.S. Congress would ratify the agreement given its dislike of the idea of differentiated responsibility (countries such as India, China, and Brazil, are regarded as competitors). Additionally, the protocol also required that the emissions of the countries that ratify the protocol represent at least 55 percent of the total for all relevant countries (Annex 1 countries) (Skolnikoff, 1999).

By 2000, under the agreement, industrialized countries had agreed to reduce their emissions of greenhouse gases to 5 percent below 1990 levels on average, during the commitment period 2008–12. Commitments differ among countries: 8 percent for the EU; 7 percent for the United States; and 6 percent for Japan. Others had committed to stay within their current emission levels and some to reduce their projected increase in emissions. Most developing countries had not tabled any commitments, fearing they would harm their development prospects (Watson et al., 1998).

The commitment refers to a "basket" of six specified greenhouse gases (Ott, 1998). The target reductions take into account "sinks" – usually forests that act to absorb greenhouse gases (Ott, 1998). This is a contentious area owing to the difficulty of calculating the effects of afforestation, reforestation, and deforestation on greenhouse gas concentrations. Recent studies by the IIASA in Austria suggest there are such major gaps in understanding of how forests influence atmospheric carbon levels that including these sinks in calculations will make the Protocol ineffective (*New Scientist*, 26/8/00, pp. 18–19).

Between 1990 and 1995, carbon dioxide emissions in Europe as a whole fell by 12 percent, largely because of the closure of large parts of Eastern European heavy industry and the switch from coal to gas for electricity generation in many countries. It is estimated that without controlling measures, by 2010 Europe will have increased its output of greenhouse gases by 5 percent from the 1990 level. The energy sector produced 35 percent of Europe's carbon dioxide emissions in 1995, but transport is a growing contributor (*Euroabstacts*, vol. 37, pp. 14–15, 5/99). The United States has by far the highest per capita emissions of carbon dioxide in the world, more than twice that of European OECD member countries (IEA/OECD, 1997).

The Kyoto Protocol, although more precise in its commitments than the UNFCCC, left substantial gaps in the design of its rules and the mechanisms for implementing them (*Responding to non-compliance under the climate change regime*, OECD Information Paper ENV/EPOC(99)21/FINAL, OECD May 1999). The protocol was considered uniquely ambitious in international environmental agreements in trying to change the way the economies of countries are powered. The mechanisms were inspired by the United States (emissions trading) and EU (the "bubble" concept) who sought to impose models with which they already had some experience. Emissions trading was based on U.S. domestic environmental policy, particularly the Acid Rain Regime (Grubb et al., 1999), but it was not clear how these concepts would operate internationally.

CLEAN DEVELOPMENT MECHANISM. The Kyoto Protocol established an innovative Clean Development Mechanism, which allowed developed countries to receive credits against their own commitments for investment to reduce greenhouse gas emissions in developing countries (e.g., energy-saving technologies) (Cosbey, 1999). In January 2000, before the protocol had specified its provisions on the Clean Development Mechanism, the World Bank launched a Prototype Carbon Fund in an attempt to begin working with market-based mechanisms for reducing carbon emissions. This fund allowed governments and private companies in richer countries to invest in developing countries and countries in transition to reduce carbon emissions or to develop reforestation schemes and to have these credited toward the richer countries' carbon emissions (Sand, 1999). The resulting emission reductions would be independently verified and the fund's contributors would be given emission reduction certificates. As of March 2000, Finland, the Netherlands, Norway, Sweden, and nine companies had agreed to participate in the fund (www.northsea.nl/jiq/news1.htm), the first project being a combined landfill and power station in Latvia where methane from decaying waste would be used to generate electricity (*Chemistry & Industr*, p. 84, 7/2/00).

INTERNATIONAL EMISSIONS TRADING. The principle of International Emissions Trading was set up to allow one country to emit more if another country agreed to emit less. Procedures included project-based flexibility mechanisms where, for example, Country 1 can enable Country 2 to move from coal to solar power as a source of energy and gain a Certified Emissions Reduction that Country 1 can offset against its commitments. Developing countries (those not undertaking any obligations under the Kyoto Protocol) could take part in project-based flexibility mechanisms. Also, under the protocol, Russia was given a generous allocation to take account of the fact that its emissions were currently low owing to the collapse of its economy, and this provided a large seller in the emissions market, which was attractive to the United States (Grubb et al., 1999).

The concept of emissions trading proved very controversial during the protocol negotiations. As a result, a minimum of enabling language was inserted in the protocol, with the details to be developed later (Grubb et al., 1999, p. 96).

The EU contribution to innovative thinking in the protocol was based on its experience of similar countries joining together in a co-operative regional system. The "bubbling" provision (Article 4) allowed a group of countries, when they ratify the protocol, to redistribute their emission commitments in ways that preserve the collective contribution. For example, some European countries such as Spain could carry a lighter burden, while others (e.g., the United Kingdom and Germany) carried a heavier burden, although access to the grouping had to be limited (e.g., the United States could not benefit from the heavier burden carried by the United Kingdom and Germany). The EU's use of this concept created resentment among other countries when it called for reductions by all other countries while some of its own members were able to increase emissions. The concept was therefore accepted as applicable to any group of countries (Grubb et al., 1999).

It is not possible to design new international agreements without considering nations' obligations under the World Trade Organization. Some measures, which may be adopted by individual nations to reduce carbon emissions, may be in conflict with WTO rules. A unilateral carbon tax, for example, would disadvantage the country implementing it, unless a similar tax could be placed on imported goods. Environmental efficiency standards on goods might also constitute technical barriers to trade (Cosbey, 1999). In addition, failure for political reasons to reach agreement within the EU on a framework for energy taxation and on a directive on renewable energy illustrated how difficult it would be to achieve targeted reductions in emissions (Hyvarinen, 2000).

POLITICAL CONSIDERATIONS. The Kyoto Protocol negotiations took place between governments, but NGOs were involved in providing information, making policy recommendations, and lobbying. Some 250 NGOs were present at Kyoto as observers. In 1989, many NGOs came together within the Climate Action Network to improve their effectiveness (Oberthür and Ott, 1999). In the Kyoto negotiations, the EU appeared weak and fragmented (Hyvarinen, 2000), and its ability to move quickly as the negotiations unfolded was hampered by the complex nature of its internal decision-making processes. Some nation states (notably Denmark) were suspicious of the negotiations carried out by representatives from Luxembourg, the Netherlands, and the United Kingdom (Oberthür and Ott, 1999). The negotiations were dominated by the United States, and its favored mechanism at that time, emissions trading, was eventually adopted. It is, therefore, ironic that the United States was one of the parties most reluctant to implement the protocol and, by 2000, had failed to make meaningful progress toward its ratification. The U.S. Congress also continued to block attempts to integrate climate change goals with energy, transport, and air pollution regulations (Kete, 1999).

Even if the provisions set out in the protocol are realized by 2012, most experts agree that there will be only marginal reductions in the degree of climate warming in this century. This raises the question whether the benefits of the Kyoto commitments are worth their cost. One view is that decisive global action to reduce poverty is a more effective use of resources than greenhouse gas reduction, especially if future victims of global warming are more wealthy and thus able to adapt to a changed climate. It has been pointed out that such questions have hardly been posed in Kyoto negotiations because of the hegemonic grip that the emissions reduction strategy has on the policy discourse (Linnerooth-Bayer, 1999). However, the time horizon of these views is very short in climate change terms, and more creative thinking may lead to solutions that can satisfy both sets of needs jointly.

The response of industry to the Kyoto Protocol has been mixed. The Carbon Club of industries launched a campaign in the United States to stop Congress from signing the protocol. However, a number of major companies have now left the Carbon Club and have signaled their intention to take the issue of climate change seriously. For example, Shell planned to reduce its own greenhouse gas emissions by 10 percent from 1990 levels by 2002 and BP by 2005 (Oberthür and Ott, 1999). Paterson (1999) suggested that, as soon as the protocol was signed, businesses began to reassess their strategies, and senior executives from companies such as BP, Amoco, Shell, and Du Pont have become outspoken advocates of changing corporate strategies to address climate issues.

The discourses surrounding the protocol produced a range of proposals for sharing the costs of greenhouse-gas reductions between developed and developing countries, reflecting a number of possible criteria for distributing emission rights and also fundamental differences over principles of fairness. It could be argued that the Kyoto Protocol made this discussion obsolete given the pragmatic political considerations that ultimately set the commitments for greenhouse-gas reduction. An important exception is the controversial principle placing responsibility for emission reductions fully on the industrialized countries. With predicted escalating emissions in developing countries, there is a gradual but significant redistribution of entitlements, but however significant this is, it will not change the disparity in per capita emissions between the United States and other industrialized countries relative to the developing countries. Issues of fair allocation are likely to remain on the international negotiating agenda.

After the U.S. presidential elections in 2001, the new administration announced that it was opposed to the Kyoto Protocol because, among other things, it set no targets for developing nations such as China and India, and the carefully negotiated compromises inherent in the protocol began to unravel. Global negotiations continued in Bonn in July 2001 and resumed again in Marrakesh in October and November 2001, despite the U.S. announcement that it would not ratify the protocol. It is likely that the numbers ratifying the protocol will exceed the required fifty-five countries. However, the absence of the United States among them and the compromises made at Bonn and Marrakesh have severely weakened the protocol's provisions (*ENDS Report*, p. 49, vol. 318; p. 53, vol. 322, 2000).

6. INTERACTIONS, LINKS, AND FUTURE ISSUES

Sections 4 and 5 have focused mainly on describing regulatory processes in place for the four types of transboundary risk identified in our framework. Some general trends emerge from this, and it is clear that there are important links between transboundary risk regulation and other international developments.

Some early developments in transboundary risk regulation were in advance of co-operative international efforts in other areas such as economic development and trade. Other transboundary risk issues have been much more resistant to international action. The difference can often be traced to the perception of where their interests lie by one or more influential industry sectors. The Montreal Protocol on Substances that Deplete the Ozone Layer was eventually so successful largely because the multinational companies that were developing CFCs could see an opportunity to sell a new and more profitable range of products than those that were the cause of the problem. The Kyoto Protocol may yet fail because the current U.S. government and a large and influential sector of American industry see it as a threat to profitability and also challenge the fairness and effectiveness of the proposed instruments. However, the introduction of the concept of tradable permits, which could be seen as shifting greenhouse gases from a "public, pervasive risk" to the status of a "traded, product-based risk," thereby creating opportunities for some industry sectors and companies to benefit from the protocol, is also changing industry attitudes in some cases.

Nations, depending on the political complexion of their governments, go through phases of being leaders and laggards, or even blockers, in international negotiations on transboundary risks. These changes can be the result of changing perceptions of where national interests lie, political international alliances driven by nonrisk related issues, or the influence of important national lobby groups such as industry or environmental NGOs.

NGOs have often had an important role at the national level in stimulating or articulating public concerns about transboundary risk issues. Now, through the globalizing potential of the internet, we are seeing highly effective international mobilization among coalitions of NGOs that seem to be capable of challenging multinational corporations or even international bodies like the WTO. However, in the past, the influence of NGOs has sometimes caused a delay in reaching agreement or implementing regulations, or led to outcomes that did not reflect their stated objectives. It is reasonable to expect that this will continue to be the case, despite their more global reach.

Scientific data and models are regularly used as weapons of argument in all these debates. Where issues are complex and data subject to high levels of uncertainty, the same data can often contribute to argument on both sides of the debate. Increasingly, there are calls for a more evidence-based approach to policy making, but it is often not clear which policy should or could emerge from the evidence.

6.1. Interactions Between Transboundary Risk Regulation and Global Trade Issues

The General Agreement on Tariffs and Trade (GATT) and its successor the World Trade Organization are primarily concerned to allow unrestricted international trade in goods and services and were not originally involved in risk regulation or environmental issues. However, they have been drawn increasingly into international risk debates. There is a tendency for some types of transboundary risk regulation to be regarded as a restriction of trade and to be challenged as such by governments and industry. On the other hand, GATT recognized that "a healthy environment and a healthy economy go hand in hand" and advocated an alliance between GATT and various environmental fora (*Chemistry and Industry*, p. 125, 17 Feb. 1992).

Following the Uruguay Round of GATT negotiations, trade ministers adopted the Decision on Trade and Environment, which anchored environment and sustainable development issues in WTO work. They set up the Committee on Trade and Environment (CTE), open to all Members of WTO, and assigned it a broad mandate, covering virtually all aspects of the trade and environment interface.

A series of agreements outlines the restrictions on trade that are regarded as legitimate by the WTO. The most relevant to risk issues is the Agreement on the Application of Sanitary and Phytosanitary Measures (SPS), which acknowledges that governments have the right to take measures to protect human, animal, or plant life or health but only to the extent necessary for this purpose, and the measures taken must be based on scientific principles (Nordstrom and Vaughan, 1999). This agreement was a major step in the globalization of food standards and set Codex standards as the benchmarks against which national regulations are evaluated. If national standards are higher than those of the Codex, the additional safeguard must be based on scientific evidence and grounded in risk assessment (Vogel, 1995, p. 188). Governments also cannot discriminate by applying different requirements to different countries without scientific justification. One result of this linkage between WTO and Codex standards is that the previously scientific Codex process is becoming politicized (Braithwaite and Drahos, 2000, p. 403). However, up to 2000, the use of the precautionary principle in such debates had not been formally recognized.

The WTO has been criticized for being secretive and not involving NGOs (Makuch, 1996), but it has since adopted a set of guidelines indicating that, although NGOs cannot be directly involved in its work or its meetings (because it is a legally binding treaty of rights as well as a forum for negotiation), there are opportunities for interactions with NGOs at the national level. The guidelines also aim to improve transparency by de-restricting documents more quickly and to organize symposia that may involve NGOs. The CTE has also agreed to extend observer status to a wide range of intergovernmental organizations (e.g., UNCTAD, World Bank, IMF, UNEP, UNDP, Commission for Sustainable Development, FAO, OECD).

In Section 2, we noted the acrimonious relationship that has developed between the WTO and NGOs around the issues of trade liberalization, its impact on transboundary environmental and health risks, and our ability to control them. Many responsible commentators share the concerns of the NGOs although they may distance themselves from the methods used to make the case. A more critical strand is beginning to emerge in some discussions of the legitimacy of NGO engagement in such debates. Patten (2000) advocated more openness in public debate at national and international levels, but noting the role of NGOs in this process, particularly those operating in the richer countries, he questioned their claim to democratic legitimacy and also in some cases the responsibility of their methods.

The interactions between the WTO and environmental NGOs at Seattle and subsequently has raised serious questions about risk regulation and the processes of decision making in modern plural democracies. We will return to this topic in Section 6.5.

6.2. Category Shifts Among Transboundary Risks

As noted in our introduction, the four categories of transboundary risk proposed here are not seen as fixed. It can be instructive in understanding the nature of a risk dispute to be aware of the ways in which protagonists in a debate are shifting the perspective on a risk from one category to another (see the following example of GM crops). A shift in categories can also be a creative way of resolving intractable problems in transboundary risk regulation. For example, considering traded risks, it has been suggested that the WTO has become a focal point for environmental disputes because it has an adjudication mechanism backed by trade sanctions as the ultimate enforcement tool, rather than the generally weaker political institutions dealing specifically with risk issues (Nordstrom and Vaughan, 1999).

This point extends beyond the issues we have classed as traded risks. For example, in the context of ozone depletion, the Montreal Protocol was easier to set up than some other protocols because the offending products were traded internationally. This aided the monitoring of compliance by making it easier to detect violations; it gave some leverage to the preferences of individual consumers who began to refuse to purchase aerosol cans (traded, product-based risk); and because of the commercial advantages to be gained by companies that were successful in developing alternative products, it gave a powerful stimulus to innovation.

In the context of global warming, the principle of International Emissions Trading incorporated into the Kyoto Protocol attempts to bring some of these incentives and potential controls into an area where they have previously been absent (i.e., converting a public risk into a traded risk, albeit one that is different from the two categories described in Section 4). It is neither product-based, not production system-based, but it is the right to pollute itself which is traded. This change has been achieved by allocating property rights in something which was previously a 'free good', or perhaps more accurately a 'free bad'.

There is already considerable interest in this approach. It is attractive to industry because it creates a tradable asset and also creates an incentive for participants to reduce emissions in the most cost-effective manner (Grubb et al., 1999, pp. 89–96). If it can be implemented effectively within the Kyoto Protocol, the use of the concept is likely to spread to other transboundary public risk areas.

Trade in food and agricultural products, in particular the export of GM food and crop commodities to Europe, has become a testing ground for the WTO and the SPS Agreement. An important aspect of this dispute is the unacknowledged shift in risk categories that has taken place as part of the debate. For many of the groups that are most active in calling for stricter regulation of GM crops, their concerns are about intensive farming systems per se, and the fact that GM crops will become an important component of intensive farming systems in the future is a major reason for their rejection (Tait, 2001). Thus, they view GM crops as a *traded, production system-based risk*. However, the only currently available basis on which to mount a challenge to GM crops is as a *traded, product-based risk*, and the legitimacy of a challenge on this basis, given the lack of scientific evidence for damage to either human health or the environment, has become a focus for vigorous debate, challenge, and counter challenge, and demands for adoption of the precautionary principle.

Viewing GM crops as a traded, production system-based risk, leading to a further round of intensification of food production systems, has strong legitimacy among the European public (National Consumer Council, 1998; ESRC Global Environmental Change Programme, 1999). European consumers have become adept in recent years in the use of their purchasing power as a political weapon. In a world of increasingly globalized trading systems and a waning of the power of national governments, the consumer boycott is becoming a new form of international governance. However, to mount an effective boycott, the product(s) in question must be identifiable, and this means that they must be labeled.

The question of labeling is extremely contentious as the producers of GM crops see no scientific evidence that would justify distinguishing them from non-GM crops. American exporting companies have resisted labeling for these reasons, although European multinational agrochemical and biotechnology companies have been prepared to label, being closer to the consumer concerns. A major, but probably not unresolvable, dilemma is emerging here. To restrict trade on the basis of no science or poor science would be a highly retrograde step; on the other hand, to deny consumers the right to chose products on the basis of their production system, just as much as any other basis, seems undemocratic.

Labeling and certification emerged much earlier in the case of tropical forest products (see Section 4.2) where a range of schemes has emerged as a way of allowing consumers to exert an influence on timber production systems in other countries (Kanowski, Sinclair, and Freeman, 2000). Markets for certified products are strongest in Western Europe and the United States, and they have not gained much market share in Japan or other Asian markets. Ironically, despite the origins of certification coming from concerns about unsustainable use of tropical timber, the vast majority of certification of wood products has taken place within Europe and North America.

6.3. The Precautionary Principle and the Risk-Based Approach

An important debate is taking place among risk analysts about the role and relevance of the precautionary principle and the risk-based approach as a basis for decision making on risk issues, nationally and internationally (Tait and Levidow, 1992).

The risk-based approach relies on quantification of probabilities and outcomes, the development of quantitative models, and the use of cost-benefit analysis as a basis for scientifically rational decision making (Dietz and Rycroft, 1987; Jasanoff, 1990, pp 248–50; Finkel and Golding, 1994; Graham and Wiener, 1997).

A contrasting feature of the precautionary principle is that, "Where there are threats of serious or irreversible damage, lack of full scientific certainty shall not be used as a reason for postponing cost-effective measures to prevent environmental degradation." (1992 Rio Conference on the Environment and Development; Rio Declaration, Principle 15). Thus, where levels of uncertainty are high, where potential impacts are very large, and/or where those impacts may be irreversible, there are grounds for adopting a precautionary approach until such time as we have sufficient scientific knowledge to make a risk-based decision (CEC, 2000).

In practice, each approach should be adopted judiciously according to the circumstances of the case, but at the extremes of the debate there is little room for such flexibility, with protagonists taking up very rigid stances. The vehemence of the debate stems from their basis in two dominant but competing paradigms in the risk field (Rosa, 1998). The risk-based approach has its roots in positivistic science including the rational actor paradigm, while the precautionary principle has strong tentacles, if not roots, in cultural theory and social constructivism. As Rosa notes, the first of these paradigms was inspired by Enlightenment thinking; the second was inspired by the need to criticize it.

In Table 10.1, we detailed the risk categories where we see each approach being generally relevant. Here we explain more fully the basis of these assumptions and qualify them to some extent where necessary.

Traded, Product-Based Risks. For most of the products traded internationally as hazardous wastes, the risks are well characterized and the impacts are predictable. In such cases, the risk-based approach would be most appropriate. However, some wastes can contain complex mixtures of products that may interact in unpredicted ways. Also, the definition of which wastes are hazardous can be arbitrary or based on political expediency.

For pesticide residues in food, the pesticides have been approved for use under national systems of registration, MRLs have been established under Codex Alimentarius standards, and detection systems are available for foods that violate the standard. Strictly interpreted, therefore, the precautionary principle is not likely to be justified. However, there are cases where a precautionary approach is demanded by individuals or pressure groups (e.g., where a pesticide is suspected to be more hazardous than shown by available scientific tests).

Also, as the case of BSE demonstrated recently, instances may arise where, with hindsight, earlier application of a precautionary approach would have been justified for international trade in some food and animal feed products.

Risk-based regulation is therefore likely to be appropriate for most traded, product-based risks, but the need for the precautionary principle should not be ruled out entirely. Also, public trust in scientists and officially convened regulatory bodies is declining, and this is exacerbated by the fact that data provided to support the case for new products are often generated mainly by the industry concerned or by scientists with strong links to industry. One outcome of this situation is an increasing number of calls for the precautionary principle to be exercised and consequent restrictions on international trade in the products involved.

GMOs provide the most conspicuous example of this kind of effect in operation. The preamble to the Cartagena Protocol reaffirms the precautionary approach as defined by the Rio Declaration. In describing the risk assessment procedure (Annex III), Article 26.1 states that "The Parties . . . may take into account . . . socio-economic considerations arising from the impact of living modified organisms on the conservation and sustainable use of biological diversity, especially with regard to the value of biological diversity to indigenous and local communities," thus extending the interpretation of "precaution" considerably beyond that described earlier.

As a result of the OECD Conference on GM food, held in Edinburgh in February 2000, the chairman of this conference (Sir John Krebs) has proposed setting up an international forum, modeled on the IPCC, to provide governments with an updated assessment of the scientific knowledge about GM technology and to set this assessment in the context of societal concerns (OECD, 2000).

A recurring problem with the precautionary principle is the extent to which it can constitute an unreasonable restriction of trade, and agreement on the Cartagena Protocol required a compromise on its relationship with WTO agreements. The protocol preamble states: "this Protocol shall not be interpreted as implying a change in the rights and obligations of a Party under any existing international agreements" and also "the above recital is not intended to subordinate this Protocol to other international agreements." An EU official noted that the effect of the two sentences was to cancel each other out (*Bridges*, vol. 4, no. 5, 8 Feb. 2000), and this will continue to be an important debating point as the implications of the protocol are clarified.

Even though there are some legitimate reasons to advocate the use of the precautionary principle in the context of GM crops, the unreasonable safeguards being demanded by some pressure groups are in danger of bringing the principle itself into disrepute and jeopardizing its adoption in circumstances where it is clearly needed (Tait, 2001).

Traded, Production System-Based Risks. For the cases in this category, formal systems of international risk regulation do not exist. Consumer pressures may result in informal agreements, as in the ITTA, but these are not very effective. The pressures that result in demands for discriminatory labeling and consumer boycotts often arise from concerns about environmental impacts within national boundaries that would, in many cases, justify the application of the precautionary principle. The consumer boycott may be effective in some cases in the short term, but it is generally a capricious and weak instrument. However, we have not yet found a way to translate public concerns of this nature into any form of international regulatory action, whether risk-based or precautionary. A major problem is that the country exploiting the resource often has very little incentive to regulate its rate of exploitation.

Public, Contained Risks. Risks in this category are likely to have their maximum impact in the country where the facility is located with often serious but diminishing impacts beyond national boundaries. Unlike traded, production system-based risks, therefore, the country where such a facility is located has every incentive to avoid an accident, and most countries have regulatory systems in place for this purpose. Concerns and uncertainties may still arise about the effectiveness of regulatory systems in some countries, but the precautionary principle would not be appropriate to deal with such issues. It is difficult to think of an example where adoption of the precautionary principle, internationally, could have prevented an accident in this category.

Public Pervasive Risks. The risks in this category often arise from complex interactions between physical and chemical processes. The diffuse or dispersed nature of pollution sources makes it difficult to identify and to control emissions before significant levels of damage have been caused. In the case of global climate change, by the time we have resolved the scientific uncertainty about its causes, it may be too late to have any influence on the impacts. This is, therefore, the risk category where the precautionary principle is most likely to be justified

Under these circumstances, it is interesting to note that NGOs are expending a very large amount of effort on promoting the use of the precautionary principle in the case of GM crops where its use may be justified to some extent, and relatively little effort in promoting its use in the case of global climate change where the potential impacts are much greater and more extensive and where there is already a considerable body of scientific evidence to support the hypothesis that the observed effects are caused by human activities. Indeed, the use of GM crops may be one relevant tactic to help mitigate the impacts of global climate change on our ability to feed the world population.

6.4. Industry Concerns and Interests in Transboundary Risk Regulation

Industry is particularly concerned about the number of different regulatory regimes with which companies have to engage, as well as the lack of international harmonization of risk regulations and risk assessment procedures. Conflicts between the EU and the United States, as well as within the EU, have demonstrated the inherent difficulties in harmonizing regulatory and administrative procedures in the face of manifestly different institutional commitments and political/administrative cultures.

Progress in transboundary risk regulation has been greater than most observers expected and has been less damaging to industry than industry expected. Industry's preferences for trade-related instruments rather than taxes seem to be gaining more widespread support among policy makers, but instruments related to trade regulation will not always be compatible with those designed to regulate pollution and to control environmentally damaging human behavior. In some cases, synergy may be possible. In others, conflicting objectives may need to be accommodated.

In considering progress to date, it has been claimed that the proliferation of new international agreements and bodies in the last decade seems to have overwhelmed most governments, resulting in a tendency to make policy as a reaction to the agenda for the next international meeting, rather than developing results-oriented long-term strategies (Hyvarinen, 2000) and the same is likely to be true of companies operating internationally. Still, as Victor and Skolnikoff (1999) show, most governments implement their agreements most of the time.

In risk regulation as in any other area, it is important to recognize the multiple perspectives and interests of different industry sectors and of different companies within sectors. This variation in response is what gives some firms a competitive advantage over others and increasingly the international operating environment of industry is being altered in favor of those companies that see risk regulation as an opportunity rather than a constraint. The idea prominent in the 1980s that tougher regulation would lead companies to relocate to "pollution havens" has been discredited by a large body of evidence (e.g., Leonard, 1988; Braithwaite and Drahos, 2000).

An interesting international dynamic is emerging whereby firms that have already improved their environmental standards because of their location in a country with strict controls then have an interest in persuading other states to follow that lead. Porter (1990) has provided several examples where a nation has gained important competitive advantage by being among the first to establish strong environmental standards.

We are now seeing the extension of that dynamic at the international level. Braithwaite and Drahos (2000, pp. 272–4) have proposed a model for this "globalization" dynamic:

- the process begins with NGO concern in key state(s) (often without public backup);
- as a result, these states then lead in promoting the case for a global regime;
- a weak framework regime is installed;
- a disaster then occurs, mobilizing concern among a global public;
- as a result, the framework is strengthened, perhaps by the development of a protocol;
- if a second disaster should strike, this will be followed by a second wave of public concern and further strengthening of the regime.

This model is applicable to *public contained risks* and to *traded, product-based risks* but seems less relevant to our other two categories. Indeed Braithwaite and Drahos note that "some environmental problems have failed to produce credible regimes because of the want of a visible disaster," and they include climate change in this category.

However, our conclusion from this review would be that, in parallel with the globalization of finance and communications, international transboundary risk regulation seems to be progressing along a trajectory of increasingly effective action.

6.5. Future Issues for Transboundary Risk Regulation

The number of issues still unresolved in transboundary risk regulation greatly outnumbers the successes. Our analysis suggests that among the most important will be questions of public involvement in risk debates and decision making, dealing equitably with wealth disparities among nations, and developing effective mechanisms to assign liability and responsibility for compensation.

Public Involvement in Risk Debates and Decision Making. It is often claimed that the key to improved risk management, globally and nationally, lies in better mutual understanding between the public or public interest groups, scientists and others working in laboratories and industry, and government policy makers and risk regulators (e.g., Tait, 1993; ESRC Global Environmental Change Programme, 1999; Renn and Klinke, 2000).

However, in complex poorly characterized areas, there may be no consensus on the relevance of particular areas of scientific expertise (Bower et al., 1998). Various scientific disciplines will compete for a voice and an influence on decision making, and policy makers will need guidance on how to discriminate among competing demands, not to exclude inputs from particular disciplines but to weight inputs according to the appropriateness of the expertise and to guard against too ready an assumption about which areas of expertise might be relevant in a particular case. In contentious areas, the agenda is often driven inappropriately by public opinion and pressure groups, but there is little critical appraisal of the methods by which public opinion is gauged. We also need to know as much about the internal and external drivers of decision making by industry and regulators as we do about subtle variations in public perceptions and values.

Paradoxically, it is easier to provide information to stakeholders and to engage in constructive participation with them where the need for it is less. Public demands for information and involvement in decision making are often loudest where there is most conflict and where uncertainty, both political and scientific, is greatest. Such cases are probably most in need of rapid resolution in transboundary risk regulation, but there is no evidence so far that greater public involvement in a situation where strong opinions and conflict are well entrenched will not merely lead to paralysis of decision making on the more urgent and contentious issues. On the other hand, sensitive public involvement in earlier stages of decision making may help to avoid the emergence of such conflicts.

Wealth Disparities Among Nations, Liability, and Cross-Border Compensation. Delays and difficulties in implementing international agreements on transboundary risk regulation are often caused by wealth disparities among nations and legitimate demands by poor nations not to be asked to pay for problems caused by richer nations. Transboundary risk regulatory mechanisms can be seen as attempts by one country or group of countries to control the economy of another.

In the case of the Forest Principles proposed at the Rio Conference, negotiators could not even reach agreement on a framework convention. Most developing countries strongly resisted a binding treaty because they saw this as an attempt by richer countries to control their economies and resources.

The need for compensatory mechanisms of this nature is increasingly being recognized, as for example in some provisions of the Kyoto Protocol, but it is not yet clear whether they can be made to work effectively.

The question of compensation is also relevant to public, contained risks where those liable for the damage could reasonably be asked to pay. In the case of pollution of the River Tisza, Hungary and Serbia, the worst-affected countries, are seeking compensation from the Romanian government or an Australian source (public or private). However, it is not clear what legal recourse they would have. The Romanian government is passing responsibility to the mine's owners, and they have gone into liquidation. The European Commission is also expected to help toward clearing up in the aftermath of the spill, but this may raise issues similar to those noted for the Ignalina plant in Section 5.1.

An important aspect of liability and compensation questions is often the prior need to assign or clarify property rights (and the responsibilities that go along with them) where none existed before. This is an area where improved international mechanisms could have a major impact on the incentives to public and private sector interests to improve their risk-related performance by strengthening transborder property rights and liabilities at the individual and corporate level (Linnerooth-Bayer, 2000) and relying to a large part on the market and courts to self-regulate (Arrow 1996; Held 1995). Transboundary liability regimes are forming, for instance, with respect to oil pollution on the seas (Brans, 1996; Brubaker, 1993) and nuclear power in Western Europe (Faure, 1995). However, for the most part there is still a general lack of state, corporate, or private liability for transboundary risk-producing agents (Bodansky, 1995; Zemanek, 1992).

Notes

1. *Bridges Weekly Trade News Digest* is published by the International Centre for Trade and Sustainable Development (ICTSD), an independent, not-for-profit organization based in Switzerland, with support from the Institute for Agriculture and Trade Policy.
2. The distinction between all GMOs, including products derived from them, and LMOs, was an important point of debate in the negotiations. LMOs are presumed capable of replication and would include seeds (e.g., maize or soya) for planting as agricultural crops or for animal feed. Ground maize or soya including material from GMOs would not be classed as an LMO.
3. The information given here is taken from the following websites: BBC news (www.news2.thls.bbc.co.uk/hi/english/world/europe/newsid/), the Hungarian Ministry of the Environment (www.ktm.hu/CIAN/angol/cyanide. htm), the World Wide Fund for Nature (WWF) (www.panda.org/news/press/ hungary.htm), and Esmeralda Exploration (www.esmeralda.com.au/3_com/ main3c3.html and /main5a.html).
4. The critical load of a soil or region is the amount of pollution that a given place can absorb before showing adverse effects.

References

Andersen, M. K. 1994. *Governance by Green Taxes. Making Pollution Prevention Pay.* Manchester University Press, Manchester.

Arrow, K. J. 1996. The theory of risk bearing: Small and great risks. *Journal of Risk and Uncertainty*, 12: 103–11.

Bodansky, D. 1995. Customary (and not so customary) international environmental law. *Indiana Journal of Global Legal Studies*, 3 (1): 105–19.

Bower, D. J., Mowatt, G., Brebner, J. A., Cairns, J. A., Grant, A. M., and McKee, L. 1998. Evaluating new and fast-changing technologies. In N. Black, J. Brazier, R. Fitzpatrick, and B. Reeves, eds., *Health Services Research Methods: A Guide to Best Practice.* pp. 226–36. BMJ Books, London.

Braithwaite, J., and Drahos, P. 2000. *Global Business Regulation*. Cambridge University Press, Cambridge.

Brans, E. H. P. 1996. Liability and compensation for natural resource damage under the international oil pollution conventions. *Review of European Community & International Environmental Law*, 5: 260–8.

Brubaker, D. 1993. *Marine Pollution and International Law: Principles and Practice*. Belhaven, London.

Carr, S. 2000. *EU Safety Regulation of Genetically Modified Crops: Summary of a Ten-Country Study Funded by DGXII under its Biotechnology Programme*. Biotechnology Policy Group, Centre for Technology Strategy, the Open University, Milton Keynes, UK.

CEC (Commission of the European Communities). 2000. *Communication from the Commission on the Precautionary Principle*, COM (2000) 1, Brussels, European Commission.

Cosbey, A. 1999. *The Kyoto Protocol and the WTO*. www.riia.org/Research/

Cosbey, A., and Burgiel, S. 2000. *The Cartagena Protocol on Biosafety: An Analysis of Results*. International Institute for Sustainable Development Briefing Note, IISD, Winnipeg, Manitoba, Canada.

Dietz, T. M., and Rycroft, R. W. 1987. *The Risk Professionals*. Russell Sage Foundation, Social Research Perspectives, New York.

ESRC Global Environmental Change Programme. 1999. *The Politics of GM Food: Risk, Science and Public Trust*, Special Briefing No. 5. University of Sussex, Brighton, UK.

Faure, M. G. 1995. Economic models of compensation for damage caused by nuclear accidents: Some lessons for the revision of the Paris and Vienna Conventions. *European Journal of Law and Economics*, 2: 21–43.

Finkel, A. M., and Golding, D. 1994. *Worst Things First? the Debate over Risk-Based National Environmental Priorities*. Resources for the Future, Washington, DC.

Giddens, A. 1999. Runaway world, Lecture 2 – Risk. *BBC Reith Lectures 1999, Hong Kong*. BBC Online Network, 07/07/99. http://news.bbc.co.uk/hi/english/static/events/reith_99/week2/week2.htm

Graham, J. D., and Wiener, J. B. 1997. *Risk vs. Risk: Tradeoffs in Protecting Health and the Environment*. Harvard University Press, Cambridge, MA.

Grubb, M., Vrolijk, C., and Brack, D. 1999. *The Kyoto Protocol: A Guide and Assessment*. The Royal Institute of International Affairs, Earthscan Publications Ltd., London.

Held, D. 1995. Democracy and the new international order. In D. Archiburgi and D. Held, eds., *Cosmopolitan Democracy: An Agenda for a New World Order*. Polity Press, Cambridge.

Hough, P. 1998. *The Global Politics of Pesticides. Forging Consensus from Conflicting Interests*. Earthscan, London.

Houghton, J. T., Meira Filho, L. G., Callander, B. A., Harris, N., Kattenberg, A., and Maskell, K., eds. 1996. *Climate Change 1955: The Science of Climate Change*. Cambridge University Press, Cambridge.

House of Lords Select Committee on Science and Technology. 1993. *Regulation of the United Kingdom Biotechnology Industry and Global Competitiveness*, 7th Report, Session 1992/93, pp. 187–96. HMSO HL Paper 80-I, London.

Hyvarinen, J. 2000. *The EU in the International Climate Negotiations – Lost and defeated?* Institute for European Environmental Policy, Jan. 2000, p. 1, www.northsea.nl/jiq/cop6jhart.htm.

IEA/OECD. 1997. *CO_2 Emissions from Fuel Combustion.* IEA/OECD, Paris.

IPCC (Intergovernmental Panel on Climate Change). 1995. *Climate Change, 1995 – The Science of Climate Change: Summary for Policy Makers and Technical Summary of the Working Group I Report.* The Met. Office Graphics Studio, 95/869, Bradenell, Berkshire, UK.

Jasanoff, S. 1990. *The Fifth Branch: Science Advisers as Policy Makers.* Harvard University Press, Cambridge, MA.

Kanowski, P., Sinclair, D., and Freeman, B. 2000. Issues in certification. *Tropical Forest Update,* ITTO Newsletter, 10 (1). www.itto.or.jp/newsletter/v10n1/0.html.

Kasperson, J. X., and Kasperson, R. 2000. Border crossings. In J. Linerooth-Bayer, R. Loefstedt, and G. Sjoestedt, eds., *Transboundary Risk Management in Europe.* Earthscan, London.

Kempel, W. 1993. Transboundary Movements of Hazardous Wastes. In Gunnar Sjostedt, ed., *International Environmental Negotiation,* pp. 48–62. Sage Publications, London.

Kete, N. 1999. Review of "The Kyoto Protocol – A Guide and Assessment" by Grubb et al., 1999. *Nature,* 402: 233–4.

Kimmins, H. 1997. *Balancing Act – Environmental Issues in Forestry.* UBC Press, Vancouver.

Krueger, J. 1999. What's to become of trade in hazardous wastes? *Environment,* 41 (9): 10–21.

Kummer, K. 1994. *Transboundary Movements of Hazardous Wastes at the Interface of Environment and Trade,* UNEP Environment and Trade Series No 7. Quoted in Krueger (1999).

Kyriakou, D. 2000. The Seattle World Trade Organisation (WTO) impasse. *IPTS Report No 43.* European Commission Joint Research Centre, Seville.

Leonard, H. J. 1988. *Pollution and the Struggle for the World Product: Multinational Corporations, Environment and International Competitive Advantage.* Cambridge University Press, New York.

Linnerooth-Bayer, J. 1999. Climate change and multiple views of fairness. In F. L. Toth, ed., *Fair Weather? Equity Concerns in Climate Change.* pp. 44–65. Earthscan, London.

Linnerooth-Bayer, J. 2000. Introduction. In J. Linnerooth-Bayer, R. Loefstedt, and G. Sjoestedt, eds., *Transboundary Risk Management in Europe.* Earthscan, London.

Linnerooth-Bayer, J., and Loefstedt, R. 1996. Transboundary environmental risk management: An overview. Paper presented at the Conference on Transboundary Environmental Risk Management (6–8 October 1996), Warsaw.

Linnerooth-Bayer, J., Loefstedt, R., and Sjoestedt, G. 2000. Transboundary risk management in Europe. In M. P. Cottam, D. W. Harvey, R. P. Pape, and J. Tait, eds., *Foresight and Precaution. Proceedings of ESREL 2000, SARS and SRA-Europe Annual Conference* (May 2000), Balkema, Rotterdam, pp. 165–70.

Loefstedt, R. E. 1994. Environmental aid to Eastern Europe: Swedish and Estonian perspectives. *Post Soviet Geography,* 35: 594–607.

Loefstedt, R., and Jankauskas, J. 2000. Swedish Aid and the Ignalina Nuclear Power Plant. In J. Linnerooth-Bayer, R. Loefstedt, and G. Sjoestedt, eds., *Transboundary Risk Management*. Earthscan, London.

Makuch, Z. 1996. The World Trade Organisation and the General Agreement on Tariffs and Trade. In J. Werksman, ed., *Greening International Institutions*. pp. 94–115. Earthscan, London.

Medvedev, Z. A. 1993. The legacy of Chernobyl: The prospects for energy in the former Soviet Union. In B. Cartledge, ed., *Energy and the Environment. The Linacre Lectures 1991–92*, pp. 127–45. Oxford University Press, Oxford.

Medvedev, Z. A. 1990. *The Legacy of Chernobyl*. Blackwell, Oxford.

National Consumer Council. 1998. *Farm Policies and our Food: The Need for Change*. London, National Consumer Council, March 1998, PD11/B2/98.

Nordstrom, H., and Vaughan, S. 1999. *Trade and Environment*, WTO Special Studies 4. World Trade Organization, Geneva, Switzerland.

Oberthür, S., and Ott, H. E. 1999. *The Kyoto Protocol. International Climate Policy for the 21st Century*. Springer, Berlin.

OECD. 2000. *GM Food Safety, uncertainties and assessment*. C(2000)86/ADD3. Organization for Economic Co-operation and Development, Paris.

O'Riordan, T., ed. 1994. *Ecotaxation*. Earthscan, London.

Ott, H. E. 1998. The Kyoto Protocol: Unfinished business. *Environment*, (July/August): 17–45.

Ovink, B. J. 1995. Transboundary shipments of toxic waste: The Basel and Bakamo Conventions: Do Third World countries have a choice? *Dicken's Journal of International Law*, 13: 281–95.

Paterson, W. 1999. *Turning Fine Words into Deeds*. www.riia.org/Research/eep/eeparticle.htm.

Patt, A. 2000. Transboundary air pollution: Lessons for useful analysis. In J. Linnerooth-Bayer, R. Loefstedt, and G. Sjoestedt, eds., *Transboundary Risk Management*. Earthscan, London.

Patten, C. 2000, *BBC Reith Lectures 2000*. news.bbc.co.uk/hi/english/static/events/reith_2000/lecture1.stm.

Pearson, M., and Smith, S. 1991. *The European Carbon Tax: An Assessment of the European Commission's Proposals*. The Institute of Fiscal Studies, London.

Porter, G., and Welsh Brown, J. 1991. *Global Environmental Politics*. Westview Press, Boulder, CO.

Porter, M. 1990. *The Competitive Advantage of Nations*. Macmillan, New York.

Rees, J. 1994. *Hostages of Each Other: The Transformation of Nuclear Safety Since Three Mile Island*. University of Chicago Press, Chicago.

Renn, O., and Klinke, A. 2000. Public participation across borders. In J. Linnerooth-Bayer, R. Loefstedt, and G. Sjoestedt, eds., *Transboundary Risk Management*. Earthscan, London.

Rosa, E. A. 1998. Metatheoretical foundations for post-normal risk. *Journal of Risk Research*, 1 (1): 15–44.

Sand, P. H. 1999. *Transnational Environmental Law: Lessons in Global Change*. Kluwer Law International, The Hague.

Sands, P., ed. 1988. *Chernobyl: Law and Communication*. Grotius Publications, Cambridge.

Selin, H., and Hjelm, O. 1999. The role of environmental science and politics in identifying persistent organic pollutants for international regulatory action. *Environmental Reviews*, 7: 61–8.

Shaw, R. W. 1993. Acid-rain negotiations in North America and Europe: A study in contrast. In G. Sjostedt ed., *International Environmental Negotiation*, pp. 84–109. Sage, London.

Sjostedt, G. 1993. Negotiations on nuclear pollution: The Vienna Conventions on notification and assistance in case of nuclear accident. In Gunnar Sjostedt, ed., *International Environmental Negotiation*, pp. 31–47. Sage, London.

Skolnikoff, E. B. 1999. The role of science in policy. The climate change debate in the United States. *Environment*, 41 (5): 18–45.

Szell, P. 1993. Negotiations on the ozone layer. In G. Sjostedt, ed., *International Environmental Negotiation*, pp. 31–47. Sage, London.

Tait, J. 1993. Written evidence on behalf of ESRC to Report of House of Lords Select Committee on Science and Technology on *Regulation of the United Kingdom Biotechnology Industry and Global Competitiveness*, 7th Report, Session 1992/93. pp. 187–96. HMSO HL Paper 80-I, London.

Tait, J. 2001. More Faust than Frankenstein: The European debate about risk regulation for genetically modified food. *Journal of Risk Research*, 4(2): 175–89.

Tait, J., and Morris, D. 2000. Sustainable development of agricultural systems: Competing objectives and critical limits. *Futures*, 32: 247–60.

Tait, J., and Levidow, L. 1992. Proactive and reactive approaches to risk regulation: The case of biotechnology. *Futures* (April): 219–31.

Tolba, M. K., and Rummel-Bulska, I. 1998. *Global Environmental Diplomacy. Negotiating Environmental Agreements for the World, 1973–1992*. The MIT Press, Cambridge, MA.

Tomlinson, N. 2000. *The Concept of Substantial Equivalence, its Historical Development and Current Use*. Joint FAO/WHO Expert Consultation on Foods Derived from Biotechnology. (29 May–2 June 2000), Geneva.

Victor, D. G., and Skolnikoff, E. B. 1999. Translating intent into action: Implementing environmental commitments. *Environment*, 42: 16–20, 39–44.

Vogel, D. 1995. *Trading Up: Consumer and Environmental Regulation in a Global Economy*. Harvard University Press, Cambridge, MA.

von Weizsäcker, E. U. 1994. *Earth Politics*. Zed Books, London.

Watson, R. T., Dixon, J. A., Hamburg, S. P., and Moss, R. H. 1998. *Protecting Our Planet Securing our Future*. World Bank Paper. http://www-esd.worldbank.org/planet/.

World Commission on Environment and Development. 1987. *Our Common Future*. Oxford University Press, Oxford.

Zemanek, K. 1992. State responsibility and liability. In W. Lang, H. Neuhold, and K. Zemanek, eds., *Environmental Protection and International Law*, pp. 187–201. Graham & Trotman, Boston.

Further Reading

The following publications, selected from the above list, give a good overview of the most important issues discussed in this paper, including a historical perspective.

Braithwaite, J., and Drahos, P. 2000. *Global Business Regulation*. Cambridge University Press, Cambridge.

CEC. 2000. *Communication from the Commission on the Precautionary Principle*, COM (2000) 1, Brussels, European Commission.

Graham, J. D., and Wiener, J. B. 1997. *Risk vs. Risk: Tradeoffs in Protecting Health and the Environment*. Harvard University Press, Cambridge, MA.

Grubb, M., Vrolijk, C., and Brack, D. 1999. *The Kyoto Protocol: A Guide and Assessment*, The Royal Institute of International Affairs. Earthscan, London.

Hough, P. 1998. *The Global Politics of Pesticides. Forging Consensus from Conflicting Interests*. Earthscan, London.

Linnerooth-Bayer, J., Loefstedt, R., and Sjoestedt, G. 2000. *Transboundary Risk Management in Europe*. Earthscan, London.

Sand, P. H. 1999. *Transnational Environmental Law: Lessons in Global Change*. Kluwer Law International, The Hague.

Tait, J. 2001. More Faust than Frankenstein: The European debate about risk regulation for genetically modified food. *Journal of Risk Research*, 4(2): 175–89.

Vogel, D. 1995. *Trading Up: Consumer and Environmental Regulation in a Global Economy*. Harvard University Press, Cambridge, MA.

World Commission on Environment and Development, 1987. *Our Common Future*. Oxford University Press, Oxford.

11

Environmental Risks and Developing Countries

An Asian Perspective

Michinori Kabuto, Saburo Ikeda, and Iwao Uchiyama

1. INTRODUCTION

Many developing countries in Asia face significant environmental problems consisting of both traditional and modern risks, and overlaps between them, resulting from a delay in their "risk transition" (Edgerton et al., 1990, Smith, 1997). To address such problems, more efficient and integrated environmental management than that historically adopted in the developed countries must be encouraged. Namely, the nations must be able to address and reduce the traditional risks through development activities without producing or worsening the modern risks of environmental pollution and also the damage to natural environments caused, for example, by deforestation and desertification. Moreover, it is likely that further

The authors heartily thank all the study subjects and also the following persons for kind cooperation in conducting the present surveys in Indonesia and China: Drs. S. Sumantri (Ministry of Health), A. T. Tugaswati (Lampung University), M. I. Z. Duki (Gunma University), S. Sudarmadi (Gunma University), N. Herawati (Gunma University), Li Wei (Beijing Normal University), Yang Zhi Min (Chengdu Institute for Environmental Protection Sciences), Chen Yu-de (Ministry of Heath & Beijing Medical College), Jin Yinlong (Institute for Environmental Health and Engineering, Chinese Academy of Preventive Medicine), and many others. Useful suggestions on the text were provided by Madhav Badami (University of British Columbia).

This series of surveys has been conducted by M. Kabuto, principal investigator, and Y. Honda (Tsukuba University) for China, Dr. S. Suzuki (Gunma University Department of Public Health) for Indonesia, and R. Ohtsuka (University of Tokyo, Department of Human Ecology) for South Asia, both of whom were chief researchers for our HDP study (1996–9) in Indonesia and South Asia, respectively, although the latter study was not described here.

This chapter was prepared based on the main part of our HDP survey for urban areas in Indonesia and China. Global environmental issues such as deforestation and desertification (water shortage and, for example, resulting arsenic poisoning) may be more serious generally in South Asia than in Indonesia or China. The results of human ecological studies on rural areas in India, Nepal, and Bangladesh (the other part or our HDP study) will be reported separately.

delay in environmental risk management would increase various types of so-called transboundary risks resulting from international movement of contaminated foods or wastes, long-range atmospheric transport of pollutants, GHG (greenhouse gas) emissions, and degradation of the global ecosystem in general (McMichael, 1999). These concerns indicate that more comprehensive risk controls from both domestic and international perspectives are urgently needed.

This chapter presents an overview of the extent to which, and how, environmental health risks have been assessed for developing Asian countries, especially China and Indonesia. Then, data collected through our Human Dimension Programme (HDP) study are reviewed indicating the level of risk knowledge, awareness, and perception of urban residents in these two countries. Finally, an international perspective is provided to help identify desirable risk research projects and approaches to sound risk management and governance.

2. MAJOR ENVIRONMENTAL HEALTH RISKS IN CHINA AND INDONESIA

When the current information on environmental health risks reported by the World Bank, the World Health Organization (WHO), the World Resources Institute (WRI), and so on are reviewed, they have focused mostly on traditional risks, especially for infant and premature adult mortality caused by poor nutrition, indoor air pollution related to biomass fuel, and natural hazards among other causes.[1] These risks are inversely correlated with the so-called human development index, incorporating life expectancy, education levels, and gross domestic product (GDP) (UNDP, 1999). Generally, the modern risks associated with industrialization, motorization, or urbanization have not been evaluated well, with the exception of air pollution, probably owing to the scarcity of monitoring and health data for estimating them. This is also the case for overlapping risks. Various risk overlap phenomena have been identified, including "risk genesis," "risk synergism," "risk mimicry," "risk competition," "risk layering," or "risk transfer" (Smith, 1997).

In this section, therefore, available information is reviewed separately for the traditional and modern risks, especially for China and Indonesia.

2.1. The Traditional Health Risks

Infectious/endemic diseases, nutritional deficiency, indoor air pollution, and an inability to deal with natural hazards still prevail, especially in the poor areas of developing countries. A new index, DALYs (disability adjusted life years) has been developed by weighting the disability effects on daily activities of each of the major diseases (Murray and Lopetz,

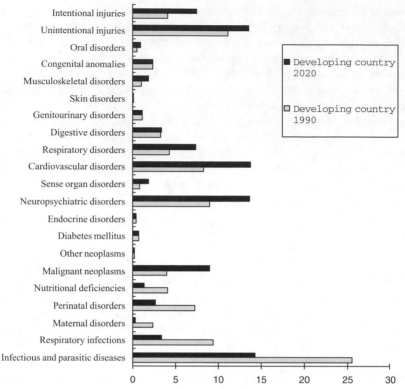

FIGURE 11.1. Percentage of DALYs reduced by each of the disease burdens in developing countries in 1990 and 2020. (*Source:* Murray and Lopetz, 1997.)

1997). While the data are not disaggregated by sex, age-group, or region (because of the general scarcity of both the systematic monitoring data on environment and health for developing countries), a general profile of the traditional risks in the developing countries as a whole is provided by the estimated percent DALYs by each of the disease burdens, as shown in Figure 11.1.

Figure 11.1 indicates that the traditional risks constitute a very high proportion of the DALY's lost in developing countries in 1990, but they are expected to contribute less in 2020 (Murray and Lopetz, 1997). It is expected according to these estimates that infectious/enparasitic diseases, maternal disorders, and nutritional deficiencies will be halved during the next years, whereas the relative contributions of respiratory, cardiovascular, and neuropsychiatric disorders as well as injuries would increase almost twofold. Further, there will likely be large differences in the future trends of DALYs between countries and also between regions within the same developing country. In the case of China, traditional risks have tended to be more serious in the socioeconomically deprived western part than in the eastern part, and also in rural areas compared to urban areas (Chen, 1997).

In Xuanwei county, Yunnan province, China, a study of the effects of indoor air pollution on lung cancer found the highest cancer mortality rates in China, and even in the world. The annual age-adjusted lung cancer rate for the period 1973–9 was 26.49/100,000, comprising 49.5 percent of total malignant neoplasms (He, 1990). Xuanwei residents, especially females, are exposed to indoor particulate concentrations exceeding 100 times the U.S. ambient 24-hour standard. The particulates from smoky coal burning are mainly smaller than 10 μm in diameter and remain longer in the air; they also are easily inhalable and therefore more effectively deposited in the lung (WHO, 1994; WRI, 1998). In rural areas of Yunnan province, indoor particulate concentrations from coal burning were found to be as much as 270–5,100 micrograms per cubic meter. These concentrations were the highest among twelve selected places throughout China (WHO, 1997).

Another by-product of indoor burning is carcinogenic polyaromatic hydrocarbons (PAH), such as benzo(a)pyrene or B(a)P, which are a major risk factor contributing to the high lung cancer prevalence in Xuanwei (Table 11.1). The B(a)P indoor concentrations during cooking are comparable to occupational exposure levels, such as those in coke oven plants (WHO, 1994; Lan, He, and Jin, 1990). Air pollution, both indoor and outdoor, was identified to be a major risk factor for respiratory diseases, the leading causes of death in China. It was estimated that 289,000 premature deaths in China as a whole could be prevented each year if air pollution concentrations were reduced to standard levels (World Bank, 1997).

Several studies provide evidence of the complex relationship between air pollution and human health in China. In a study in Beijing, it was found that the risk of mortality from COPD (chronic obstructive pulmonary disease) increased 38 percent with the doubling of particulate emissions (Xu et al., 1994). The study of the relationship between air pollution and chronic bronchitis and asthma in Shenyang showed that the prevalence

TABLE 11.1. *Age-Adjusted Mortality of Lung Cancer in Xuanwei, Yunnan Province, China*

Xuanwei Incidence Areas	Mortality (per 100,000)	National Lung Cancer Mortality (per 100,000)
Low	5.98	
Medium	20.90	Male, 6.28
High	126.06	Female, 3.20

Source: He (1990).

TABLE 11.2. *Prevalence of Respiratory Diseases of Residents Owing to Air Pollution, Smoking, and Coal-Burning in Shenyang, 1993 (standard error in parenthesis)*

Air Pollution	Smoking	Coal Burning	Disease	Prevalence (%)			P^a
				Control	Medium Polluted Area	Heavily Polluted Area	
+	−	−	Chronic bronchitis	2.3 (1.00)	7.0 (3.04)	24.4 (10.61)	<.01
+	+	+	Chronic bronchitis	6.6 (2.87)	19.4 (8.43)	50.0 (21.74)	<.01
+	−	−	Asthma	1.9 (1.00)	2.8 (1.47)	13.6 (7.16)	<.01
+	+	+	Asthma	5.2 (2.74)	8.7 (4.58)	24.6 (12.95)	<.01

[a] P for statistical significance with Chi-square test.

Source: Yin (1993).

TABLE 11.3. *Survey of Respiratory Disease and Symptoms Among Adults in Three Sections of Taiyuan City, Shanxi Province, 1998–2000*

	Comparison of Prevalence Rate (%) of Respiratory Tract Symptoms and Diseases in Three Different Areas			
Symptom	Shanglan (Low Pollution)	Yingze (Medium Pollution)	Yidian (Heavy Pollution)	χ^2
Cough	7.33	13.33	15.75	99.879
Sputum	9.19	16.22	19.66	116.943
Wheezing	2.26	2.22	3.18	10.914
Dyspnea	20.15	33.66	35.75	169.197
Bronchitis	5.42	12.91	15.33	96.492
Pneumonia	2.92	7.08	8.31	49.28
Chronic bronchitis	2.98	6.39	8.28	50.581
Emphysema	0.54	0.41	0.6	1.394
Asthma	1.29	1.90	2.75	13.663
COPD	3.73	7.27	9.39	51.352

Source: Jin et al. (2000).

rates of chronic bronchitis and asthma were 25 and 14 percent, respectively, in heavily polluted regions (Yin, 1993.) (See Table 11.2.) Another compelling piece of evidence is the high incidence of respiratory disease in Taiyuan city, Shanxi province, which currently has the highest level of air pollution in China. As shown in Table 11.3, a comparison of prevalence rates of respiratory diseases and respiratory tract symptoms identified bronchitis and COPD as being predominant (Jin, 2000).

Other important traditional risks prevalent in large areas in China are fluorosis (dental and skeletal) and arsenism (arsenic poisoning), both associated largely with drinking-water pollution. A portion of the fluorosis cases have also been associated with indoor air pollution caused by the burning of coal containing fluoride for cooking and heating.

It is known that drinking water with a high level of fluoride can cause endemic dental and skeletal fluorosis. In China, fluoride levels as high as five times the WHO guideline (1 mg/L) have been found in the northeast and central parts of the country, except in Shanghai, Hainan, Guizhou, and Ningxia. In 1999, the Chinese Health Statistics Yearbook reported a national total of 24,074,335 cases of dental fluorosis and 1,143,620 cases of skeletal fluorosis from drinking water sources. An earlier nationwide survey carried out by the Institute of Environmental Health and Engineering of the Chinese Academy of Preventive Medicine in the early 1980s found that endemic fluorosis was prevalent in 238 cities, 1,095 districts/counties, and 125,817 villages with a total affected population of 85,610,000. Among 41,350,000 people who were investigated, 18,120,000 (44%) suffered dental fluorosis, and 830,000 (2%) suffered skeletal fluorosis. In Jiangsu in 1984, a correlation study was carried out between the fluoride concentration in drinking water in some regions and the prevalence of dental and skeletal fluorosis. In Table 11.4, significant correlations with fluoride concentration in drinking water were found for dental fluorosis (correlation coefficient = 0.974) and for skeletal fluorosis (correlation coefficient = 0.829) (Chinese Academy of Preventive Medicine, 1988).

Endemic arsenism is a systemic disease caused by drinking water with high arsenic concentration and manifested as skin hyperpigmentation, hypopigmentation, and hyperkeratosis. The Institute of Xinjiang Endemic Disease Preventive Research studied the relationship between arsenic concentration in drinking water (more than 0.1 mg/L) and the prevalence of arsenism in fifteen regions. They found a significant positive correlation between the arsenic concentration and disease prevalence, as shown in Figure 11.2. Further, the mortality rate for malignant tumors was significantly correlated with the arsenic concentration in drinking water (Wang, Sun, and Feng, 1996). In Hangmianhouqi region, Inner Mongolia, the prevalence rate for arsenism is 49.3 percent (where the arsenic concentration was above 0.65 mg/L or 65 times the WHO guideline) (Niu, 1996).

Endemic arsenism is clearly a serious problem mainly in Xinjiang, Inner Mongolia, Shanxi, Guizhou, and Taiwan.[2] In Xinjiang alone, some 2,000 cases have been found, and 100,000 people are at risk of exposure to high levels of arsenic in drinking water. In Inner Mongolia and Shanxi, there is a high-arsenic region that is more than 1,000 kilometers long and about 10 to 40 kilometers wide. This region includes 5 cities, 11 counties, and 628 villages (farms) in Inner Mongolia with reported cases of 2,455 patients and 2,620,000 people at risk (Wang, 1997). In 1988, data from the "National

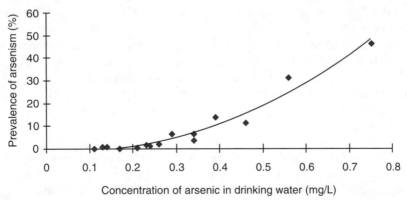

FIGURE 11.2. The exposure-response relationship between arsenic and arsenism in Xinjiang, 1992. (*Source:* Wang et al., 1996.)

TABLE 11.4. *Relationship Between the Fluoride Concentration in Drinking Water in Jiangsu and the Prevalence of Fluorosis, 1984*

Fluoride Concentration (mg/L)	Population Investigated	Dental Fluorosis		Skeletal Fluorosis		Total	
		Cases	Prevalence (%)	Cases	Prevalence (%)	Cases	Prevalence (%)
<0.5	140,00_	4,929	3.5	35	0.02	4,964	3.6
0.5–1.0	35,33_	4,066	11.5	54	0.15	4,120	11.7
1.0–2.0	94,05_	43,658	46.4	225	0.24	43,883	46.7
2.0–4.0	43,97_	26,069	50.3	538	1.22	26,607	60.5
4.0–5.0	6,02_	4,456	73.9	151	2.50	4,607	76.4
>5.0	10,00_	8,379	83.8	46	0.46	8,425	84.2
TOTAL	329,40_	91,557	27.8	1,049	0.32	92,606	28.1

Source: Ministry of Health (1984a).

Survey on Drinking Water Quality and Waterborne Diseases" conducted by the Chinese Academy of Preventive Medicine indicated that around 14.7 million people in China drank water with arsenic concentration of more than 0.03 mg/L (the WHO guideline value is 0.01 mg/L). The main source of arsenic was groundwater wells (Zhang and Chen, 1997). Most of the malignant neoplasms induced by arsenic are tumors affecting the epithelial tissues (skin, respiratory, digestive, and urinary systems) which account for 80 percent of total cases. Among all cases, skin cancer is the most prevalent (Ministry of Health, 1988).

Especially in rural areas in China, severe biological pollution of drinking water sources has been found to cause both liver cancer and intestinal infectious diseases. The incidence of intestinal infectious diseases, such as hepatitis A, typhoid fever, bacillary dysentery, diarrhea, and cryptosporidiosis in China has been observed to be 10–100 times higher than in developed countries (Zeng et al., 1997). Domestic sewage and excreta as well as agricultural runoff find their way into surface waters, such as rivers, lakes, and ponds, which are the sources of drinking water for 25 percent of the population. Eutrophication of these bodies of surface water supports the growth of cyanobacteria especially from July to September. Cyanobacteria can produce toxins such as microcystins, which are potent liver cancer promoters and are directly hepatotoxic to poultry, livestock, animals, and humans. Microcystins in drinking water cannot be completely removed by common disinfection nor heating (Wang et al., 1995; Ling, 1999).

A compelling piece of evidence showing the correlation between drinking water contamination and health is the liver cancer prevalence in Qidong county, Jiangsu province. Qidong county has the largest number of cases of liver cancer morbidity with about 50/100,000 for many years. Since 1972, several investigations of the relationship between liver cancer morbidity among Qidong residents and different sources of drinking water have been performed. The results consistently show that pond/lake waters are the most polluted and cause the highest incidence of liver cancer morbidity, as shown in Table 11.5 (Gu, 1988; Su, 1980).

There are also data showing the prevalence of parasitic and vector-borne diseases, although they are not mentioned here. It is suggested that all these available data should be combined on a regional basis to estimate the total extent of traditional risks. Such a database may be used, for example, to predict the health impacts of global climate change, if reliable information on their future trends becomes available. In relation to global climate change, information regarding the other aspects of environmental risks, or ecological risks and QOL (Quality of Life) risks should also be combined and considered simultaneously. These issues have been considered in our subsequent HDP study (1997–99) and will be addressed in the near future.

TABLE 11.5. *Liver Cancer Morbidity Rates Versus Different Sources of Drinking Water in Qidong County, Jiangsu Province*

Source of Drinking Water	Morbidity (per 100,000)	
	Gu (1988) 1971–81	Su (1980) 1971–72
Pond/lake	141.40	101.35
Irrigation	72.32	64.57
River	43.45	42.64
Shallow well	22.26	0
Deep well	11.70	0

2.2. The Modern Health Risks

Modern risks arise from more recent industrial activity and emissions. These have been investigated mostly for air pollution from industries and automobiles. Generally, there are no sufficient systematic and reliable data on exposure to chemical pollution in water and soil, or to industrial and domestic waste, although the public is becoming more sensitive than ever to these environmental issues. Available information for a set of these risks is considered.

Since the late 1980s, several studies of urban air pollution and its health risks have been conducted in the developing countries, such as those by WHO/UNEP(1992) and the World Bank (1992). These studies suggest that air pollution concentrations could increase (by several fold or over tenfold) over the next several decades as a result of motorization, increases in energy consumption, and other activities. According to these estimates, it is also expected that around 300,000 to 700,000 infant deaths in developing countries could be prevented, particularly in India and China, if ambient SPM (suspended particulate matter) concentrations are reduced to the level of the WHO guideline. It has been estimated that serious indoor pollution owing to biomass fuel use causes four million infant deaths annually. However, the scientific evidence indicating such specific cause-effect relationships between air pollutions and infant mortality is very limited. SPM, especially $PM_{2.5}$ (particulate matter with a diameter less than 2.5 μm) is also related to the mortality in aged people, especially as a result of cardiopulmonary disease.

Based on dose-response relationships derived from the preceding epidemiologic studies, World Bank (1997) estimated that 178,000 infant deaths, 66,000 child asthma cases, various respiratory symptom cases of more than the number of the total population, more than 346,000 hospital admissions and 6.8 million emergency cases could be avoided in China as a whole if SPM concentrations were reduced to 60 μg/m^3 or below. It should be noted that these risk reduction estimates are based on average air pollution levels so they should be interpreted carefully. Moreover, domestic fuels have been changing rapidly from coal to natural gas, and drinking water from untreated to treated in most houses, especially in major cities. On the other hand, environmental pollution as well as waste from industries and motor vehicles appears to be increasing, causing many complaints among the public especially around urban areas.

It should also be noted that in these health risk estimates for air pollution in the major cities, there are large uncertainties. According to the official data, daily average ambient SO_2 levels, for example, have ranged from 300 to 400 $\mu g/m^3$ in Chongqing for the last fifteen years. However, our spot measurements of SO_2 at several sites in Chongqing did not show such high levels, although the rank order of air pollution severity among Chongqing, Chengdu, Shanghai, and Dalian were the same. Thus, it is suggested that precise exposure estimates based on actual measurements are needed for epidemiological studies and health risk assessments.

In China, chronic obstructive pulmonary diseases induced by air pollution was found to be a major cause of illness. Many studies have confirmed that the prevalence of chronic respiratory problems, such as chronic bronchitis, increases in the heavily polluted regions. A survey carried out on the population above 40 years old in Shenyang showed that the morbidity resulting from COPD, especially chronic bronchitis and asthma, markedly increased with the degree of air pollution (Figure 11.3) (Jin, 2000). In Shanghai, it was reported that given a constant concentration of SO_2 in ambient air, any increase of TSP by 0.1 mg/m^3 is associated with a corresponding increase in the likelihood of coughing (1.20), vomiting (1.23), stifling (1.13), chronic bronchitis (1.29) and pulmonary emphysema (1.59) (Hong, 1991).

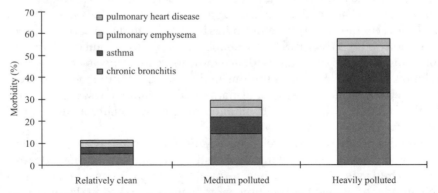

FIGURE 11.3. Morbidity owing to various respiratory conditions in population above 40 years old in Shenyang, 1993. (*Source:* Jin, 2000.)

Lead is a known toxicant causing neurological, reproductive, and carcinogenic effects. According to a three-year national survey of twenty-eight selected cities in China, blood lead levels of more than half of Chinese urban children exceeded the WHO standard of 100 $\mu g/L$. Detailed findings are shown in Table 11.6. WHO has suggested that learning impairment as well as behavioral abnormality can happen following postnatal exposures that result in blood lead levels of 109 to 330 $\mu g/L$.

Town and village enterprises (TVEs) and other private sector units are playing a larger role in the economic development of China (Zhao, 1999). TVEs are enterprises established outside of urban areas with the majority of investment (>50 percent) from rural collective organizations or farmers. There are around seven million TVEs, which account for more than 60 percent of the total industrial GDP. According to the China Environment Yearbook, TVEs discharge about 50 percent of the industrial wastewater and air pollution. Most TVEs have no wastewater or hazardous waste treatment facilities (Wu et al., 1999; WRI, 1999). Although there is little information on the health effects due to TVEs, it is likely that they have become a major risk factor especially in rural areas.

In Jakarta, Indonesia, the World Bank (1994) has conducted a preliminary assessment of health risks due to environmental pollution from urban, industrial, and other sources including human wastes (fecal coliform). The estimated risks in terms of number of mortalities and morbidities, hospital admissions, restricted days and activities owing to air pollution (suspended particulate matter (SPM), lead, nitrogen oxide) were converted to monetary costs and are shown in Table 11.7. The central estimate of the monetary costs of health damage caused by air pollution was more than U.S. $200 million per year. Similarly, the health damages caused by unsafe water was expected to be more than U.S. $300 million per year. The World Bank also predicted that air pollution in terms of SPM would continue to increase rapidly in Jakarta and other cities in Indonesia, with corresponding increases in health damage costs.

Regarding air pollution in Jakarta, there are also some reports on the personal exposure to lead and nitrogen dioxide (NO_2). Lead concentrations and related substances in the blood and urine as well as respiratory symptoms for the last several years have been examined in Tri-Tugaswati and Kiryu (1996). Air lead concentrations along major roads ranged from 1 to 3 $\mu g/m^3$ or 10- to 30-fold higher than those in Japan. A significant temporal increase in urinary lead was observed in people living in heavy traffic areas. Concentrations of NO_2 ranged from 20 to 30 ppb in the highest cases in central Jakarta. Although these values do not exceed existing standards, there was found to be a significant correlation between those values and the prevalence of respiratory symptoms examined using the ATS-DLD questionnaire. According to them, indoor air pollution was closely correlated

TABLE 11.6. *Survey of Lead Concentrations in Blood Among Children in Some Cities of China 1997–2000*

Year	Monitoring Sites	Number of Samples, n	Age (year)	Lead Concentration in Blood (µg/L) mean (SD)	% exceeding 100 µg/L
1997	Wuxi	1122	1–5	86.9 (31.3)	27.5
1997	Jiangsu	1101	2–6	85.9 (44.9)	31.5
1997	Shanghai	1967	3 mo–6	96	37.8
1997	Henan (Kaifeng)	1013	2–6	94 (51)	52.8
1998	Beijing	246	1–6	–	68.7
1998	Lanzhou	808	3–7	101.6 (48.8)	39.1
1999	Yunnan:	1004	3–12		50.4
	Kunming			125.2 (48)	
	Jianchuan			81.5 (32)	
	Lijiang			125.8 (54.7)	
1999	Taiyuan	845	3–6	100.8 (28.9)	44.3
1999	Henan	1085	0–9	135.8 (10.0)	57.94 (rural)
					56.38 (rural)
2000	Beijing	932	6 mo–12	158 (74)	77.5

Source: Wu (2000).

TABLE 11.7. *Costs of Health Damage from Air Pollution in Jakarta, 1990*

Pollutant	Quantity (units/yr)	Unit Value ($US)	Total Value Estimates (U.S.$ million/yr) Low	Central	High
Suspended particulate matter					
Avoidable mortality (cases)	1,500	75,000	15	112.5	262.5
Restricted activity days	6,200,000	3.1	19.2	19.2	19.2
Outpatient visits	41,000	16	0.3	0.7	1.1
Hospital admissions	18,000	260	1.6	4.7	24.7
Respiratory illness (children)	104,000	5.6	0.6	0.6	0.6
Asthma attacks	303,000	2.5	0.8	0.8	0.8
Respiratory symptoms (days)	46,000,00	0.4	18.4	18.4	18.4
SUBTOTAL			55.9	156.9	327.3
Lead					
Avoidable mortality (cases)	340	75,000	3.4	25.5	59.5
Hypertension (cases)	62,000	5.5	0.3	0.3	0.3
Myocardial infarction (cases)	350	1,230	0.4	0.4	0.4
IQ loss in children (points)	300,000	115	36.1	36.1	36.1
SUBTOTAL			40.2	62.3	96.3
Nitrogen dioxide					
Respiratory symptoms (days)	1,800,000	0.4	0.7	0.7	0.7
TOTAL			96.8	219.9	424.3

Source: World Bank (1994).

with outdoor air pollution levels especially in suburban and rural areas, and in the subjects' houses fuels were mostly kerosene.

As mentioned earlier, even when only the modern health risks out of the total environmental risks are overviewed briefly, it is obvious that significant risks are already associated with environmental pollution, and these would be worsened in various regions under the business-as-usual scenarios. In the case of modern risks at the regional level, it is apparent that those risks could be reduced through regulatory actions and emission controls and by changing the principal energy source from coal to natural gas. Some of these measures have already been taken especially in urban areas, but significant effects (i.e., reduced concentrating) have not yet been observed in the official data that we reviewed. It is apparent that more attention than ever should be paid to environmental pollution produced by the TVEs, for example.

3. PUBLIC PERCEPTION OF ENVIRONMENTAL HEALTH RISKS IN CHINA AND INDONESIA

As a main part of our HDP study, awareness and perception of environmental health risks were surveyed using a questionnaire developed by the authors and administered in three cities in Indonesia and five cities in China (Figure 11.4). The subjects, who were randomly selected, were basically community (lay) people. The sample sizes were 1,002 in Jakarta, 1,190 in Bandung, 719 in Bandar Lampung, 1,116 in Beijing, 957 in Chongqing, 628 in Chengdu, 943 in Shanghai, and 972 in Dalian. In Jakarta, an additional group of 430 university graduates was also surveyed.

In the case of China, subjects were randomly selected from three areas in the central part of each of the cities based on land use or distance from major roads, considering especially pollution from industries and/or motor vehicles. Each of the subjects was interviewed with a questionnaire developed by Kabuto and Honda, National Institute for Environmental Studies, and translated into Chinese (Kabuto et al., 1998; Honda et al., 1998; Kabuto and Honda, 1999). As shown in Table 11.8, to the question "Is there any industrial pollution ?," the percentage of "yes" responses in China was the highest in Chongqing (79%), followed by Chengdu (71%), Beijing (56.8%), Dalian (54%), and Shanghai (20%). The positive response rate in Indonesia was lower: Jakarta (16%), Bandung (22%), and Lampung (15%). Although data are not shown, the response rate was higher in the educated group in Jakarta (30%). As for global environmental issues, the positive responses were generally more than half, showing much different trends compared to those for the regional industrial pollution issue. This rate was the highest in Shanghai (96% for global warming), as shown in Table 11.9.

Moreover, as illustrated in Figures 11.5–11.7, which depict the results of the data in China, the odds ratios of a positive response for each of the

TABLE 11.8. *Responses to the Question: "Is there any industrial pollution?" in Five Cities in China [number, (percent yes)]*

	Beijing	Chengdu	Chongqing	Dalian	Shanghai
No	482	282	239	452	634
Yes	634 (56.8)	687 (70.9)	906 (79.1)	520 (53.5)	158 (19.9)
TOTAL	1116	969	1145	972	792

Source: Kabuto and Honda (2001).

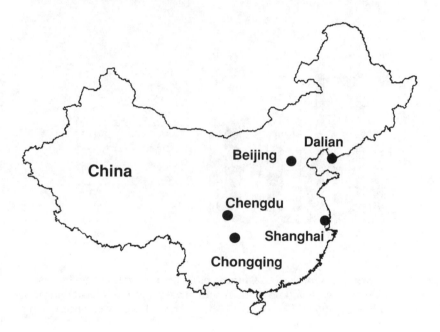

FIGURE 11.4. Locations of the eight subject cities in Indonesia and China in our HDP study (1997–9).

TABLE 11.9. *Responses to the Question: "Do you know about global warming?" in Five Cities in China [number, (percent yes)]*

	Beijing	Chengdu	Chongqing	Dalian	Shanghai
No	288	382	319	111	33
Yes	828 (74.2)	587 (60.6)	826 (72.1)	861 (88.6)	759 (95.8)
TOTAL	1116	969	1145	972	792

Source: Kabuto and Honda (2001).

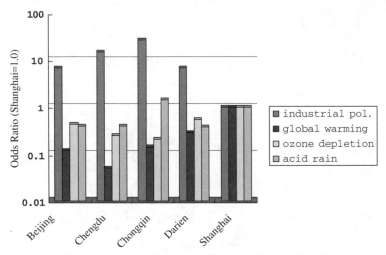

FIGURE 11.5. Odds ratios for various Chinese cities of "yes" response when asked if respondents were aware of/know about industrial pollution, global warming, ozone layer depletion, and acid rain, with Shanghai treated as the reference. (*Source:* Kabuto and Honda, in press.)

cities were roughly proportional to the actual pollution situation, as far as industrial pollution is concerned. It may be surmised that the pollution level within each of the cities contributes significantly to the positive response. This was also the case for the questions of "Is your own health injured by industrial pollution?" and "Are counteractions needed for industrial pollution?" In Chongqing, 96 percent of people answered "yes" to the question "Is your health injured by industrial pollution?" especially in the industrialized area of the city. However, no significant odds ratios for the positive response for age groups or education levels were found, suggesting that people are aware of and perceive the existence and possible risks of industrial pollution, regardless of age group or education level. As for global environmental issues (global warming, ozone layer depletion, or acid rain), however, the odds ratios of each of the studied cities

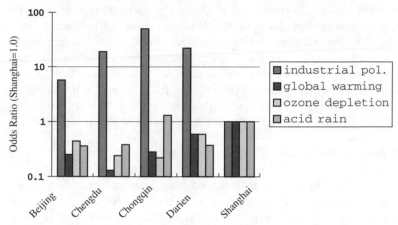

FIGURE 11.6. Odds ratios for various cities of "yes" response to questions that asked if respondents think their own health is/would be injured by industrial pollution, global warming, ozone layer depletion, and acid rain, with Shanghai treated as the reference. (*Source:* Kabuto and Honda, in press.)

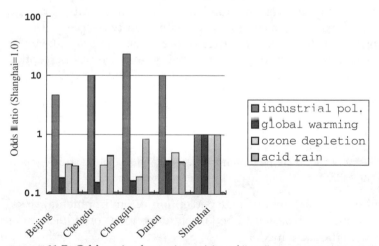

FIGURE 11.7. Odds ratios for various cities of "yes" response to questions that asked respondents whether actions are needed for industrial pollution, global warming, ozone layer depletion, and acid rain, with Shanghai treated as the reference. (*Source:* Kabuto and Honda, in press.)

showed reversed trends compared to those for regional issues, with the positive response rate in Shanghai much higher than in the other cities. It was also shown that risk perception regarding global issues is much higher among educated people as well as among younger people, suggesting their awareness and perception of those issues are influenced by education and possibly the risk information received through the mass media including TV/radio, since most of the subjects have their own TV/radio (excluding the temporary residents from rural areas, who were excluded from the present analyses).

In Indonesia, more marked differences in responses were observed between educated people and community people for the same questions, with similarly high percentages in the educated groups but much lower ones in the community group, compared to the people in Chinese cities. However, in both China and Indonesia, most subjects were not aware of the toxic health effects of mercury, arsenic, PCB, dioxins, and so on with the exception of AIDS, about which almost everyone was very aware and anxious.

It may be expected that education level affects awareness of pollution and its sources, in addition to risk perception. The markedly low frequencies of positive responses among the community people in Bandung and Lampung could be attributed mostly to their low awareness associated with low education level. It should be noted that there were also marked differences in socioeconomic status as well as lifestyles between the educated and general community people. Most of the educated people surveyed read newspapers every day and own cars, air conditioners, and refrigerators, for example, in contrast to the community people.

Regarding the attitudes of the public to the risks, Hirose et al. (1998), who summarized the results of a series of their studies on risk perception among Japanese subjects from 1992 through 1998, stated that

social awareness is conservative and there are cases where perceptions are not changed easily despite changes in social reality. To the contrary, in a society where the influence of media is strong, there are cases where social awareness goes first like a precursor before social reality is changing. It seems certain that an idea to control risk has now begun taking root in among the Japanese mind. True enough, awareness is a tool for adaptation to environment.

Thus, knowledge, awareness, and perception of environmental risks[3] among the public are expected to be largely dependent on education as well as access to the mass media.

Even if awareness of risks is high, risk perception can vary widely, depending on the nature of risks, experiences of risk-related accidents, cultural background, benefit to the individual, and other factors. According to the Third Annual Environmental Monitor report (Environics, 1999), which presents the results of surveys of 1,000 citizens in each of twenty-seven countries on a variety of environmental and natural resource issues, the percentage of people who believe that their health is affected by environmental pollution a great deal or a fair amount was highest in China (98 percent) and the lowest in Germany (39 percent). People living in affluent countries appear to be much less likely to blame environmental pollution for seriously affecting their health, whereas the majority in developing countries believe the opposite. Moreover, according to a public opinion survey of the citizens in seventeen countries including Europe, the United States, and Japan by M. Aoyagi (personal communication), attitude formation toward actions that counteract global warming tends to be negative when individuals think that developing countries are responsible for it, and vice versa, suggesting that awareness and perception could be modified by attitudes or background factors if they know the associated risks.

Thus, as has been clearly demonstrated here, risk awareness and perception have already been enhanced among highly educated people in Indonesia, and even among community people in China. Accordingly, it is hypothesized that risk awareness and perception among the public would be elevated proportionally with the severity of the modern risks at the regional level, once they know how hazardous these risks are in their daily lives. However, awareness and perception of the risks at the global environmental level may be dependent on various factors including education and other socioeconomic factors, age, sex, and even opinions/attitudes. The latter tendency may be the case for the risks of small amounts of exposures to toxic chemicals.

4. RISK AWARENESS AND PERCEPTION AMONG EXPERTS ON PUBLIC HEALTH AND ENVIRONMENTAL SCIENCES

Apart from the previously mentioned survey of the public, experts in the public health and environmental preservation sectors in Indonesia, China, and some other Asian countries were also surveyed in order to understand their knowledge, awareness, and perception of environmental risks, and also to get a sense of the state of environmental education especially on risk assessment. A questionnaire developed by the authors was distributed to various persons who were on a list of members included in the Asian network of Japan NIPH as well as the collaborators in our HDP study in Indonesia and China. In this section, responses from fifty-three participants from whom completed answers were obtained are analyzed.

The subjects included fourteen Indonesians and nine Chinese. Other respondents were from India, Bangladesh, Singapore, Malaysia, Australia, Mongolia, and also Japan. Respondents' ages ranged from 24 to 67 years old. Ten subjects were women and others were men.

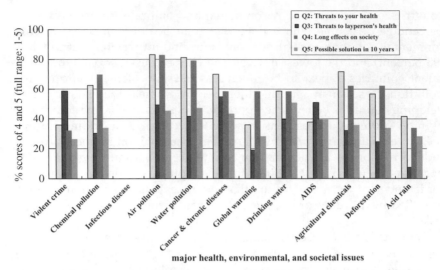

FIGURE 11.8. Percentage of the fifty-three experts in Indonesia, China, and several other Asian countries who scored 4 and 5 on a scale from 1 to 5 in response to the questions regarding the impacts of each of the major environment, health and societal issues in the developing countries in terms of "threat for your health," "threat for laypersons' health," "long effects on your society," and "possible solution in 10 years." The figure shows that the majority of the experts of environmental sciences and public health feel air and water pollution are the most important issues compared to crime, STD (sexually transmitted diseases), and acid rain. (*Source:* Kabuto and Honda, unpublished data).

As the most important societal goal, more than 70 percent of the experts rated education, environmental preservation, parks and forests, budget, and reducing poor and infant mortality. Their responses to several questions are summarized in Figure 11.8, showing that air and water pollution, cancer and chronic diseases, and chemical pollution were their primary concerns in terms of threat to their health as well as long-term effects on society, while they were not concerned with infectious diseases at all. Moreover, they evaluated threats of those major health, environmental, and societal issues to be much less to lay persons compared to those to themselves, suggesting that, they consider, that lay persons are not so vulnerable to those threats.

Regarding lectures on various risk assessment methods in their institutes or universities, the highest frequency of around 60 percent was observed for carcinogenic assessment, then 40 percent for toxicological assessment, and lower percentages for life cycle assessment and ecological assessment (Figure 11.9). It was not certain, however, whether these lectures are meaningful for policy making at the local and national government levels.

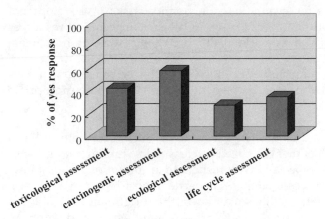

Lectures related to risk assessment

FIGURE 11.9. Percentage of experts indicating availability of lectures or related information on indicated subject at their institution. (*Source:* the same as in Figure 11.8)

The public health and environmental experts and medical university staff that were interviewed have knowledge/awareness of environmental risks, risk assessment, and so on compared to the community people, although most of the experts did not know about the Society for Risk Analysis (only six knew about it well), and they generally had only weak connection with international organizations including the UN (data not shown).

5. SCOPE OF ENVIRONMENTAL HEALTH RISK MANAGEMENT AND GOVERNANCE

It was in 1988 when major research priorities related to environmental risk in the developing countries were identified in a comprehensive way at a workshop held at the East-West Center (Edgerton et al., 1990). These priorities included:

1. conceptual and theoretical study of risk and development (such as risk transition theory);
2. environmental considerations in development policy (risk assessment as part of environmental impact assessment and development planning, comparative risk and cost assessments of environmental standards, international risk/hazard transfer);
3. methodologies for evaluating chemical risks in development (the use of hazard indicators and hazard screening methods, methods and databases for evaluating models, the role of total exposure, and specific source assessment in evaluating risk);

4. methodologies for evaluating nonchemical risks in development (risks from natural hazards, risks to the sustainability of resources and biodiversity, risks of global climate change through human activity), and
5. risk communication and education (including cross-cultural and intracultural differences of risk perception for setting more appropriate risk assessment methods).

All of these risk research priorities seem to be still appropriate and necessary, although each of them has been addressed in a preliminary manner for the last ten years, especially with respect to air pollution issues. However, as far as health risk estimates are concerned, those reported by several international organizations focus mainly on the traditional risks and are usually too highly aggregated regarding sex, age, region, and so on. In order to estimate health risks in the developing countries, as listed here, there could be various approaches including health impact analysis, risk assessment methods, and monitoring of specific diseases in relation to corresponding chemical pollution. Reliable monitoring data on health and the environment are essential. Generally, evidence from epidemiological studies is preferred. Basic health endpoint data availability has recently been increasing in China, which in turn will facilitate health risk estimation. As an example, in the disease surveillance project covering 145 sampling sites over the nation, which has been conducted by the Chinese Academy of Preventive Medicine with support from the World Bank for the last ten years, quality controlled cause-specific mortality data have been collected.

As for the concept of risk and development, risk transition theory, which is based on a classification of the traditional risks and modern risks, should be slightly modified as shown in Table 11.10, since global climate change and transboundary risk issues should also be included in the modern risks. As mentioned earlier, the traditional risks are expected basically to be a function of development levels of the society as a whole, whereas the modern risks at the regional level and also at the global level are associated in different ways with those development activities through industrialization, motorization, and also urbanization, which should be managed with various regulatory approaches. Thus, it seems necessary to divide the modern risks at the regional and global levels in various catagories of risk assessment and management. It is also apparent that integrated and efficient methods to regulate the modern risks while balancing developmental activities should be considered for each of the regions according to the three types of health risks as shown in Table 11.10. Disaggregating environmental risk estimates by sex, age groups, and local level, transregional level, and global level are also essential.

The experience in the Asian developed countries has shown that leadership by the central government and risk researchers at the national level

TABLE 11.10. *Classification of Environmental Risks*

1. Traditional risks	1. Traditional risks at community/local and regional level	Indoor air pollution (coal, biomass fuels) Infectious disease Malnutrition Endemic diseases Natural hazards
2. Modern risks	2. Modern risks at the local level	Air and water pollution Waste disposal
	3. Modern risks at transregional level	Deforestation, acid rain, coastal eutrophication, desertification, etc.
	at global level	Global warming, ozone layer depletion and UV increase etc. Dioxins/EDCs/POPs etc.

are critically important for promoting sound risk management. According to the authors' private communications, the Chinese government is now discussing risk assessment methods as part of their official procedures. It is also evident as mentioned previously, that public awareness of the need for government actions to address modern health risks at the regional level has been enhanced, at least in urban areas.

When these situations of the government, the public, and the experts in terms of risk, awareness, and perception as well as the overlaps of traditional and modern risks are considered, the idea of a dual-benefit approach that addresses both types of risk as proposed by Wang and Smith (1999) and WHO (1999) seems timely. For example, changing electric energy sources from coal to natural gas or hydropower may provide near-term reductions in health risks at the regional level and also long-term GHG emission reductions. This type of approach would be considered and developed much more if international collaborative work on this aspect were strengthened.

In conclusion, environmental management based on health risk is to be promoted in the developing countries by strengthening risk research in general and improving environmental education and access to risk information among the public. At various steps of this process, international collaboration seems essential to address the current risk overlapping issues as early as possible, which may eventually give way to international governance of global environmental and transboundary risk issues (Zhai and Ikeda, 1998).

To this end, the first China and Japan Joint Conference on Risk Assessment and Management (CJCRAM98) was held in Beijing in 1998, with the second in 2001. Through our HDP study, networks for cooperative studies have been formed between the Japan National Institute for Environmental Sciences (NIES), the WHO, and various institutes in China, Indonesia, and the South Asian countries. Moreover, the Asia Pacific Network for Global Change Research (APN homepage: http://www.rim.or.jp/apn), an intergovernmental organization created by the Japan Ministry of Environment supports global environmental research in the Asia-Pacific region. Many other cooperative plans have been made by the Center for International Environmental Cooperation in Japan. As for monitoring transboundary risks, a multilateral agreement was signed in 1999 for exchanging data in a unified context among North-Asian countries.

Notes

1. Globally, about 43 percent of the total burden of disease due to environmental risks affects children under 5 years old, even though they constitute only 12 percent of the total population (Smith, Corvalan, and Kjellström, 1999).
2. Elevated arsenic concentrations in drinking water and resulting severe arsenic poisoning now plagues much of Asia (see, for example, Berg et al., 2001; and the website at http://www.unu.edu/env/Arsenic/Index.htm). Ironically, in many areas, wells were dug into arsenic-rich (but unmonitored) groundwater aquifers to obtain a water supply that would be free from the sewage/microbial contamination (discussed in Section 2.2) that plagued surface-water supplies in these localities. This tragedy is characteristic of many risk-risk trade-offs that occur when a new technology is introduced to address an existing risk, but with inadequate characterization and safeguards.
3. In this chapter, knowledge, awareness, and perception of risks are not strictly differentiated, since they are closely interrelated with each other. There are people who know a risk but are not aware of and perceiving it, whereas other people are not aware of it due to a lack of knowledge.

References

Berg, M., Tran, H. C., Nguyen, T. C., Pham, H. V., Schertenleib, R., and Giger, W. 2001. Arsenic contamination of groundwater and drinking water in Vietnam: A human health threat. *Environmental Science & Technology*, 35 (13): 2621–26.

Chen, M., ed. 1997. *Year Book of Health in the People's Republic of China 1997*. People's Medical Publishing House, Beijing.

Chinese Academy of Preventive Medicine. 1988. *Report of Nationwide Survey on Quality of Drinking Water and Water Borne Diseases*. Institute of Environmental Health and Engineering, Beijing.

Edgerton, S. A., Smith, K. R., Carpenter, R. A., et al. 1990. Priority topics in the study of environmental risk in developing countries: Report on a workshop held at the East-West Center, August, 1988. *Risk Analysis*, 10 (2): 273–83.

Environics International Ltd. (1999). *The Environmental Monitor-Global Public Opinion on the Environment-1999 International Report*. Environics International Ltd., Toronto.

Gu, G. W. 1988. Progress of study on drinking water and liver cancer, *Journal Environment and Health*, 5 (1): 44–6.

He, X. Z. 1990. *Lung Cancer and Indoor Air Pollution From Coal Burning*. Yunnan Sci-Tech Publishing Company, Yunnan.

Hirose, H., Fuwa, H., Uchiyama, I., Murayama, R., Ishizuka, T., Tsuchida, S., and Nakaune, N. 1998. Empirical study on perceived environmental risks of the Tokyo Metropolitan residents. In *Risk Research and Management in Asian Perspective: Proceedings of the First China-Japan Conference on Risk Assessment and Management*, edited by Beijing Normal University, Society for Risk Analysis, Japan Section, National Natural Science Foundation of China, pp. 433–9. International Academic Publishers, Beijing.

Honda, Y., and Kabuto, M. 1998. Questionnaire surveys in Beijing and Chengdu, China-Environmental health risk perception level and its relation to socioeconomic factors. In *Risk Research and Management in Asian Perspective: Proceedings of the First China-Japan Conference on Risk Assessment and Management*, edited by Beijing Normal University, Society for Risk Analysis, Japan Section, National Natural Science Foundation of China, pp. 286–90. International Academic Publishers, Beijing.

Hong, C. J. 1991. Study on the effect of air pollution on human health in Shanghai city. *The Proceedings of Japan-China Medical Communication Conference*, p. 172.

Jin, Y. L. 2000. *Quantitative Study on the Effect of Air Pollution by Coal Burning on Human Health in China*. The 9th Five-Year Plan. Ministry of Science and Technique.

Jin. Y. L. 2000. *The Challenge of Environmental Health in China*. Proceedings of the 9th Conference of World Federation of Public Health Association, (September 2000), Beijing Convention Center.

Kabuto, M., and Honda, Y. 1998. Risk awareness and perception in Asian developing countries as a function of environmental risk transition: A HDP (Human Dimensions Programme) study plan. In ed., Beijing Normal University, Society for Risk Analysis, Japan Section, National Natural Science Foundation of China, *Risk Research and Management in Asian Perspective: Proceedings of the First China-Japan Conference on Risk Assessment and Management*, pp. 595–600. International Academic Publishers.

Kabuto, M., and Honda, Y. 1999. Status and perspectives towards the early 21st century of the "environmental risks transition" in relation to awareness and perception of the environmental risks in Asian countries: An overview of our HDP approach. *Proceedings of 1999 Open Meeting of the Human Dimensions of Global Environmental Change* (June 24–26, 1999), Kanagawa.

Kabuto, M., and Honda, Y. 2001. Awareness and perception of health risks associated with regional and global environmental issues among community people in 5 major cities, China. *Global Environmental Research* 5 (1): 85–95.

Lan, Q., He, X. Z., and Jin, Z. Y. 1990. *Retrospective Cohort Study on Lung Cancer and Indoor Air Pollution in Xuanwei: Lung Cancer and Indoor Air Pollution from Coal Burning*. Yunnan Sci-tech Publishing Company, Yunnan.

Ling, B. 1999. Health impairments arising from drinking water polluted with domestic sewage and excreta in China. *Water, Sanitation and Health*. WHO, Geneva.

McMichael, A. J. 1999. Epidemiology and society: A forum on epidemiology and global health. *Epidemiology* 10 (4): 460–4.

Ministry of Health. 1984. *Summary of the Nationwide Water Investigation*. Peoples Republic of China, Ministry of Health, Beijing.

Ministry of Health. 1988. *Drinking Water Atlas of China*. China Cartographic Publishing House, Beijing.

Murray, C. J., and Lopetz, A. D. 1997. Alternative projections of mortality and disability by cause 1900–2020: Global burden of disease study. *Lancet*, 349: 1498–504.

Niu, S. R. 1996. The status of arsenism and its prevention in Chinese mainland. *ZhongHua Weizhi (Taiwan)*, 15: 1–5.

Smith, K. R. 1997. Development, health and the environmental risk transition. In G. S. Shahi et al., eds., *International Perspectives on Environment, Development and Health: Toward a Sustainable World*, pp. 51–62. Springer, New York.

Smith, K. R., Corvalan, C. F., and Kjellström, T. 1999. How much global ill health is attributable to environmental factors? *Epidemiology*, 10 (5): 573–84.

Su Delong. 1980. Drinking water and liver cancer. *Chinese Journal of Preventive Medicine*, 14 (2): 65.

Tri-Tugaswati, A., and Kiryu, Y. 1996. Effect of pollution on respiratory symptoms of junior high school students in Indonesia. *Southeast Asian Journal of Tropical Medicine and Public Health*, 27: 792–800.

UNDP. 1999. *China Human Development Report – Transition and the State*. United Nations Development Programme.

Wang, H. 1995. Study on seasonal dynamics of algae and microcystins in a lake. *Journal of Environment & Health*, 12 (5): 196.

Wang, L. F. 1997. *Endemic Arsenism and Blackfoot Disease*. Xinjiang Sci-tech and Sanitation Publishing Company, Urumuqi.

Wang, L. F., Sun, X. Z., and Feng, Z. Y. 1996. The relationship between arsenism and arsenic in ground water. *Survey Disease* (in Chinese), 11 (3): 54–5.

Wang, X., and Smith, K. R. 1999. Secondary benefits of greenhouse gas control: Health impacts in China. *Environmental Science and Technology*, 33 (18): 3056–61.

WHO and UNEP. 1992. *Urban Air Pollution in Megacities of the World*. Blackwell, Oxford.

WHO. 1994. Lung cancer and indoor air pollution in China. *Women, Health & Environment: An Anthology*. World Health Organization, Geneva.

WHO. 1997. *Health and Environment in Sustainable Development: Five Years after the Earth Summit*. World Health Organization, Geneva.

World Bank. 1992. China: Environmental Strategy Paper. China and Mongolia Department, Washington, DC.

World Bank. 1994. *World Population Projections. 1994–95 Edition. Estimates and Projections with Related Demographic Statistics*. The World Bank, Washington, DC.

World Bank. 1997. *Clear Water, Blue Skies – China's Environment in the New Century: China 2020*. The World Bank, Washington, DC.

WRI, UNEP, UNDP, and World Bank. 1999. *World Resources 1998–99 – A Guide to the Global Environment*. WRI, New York.

Wu, C., Maurer, C., Wang, Y., Xue, S., and Davis, D. L. 1999. Water pollution and human health in China. *Environmental Health Perspectives*, 107 (4): 251–6.

Wu, Y. 2000. *The Accumulation of Harmful Substance in Human Body – The National Dynamic Study*. Institute of Environmental Monitoring, Chinese Academy of Preventive Medicine, Beijing.

Xu, X., Gao, J. J., Dockery, D. W., and Chen, Y. 1994. Air pollution and daily mortality in residential area of Beijing, China. *Archives of Environmental Health*, 49 (4): 216–22.

Yin, X. R. 1993. Effects of air pollution on human health. *Proceedings of the Conference of Epidemiologist on Air Pollution*, Beijing.

Zhai, G., and Ikeda, S. 1998. A transfrontier risk profile in Northeast Asia. *Proceedings of the First China-Japan Conference on Risk Assessment and Management*, pp. 609–16. Beijing Normal University, Society for Risk Analysis, Japan Section, National Natural Science Foundation of China. International Academic Publishers, Beijing.

Zeng, G. 1997. Infectious diseases control and prevention in China for the 21st century. *Chinese Journal of Epidemiology*, 18 (2): 67.

Zhang, L., and Chen, C. 1997. Geographic distribution and exposure population of drinking water with high concentration of arsenic in China. *J Hygiene Research* (in Chinese), 26 (5): 310–13.

Zhao, Q. 1999. *Control Engineering for Occupational Injuries in Xiang Zhen Industries*. China Labor Press, Beijing (in Chinese).

SUMMARY AND FUTURE DIRECTIONS

12

State of the Art and New Directions in Risk Assessment and Risk Management

Fundamental Issues of Measurement and Management

Rae Zimmerman and Robin Cantor

1. INTRODUCTION

Risk assessment and risk management have grown out of many different intellectual traditions, with many approaches to practice within an array of governance contexts. Over the past few decades, the analytical and managerial approaches to risk have evolved substantially, helping to inspire new methods and directions for future applications. Many views and perspectives are needed to describe risk adequately. Thus, we need common frameworks that will allow comparisons and judgments to be made regarding risk management trade-offs. Changes in society, including the growing complexity of the population and its diversity, the speed with which new technologies are introduced, and the emergence of regional and global economies and new modes of international governance have also altered the ways in which we view risk.

The preceding chapters highlight many conceptual and contextual aspects of risk problems that emerge from the intersection of uncertain outcomes and consequences, and the need to govern in the face of these. In democratic societies, risk problems place particular demands on governance that have substantial implications for how we allocate, use, and manage available resources, products, technologies, and infrastructure. In this concluding chapter, we summarize the relationship between these special demands on governance and some of the enduring themes that have influenced the development of risk analysis as a discipline. We consider as well how these demands have evolved and therefore continue to challenge our core risk concepts, general analytical approaches, and risk management strategies.

As discussed in the introduction, a general objective of this book was to assess the state of the art of risk analysis, and authors were encouraged to emphasize where consensus, controversy, and uncertainties exist in our concepts and approaches. As a result, the chapters both identify major cornerstones of the current state of the field and important directions for future efforts that would address the uncertainties and controversies.

Discussions during the Airlie House symposium held in Warrenton, Virginia, in June 2001 also revealed that while risk analysis has matured as a field of study, the portfolios of risk problems are varied and changing. Speakers from different disciplines (primarily the social sciences, engineering, environmental science, and health) and different international perspectives (primarily North America, Europe, and Asia) set forth how meeting these new needs would challenge our core concepts and current worldwide analytical capabilities. Sessions organized around the issues of equity, efficiency, and the integration of analysis and deliberation reinforced the fact that even though we have made much progress, especially in the technical field of risk assessment, the intersection of risk and governance continues to foster evolution and controversy in the science and practice of risk analysis.

The introductory chapter, and the overall organization of the book, is divided into three themes: (1) the character of risk, (2) advances in the methodology for risk assessment and analysis, and (3) the needs and demands for effective risk management. Accordingly, this concluding chapter follows this frame and provides the beginning of a roadmap for the evolving state of the art and future directions for risk assessment and management.

2. THE CHARACTER OF RISK

Major advances have been made in how risk is characterized. Several of the chapters in this volume and the symposium discussions emphasized that risk concepts have evolved substantially over time. In the past, risk analysis largely emphasized traditional statistical approaches to evaluate consequences as well as their probabilities and expected values. There is now a growing recognition that a much richer set of characteristics is needed to describe risks.

The current state of the art continues to focus on the uncertainty and variability of risks, as represented by Chapters 2–5. Aspects of Chapter 6 bear upon these issues as well. The distinction between variability and uncertainty provides an important foundation for managing risks that affect members of the population in different ways: variations in both exposure and susceptibility, and our knowledge and characterization of these differences is highly limited. As described in Chapter 2, differentiating between the concepts of variability and uncertainty eliminates much of the confusion that was persistent in the risk field, especially in the way in which the two concepts were quantified. Consideration of variability as a management tool enables us to identify differences in the characteristics of populations as a basis for understanding the differences in the risks to which they are exposed. Consideration of uncertainty allows us to direct research and data collection to most effectively improve this characterization, and thereby more effectively design risk management programs to reduce or eliminate the most severe problems.

While variability in exposure and response is clearly important across the different members of a population, as Rhomberg notes in Chapter 3, this variability can be identified at many scales of biological organization, including the organs, tissues, and cells of a given organism. Rhomberg thus extends the concept of variability to compare and synthesize alternative approaches to dose-response modeling. He particularly notes the similarity in how variability applies to cancer and noncancer risks at different levels of organization in biological systems. The chapters in Part 1 underscore a fundamental fact: without uncertainty there would be no need for risk assessment, only the need for management or adaptation to known outcomes. Uncertainty, thus, continues as an important focus of risk assessment research and practice.

Given the different implications of variability and uncertainty for risk management, Cullen and Small in Chapter 6 underscore the need to portray attributes of risk as distributions rather than as point estimates. This need arises over the continuum of the many phases of decision making from hazard identification through exposure assessment to dose-response and risk characterization. The more subtle issue is how the uncertainty and variability profiles look and interact as one advances through the various steps of risk assessment. In addition, if we rely on judgment in conjunction with quantitative analytical techniques to assess uncertainty (as inevitably we must), a whole array of biases that have been identified in the literature on risk perception and judgment come into play. Furthermore, the very framing and formulation of quantitative techniques often imply their own sets of judgments and biases.

Concern for extreme, catastrophic, and rare events have recently become more important in how we view and manage risk. For example, in contrast to relying on an expected value to characterize uncertainty, Bier et al. in Chapter 4 emphasize that catastrophic risks require a different, and often a more creative, approach to risk characterization. Thus, the extent to which a risk has a great potential for surprise outcomes is now more likely to influence its characterization and, consequently, the design of a risk management strategy. The expanding focus on extreme events and surprises is the result, in part, of our new tools and parameters to address these risks. It is also the result of a sense that extreme, catastrophic, or surprise events may be more important risks today than in the past, as a result of various new and emerging threats such as global change, terrorism, and illegal access to information and computer systems. Rather than being led by the law of averages, societies tend to expect (and fear) extremes. As Chapter 4 points out, a focus on extremes not only affects how we use experts in the risk management process but also how we communicate risk. Whether a focus on extremes is good governance is, of course, an issue of public policy and debate.

A fourth perspective in defining the character of risk is addressed by English in Chapter 5: the convergence of the environmental risk and justice movements. This convergence has had a profound effect upon how risk is viewed, combining the notion of variability of risk across populations with recognition of its implications for social justice. Concepts of justice address distributional issues across populations with different resources and capacities to deal with risk. Hence, English addresses risk across spatial and temporal distributional dimensions as well as cultural differences in how risks are interpreted. It has become clear from this perspective that risk has a subjective element that derives from its cultural context. English combines risk management and justice by identifying six different dimensions of environmental decision making. These depend on "who makes the decision, under what conditions, and with what consequences." English concludes that the justice movement has changed the way in which exposures are analyzed, the nature of the measurement instruments, and the models and metrics used for choosing among alternatives for risk management. The key challenges from an analytical point of view are fitting risk into the definition of social community, within spatial units of analysis and within geographic boundaries for study areas.

3. ADVANCES IN RISK ASSESSMENT AND ANALYSIS

After reviewing the state of the art for defining the character of risk and how this has changed to incorporate evolving notions of uncertainty, variability, and justice, Part 2 addresses methods for measuring and valuing risk. Chapter 7 emphasizes that the choice of a risk measure often influences the identification of risk problems and the consequent selection of policy prescriptions. This same viewpoint was expressed frequently in the symposium discussions. Three principal themes are emphasized in defining current methodology for risk measurement and valuation. The first emphasizes the need to value risky outcomes in order to compare alternative management options. This has transformed the way in which risks are analyzed and involves a synthesis of the environmental valuation and risk literature. In Chapter 7, Gregory summarizes the state of the art for valuing risk and risk consequences, and how this influences choices among risk management options, since valuation is a way of weighing risks to set priorities. Moreover, risk valuation incorporates risk perception directly into the decision-making framework. At the broadest level, increasing attention to the factors associated with perceptions of risk and differences in lay and expert assessments of risk have driven risk analysts to use a multidimensional construct for risk attributes and their measures. This construct includes both consequences and their probabilities, as well as such factors as the voluntary nature of risk, whether it affects children or adults, risk latency, and the potential for catastrophic consequences. However, as Gregory notes, an increasing attention to risk perception will often lead to additional complexities for the measurement of risk, as analysts struggle with reliable interpretations of these measures.

The second theme in risk measurement and valuation is addressed by Graham, Johansson, and Nakanishi who discuss efficiency analysis as a means of considering trade-offs and making resource allocation decisions to manage risk. They adapt the more conventional methods of efficiency analysis to the specific needs of risk management indicating that "efficiency [in the conventional sense] is not a necessary or sufficient condition for good (societal) risk management." Differences from traditional approaches to economic efficiency arise because there are concerns about the fairness of the process and the equity of the method by which gains and losses are distributed. The informational requirements for applying efficiency to risk management are complex, involving measurements associated with willingness to pay, discounting, and other issues. The key managerial tools discussed include cost-benefit, cost-effectiveness, and value-of-information analysis. The authors note in Chapter 8 that cost-effectiveness is particularly appealing in health risk assessment when measures are used that incorporate mortality and morbidity information into effectiveness metrics, such as quality adjusted life years. The preferred technique depends upon the extent to which risks can be monetized and how certain the risks are. Thus, conventional economic tools must be adapted for assessing risks where nonmonetary and intangible factors are also important.

The third theme is the expanding scope of contextual, spatial, and temporal factors that are measured in a risk analysis. Perhaps the best example of this theme is found in Chapter 10 with its emphasis on transboundary risks and the subsequent demands these place for large-scale analysis and the incorporation of intergovernmental legal and institutional issues. Chapters 2 and 5 raise related examples. They focus on the expanding contextual scope for risk measurement and argue that the environmental justice movement has helped to focus attention on risks that can arise from chronic, multiple, synergistic hazards over extended periods and over entire communities – not just from the single-source, acute exposures that are analyzed using the traditional, more limited perspectives employed in the past. With this broader perspective, the severity of risk is seen to depend on the prior exposure and health histories and sensitivities of affected communities. Similarly, the environmental justice movement has emphasized that some environmental risks are not exclusively (or even primarily) risks to health; they may also involve risks to economic, social, or psychological well-being. In Chapter 5, English further suggests a more ecocentric conception of environmental justice, which could involve analyzing nonhuman, or intrinsic, sources of value resulting from environmental changes.

4. APPROACHES AND NEEDS FOR RISK MANAGEMENT

Risk management approaches are now seen as being fully integrated with risk analysis. That is, the two processes occur jointly, rather than sequentially, that is, one being an input to the other (NRC, 1994; Stern and Feinberg, 1996; U.S. EPA, 1996). Several of the chapters provide different perspectives on risk management and its integration with risk analysis.

Chapter 9 sets the stage for a typology of management styles drawn from many disciplines and perspectives and suggests a model for synthesis, termed cooperative discourse. The aim of Renn's new model is to be "effective, efficient, legitimate, and socially acceptable." He argues that knowledge and values lead to risk problems that in turn lead to a need to address questions of trade-offs and lifestyle issues. Risk, according to Renn, introduces uncertainty, complexity, and ambiguity. Ambiguity leads to social conflict about the means for managing risks that in turn leads to a need for deliberative processes. Renn presents an array of perspectives from which participation and deliberation have been viewed and the ends to which these processes strive (i.e., conflict resolution, consensus building, policy formulation, information exchange, enhanced understanding, learning and communication). These ends are representative of a wide range of such approaches, recognizing that many others have been explored by political and policy science researchers and practitioners.

Chapter 10 examines the international institutions that address risks and their management by virtue of scale, scope, origin, role, science and technology, complexity, irreversibility, accountability, and scientific evidence. The organizational arrangements they discuss are parallel to the nature of risk management concerns. The management concerns they see as paramount are those that diminish the democratic process for individuals and for local and national governments. The state of the art is addressed in terms of the range of international controls that exist for a "representative set of risks" and their level of development. Although Tait and Bruce do not set up a typology for risk management of transboundary risks in Chapter 10, they offer a set of categories that are applicable to problems beyond those that they address, depending on whether they involve traded risks (e.g., product testing and labeling) or public risks (e.g., legal rights, international agreements). Directions for future international risk management policies are suggested, taking into account their ease of implementation, their stage of development, the degree of consensus, and other factors.

In Chapter 11, Kabuto, Ikeda, and Uchiyama examine the manner in which risk perception, knowledge, and awareness have influenced environmental risk management in a number of Asian countries. They provide an instrument and application as a basis for studies of different risks and how their management can take into account the cultures of different nations. Their evaluation demonstrates the importance of risk assessment in the developing world, the emergence of scientific and management capabilities for implementing such assessments, and the need for the sharing of methods, results, and insights across both national and disciplinary boundaries. We expect that risk assessors from these emerging centers of expertise will soon be leaders in identifying new and important risk problems, as well developing the new methods necessary for their evaluation and management.

5. FINAL OBSERVATIONS

Risk assessment and risk management as an integrated discipline focus on the multiple ways of making decisions in the face of uncertainty and variability and the decision rules for choosing one option over another. The field of risk assessment has broadened to include risk decision making, considering inputs from many different societal and disciplinary perspectives. As we adapt to a smaller world, with unprecedented access to information and cross-cultural communication, only a few keystrokes or mouse clicks away for a growing portion of the world's population, the basic problems of economic development, human health, environmental quality, infrastructure safety, national security, and democratic governance continue to challenge our scientific and management capabilities. We must ensure that the perspectives, frameworks, and methods for risk analysis continue to be advanced and adapted to meet these challenges.

Additional References

National Research Council (NRC). 1994. *Science and Judgment in Risk Assessment,* NRC Committee on Risk Assessment of Hazardous Air Pollutants, National Academy Press, Washington, DC.

Stern, P. C., and Fineberg, H. V., eds. 1996. *Understanding Risk: Informing Decisions in a Democratic Society*. National Research Council Committee on Risk Characterization, National Academy Press, Washington, DC.

U.S. Environmental Protection Agency (EPA). 1996. *Strategic Plan for the Office of Research and Development*. U.S. EPA, Office of Research and Development. EPA/600/R-96/059. U.S. EPA, Washington, DC.

Index

CL

363.
102
RIS

6000612257